Function Spaces, Entropy Numbers, Differential Operators

Function Spaces, Entropy Numbers, Differential Operators

D.E. Edmunds
University of Sussex

H. Triebel
University of Jena

CAMBRIDGE UNIVERSITY PRESS
Cambridge, New York, Melbourne, Madrid, Cape Town, Singapore, São Paulo

Cambridge University Press
The Edinburgh Building, Cambridge CB2 8RU, UK

Published in the United States of America by Cambridge University Press, New York

www.cambridge.org
Information on this title: www.cambridge.org/9780521560368

First published 1996
This digitally printed version 2008

A catalogue record for this publication is available from the British Library

Library of Congress Cataloguing in Publication data

Edmunds, D. E. (David Eric)
Function spaces, entropy numbers, differential operators / D.E. Edmunds, H. Triebel.
p.   cm. – (Cambridge tracts in mathematics; 120)
Includes bibliographical references and index.
ISBN 0 521 56036 5 (hc)
1. Function spaces.   2. Entropy (Functional analysis)   3. Differential operators.
I. Triebel, Hans.   II. Title.   III. Series.
QA323.E26   1996
515′.73–dc20   95-25362   CIP

ISBN 978-0-521-56036-8 hardback
ISBN 978-0-521-05975-6 paperback

Contents

Preface

This book deals with the symbiotic relationship between

(i) function spaces on $\mathbf{R}^n$ and in domains,
(ii) entropy numbers in quasi-Banach spaces, and
(iii) distributions of eigenvalues of degenerate elliptic differential and pseudodifferential operators,

as it has evolved in recent years.

We are mainly interested in the two scales of function spaces B^s_{pq} and F^s_{pq} with $s \in \mathbf{R}, 0 < p \le \infty, 0 < q \le \infty$, which cover many well-known classical spaces such as (fractional) Sobolev spaces, Hölder–Zygmund spaces, Besov spaces and (inhomogeneous) Hardy spaces. The theory of these spaces has been developed in its full extent in [Triα], [Triβ] and [Triγ]. Here we also deal with some recent modifications and refinements connected with spaces of Orlicz type and logarithmic Sobolev spaces.

Let $B^s_{pq}(\Omega)$ be the corresponding spaces on an (arbitrary) bounded domain Ω in $\mathbf{R}^n$. Then the embedding

$$\text{id} : B^{s_1}_{p_1 q_1}(\Omega) \to B^{s_2}_{p_2 q_2}(\Omega) \tag{1}$$

is compact if

$$s_1 - s_2 > n \left(\frac{1}{p_1} - \frac{1}{p_2} \right)_+, \quad 0 < q_1 \le \infty, 0 < q_2 \le \infty. \tag{2}$$

Let $e_k(\text{id})$ be the corresponding entropy numbers. Then there exist two positive numbers c_1 and c_2 such that

$$c_1 k^{-(s_1 - s_2)/n} \le e_k(\text{id}) \le c_2 k^{-(s_1 - s_2)/n}, \; k \in \mathbf{N}. \tag{3}$$

The history of assertions of this type begins in 1967 when M.S.Birman and M.Z.Solomyak [BiS1] proved (3) for the embedding of the Sobolev

(–Besov) spaces $W_p^s(\Omega)$ in $L_q(\Omega)$, their proof being based on the method of piecewise-polynomial approximations. Our method for proving (3) in its full extent relies on Fourier-analytical techniques and has been developed in the last few years in [ET1], [ET2], [Tri3] and [ET4].

The connection between (ii) and (iii) comes from Carl's observation (1980) that

$$|\mu_k| \le \sqrt{2}e_k, \ k \in \mathbf{N}, \tag{4}$$

where μ_k and e_k are respectively the eigenvalues (counted according to their algebraic multiplicities and ordered by decreasing modulus) and the entropy numbers of a compact operator acting in a given (quasi-) Banach space; see [Carl1], [CaT]. It is the main aim of this book to combine observations of type (3) and (4), and to apply them in order to study eigenvalue distributions of degenerate elliptic differential and pseudodifferential operators and their inverses on the basis of some recent progress made in the theory of spaces of B_{pq}^s and F_{pq}^s type.

This book may be considered as a research report mostly based on results of the authors and their co-workers obtained in the last few years. On the other hand, we review the basic material which is needed and give proofs of new results and of assertions not available in relevant books. In this sense we have tried to present a self-contained treatment, accessible to non-specialists.

There are five chapters. Chapter 1 contains elements of a spectral theory in quasi-Banach spaces. We also introduce entropy and approximation numbers and establish some of their basic properties in the context of quasi-Banach spaces, including certain results about the behaviour under interpolation of entropy numbers. Although we focus mainly on entropy numbers in this book, it is helpful to have simultaneously a close look at approximation numbers of abstract and concrete compact operators. Chapter 2 deals with function spaces of type B_{pq}^s and F_{pq}^s. In addition to providing a description of the basic notation and facts, we prove some specific assertions needed in the following chapters. Thus the chapter may be considered as a complement to [Triα], [Triβ] and [Triγ]. In Chapter 3 we calculate the entropy and approximation numbers of compact embeddings between the spaces B_{pq}^s and F_{pq}^s on bounded domains. Chapter 4 concentrates on corresponding problems for weighted spaces of type B_{pq}^s and F_{pq}^s on $\mathbf{R}^n$. Finally, Chapter 5 is devoted to applications of all these results to the distribution of eigenvalues of degenerate elliptic differential and pseudodifferential operators (and their inverses) with non-smooth coefficients.

The book is organised by the decimal system: "n.k.m" refers to subsection n.k.m, "Theorem n.k.m/l" means Theorem 1 in n.k.m, etc. All unimportant positive numbers will be denoted by c (with additional indices if there are several cs in the same formula).

Brighton and Jena
Winter, 1994.

1

The Abstract Background

1.1 Introduction

This chapter is devoted to the general functional-analytic preliminaries which will be needed throughout the book. A good deal of what is presented may be familiar to the reader in a Banach space setting, but in view of later applications we give the theory in the more general context of quasi-Banach spaces, omitting details when they are too similar to those in Banach spaces. In particular, we present a Riesz theory for compact linear operators acting in a quasi-Banach space, detailing the nature of the spectrum.

The bulk of the chapter concerns the basic properties of entropy and approximation numbers of bounded linear operators. Of especial importance is the connection between eigenvalues and entropy numbers which is established in 1.3.4. In a Banach space, this was proved by Carl [Carl1]; the quasi-Banach version given here extends the method in [CaT]. This connection plays a fundamental role in the estimates of eigenvalues of (degenerate) elliptic (pseudo)differential operators which are given in Chapter 5.

1.2 Spectral theory in quasi-Banach spaces

We begin by recalling some basic facts about quasi-normed spaces. A *quasi-norm* on a complex linear space B is a map $\| \cdot |B \|$ from B to the non-negative reals $\mathbf{R}_+$ such that

(i) $\| x |B \| = 0$ if, and only if, $x = 0$,

(ii) $\| \lambda x |B \| = |\lambda| \, \| x |B \|$ for all scalars $\lambda \in \mathbf{C}$ and all $x \in B$,

(iii) there is a constant C such that for all $x, y \in B$,

$$\| x + y |B \| \leq C \left(\| x |B \| + \| y |B \| \right).$$

1

Plainly $C \geq 1$; if $C = 1$ is allowed then $\|\cdot|B\|$ is a norm on B. Each quasi-norm defines a topology on B, compatible with the linear structure of B and with a basis of (not necessarily open) neighbourhoods of any point $x \in B$ given by the sets $\{y \in B : \|x - y|B\| < 1/n\}$, $(n \in \mathbf{N})$. The pair $\left(B, \|\cdot|B\|\right)$ is called a *quasi-normed space*; it is a particular type of (metrisable) topological vector space. Cauchy sequences in a quasi-normed space B are defined in the obvious way; if every Cauchy sequence in B converges (to a point of B), we call B a *quasi-Banach* space.

Given any $p \in (0, 1]$, a *p-norm* on a linear space B is a map $\|\cdot|B\| \to \mathbf{R}_+$ which satisfies conditions (i) and (ii) above and instead of (iii) satisfies

(iii$'$) $\|x + y|B\|^p \leq \|x|B\|^p + \|y|B\|^p$ for all $x, y \in B$.

Two quasi-norms or p-norms $\|\cdot|B\|_1$ and $\|\cdot|B\|_2$ are said to be *equivalent* if there is a constant $c \geq 1$ such that for all $x \in B$,

$$c^{-1}\|x|B\|_1 \leq \|x|B\|_2 \leq c\|x|B\|_1.$$

It can be shown that (see [Kön], p. 47 or [DeVL], p. 20) if $\|\cdot|B\|_1$ is a quasi-norm on B, then there exist $p \in (0, 1]$ and a p-norm $\|\cdot|B\|_2$ on B which is equivalent to $\|\cdot|B\|_1$; the connection between p and the constant C which appears in (iii) above is that C can be taken to be $2^{\frac{1}{p}-1}$. Conversely, any p-norm is a quasi-norm with $C = 2^{\frac{1}{p}-1}$.

Let $0 < q \leq \infty$ and let ℓ_q be the set of all complex sequences $b = (b_k)_{k \in \mathbf{N}}$ of scalars such that

$$\|b|\ell_q\| := \begin{cases} \left(\sum_{k=1}^{\infty} |b_k|^q\right)^{1/q}, & q < \infty, \\ \sup_{k \in \mathbf{N}} |b_k|, & q = \infty, \end{cases}$$

is finite. Then it is easy to see that ℓ_q is a quasi-Banach space, and even a Banach space if $q \geq 1$. In the same way, it can be verified that the L_q spaces are quasi-Banach spaces, and even Banach spaces if $q \geq 1$, when endowed with the obvious (quasi-)norm.

Let A, B be quasi-Banach spaces and let $T : A \to B$ be linear. Just as for the Banach space case, T will be called *bounded* or *continuous* if

$$\|T\| := \sup\{\|Ta|B\| : a \in A, \|a|A\| \leq 1\} < \infty.$$

The family of all such T will be denoted by $L(A, B)$, or $L(A)$ if $A = B$. For the most part, terminology which is standard in the context of Banach spaces will be taken over without special comment to the quasi-Banach situation. We shall, however, be explicit about certain spectral matters to which we now turn.

Let B be a (complex) quasi-Banach space and let $\mathscr{T}$ be the family of all closed, linear, densely defined operators in B, so that any $T \in \mathscr{T}$ has domain $\mathrm{dom}(T)$ which is a dense linear subspace of B mapped into B by T. Given any $T \in \mathscr{T}$, the *resolvent set* of T is

$$\rho(T) = \left\{ \lambda \in \mathbf{C} : (T - \lambda \mathrm{id})^{-1} \text{ exists and belongs to } L(B) \right\}.$$

Here id stands for the identity map of B to itself. The *spectrum* of T is $\sigma(T) = \mathbf{C} \backslash \rho(T)$, and we distinguish two subsets of it, the *point spectrum* $\sigma_p(T)$ and the *essential spectrum* $\sigma_e(T)$. By $\sigma_p(T)$ we simply mean the set of all eigenvalues of T, so that $\lambda \in \sigma_p(T)$ if, and only if, $\lambda \in \mathbf{C}$ and there exists $u \in B \backslash \{0\}$ such that $Tu = \lambda u$. We choose to define $\sigma_e(T)$ by means of *Weyl sequences* (sometimes called *singular sequences*): a sequence $\{u_j\}$ in $\mathrm{dom}(T)$ is called a Weyl sequence of T corresponding to $\lambda \in \mathbf{C}$ if it does not contain a convergent subsequence and $\|u_j\| = 1$, $j \in \mathbf{N}$, $Tu_j - \lambda u_j \to 0$ as $j \to \infty$. Now we can define

$$\sigma_e(T) = \{\lambda \in \mathbf{C} : \text{there is a Weyl sequence of } T \text{ corresponding to } \lambda\}.$$

For information about the essential spectrum in a Banach space context we refer to [EEv], especially Chapter IX.

If $T \in \mathscr{T}$ and $\lambda \in \rho(T)$, then of course there exists $c > 0$ such that

$$\|Tu - \lambda u\| \geq c \|u\| \text{ for all } u \in \mathrm{dom}(T). \tag{1}$$

By analogy with self-adjoint or normal operators in a Hilbert space, we introduce a class of operators characterised by property (1). We define

$$\mathscr{S} = \{T \in \mathscr{T} : \lambda \in \rho(T) \text{ if, and only if, (1) holds for some } c > 0\}.$$

With operators having pure point spectrum in mind (see [Tri1], 4.5.1, p. 254) we define

$$\mathscr{S}_0 = \{T \in \mathscr{S} : \sigma_e(T) = \emptyset\}. \tag{2}$$

Two further classes of operators will be useful:

$$\mathscr{K} = \{T \in L(B) : T \text{ is compact}\} \tag{3}$$

and

$$\mathscr{S}_1 = \left\{ T \in \mathscr{T} : \rho(T) \neq \emptyset, (T - \lambda \mathrm{id})^{-1} \in \mathscr{K} \text{ for some } \lambda \in \rho(T) \right\}. \tag{4}$$

Proposition 1 *Let $T \in \mathscr{T}$. Then the following three assertions are equivalent:*

(i) $T \in \mathscr{S}$;

(ii) $\sigma(T) = \sigma_e(T) \cup \sigma_p(T)$;

(iii) $\sigma(T) = \{\lambda \in \mathbf{C} :$ there exists a sequence $\{u_j\}$ in dom(T) with $\|u_j\| = 1$, $j \in \mathbf{N}$, and $Tu_j - \lambda u_j \to 0$ as $j \to \infty\}$.

Proof (i) $\Longleftrightarrow$ (iii). This follows immediately from (1).

(ii) $\Longleftrightarrow$ (iii). Let M be the right-hand side of (iii): we have to prove that $\sigma_e(T) \cup \sigma_p(T) = M$. Let $\lambda \in M$. If there is a sequence $\{u_j\}$ in dom(T) with no convergent subsequence, and with $\|u_j\| = 1$ and $Tu_j - \lambda u_j \to 0$, then by definition, $\lambda \in \sigma_e(T)$. On the other hand, if there is a sequence $\{u_j\}$ with the same properties save that it is convergent, to u say, then $\|u\| = 1$ and $Tu_j \to \lambda u$. Since T is closed it follows that $u \in $ dom(T) and $Tu = \lambda u$, so that $\lambda \in \sigma_p(T)$. Hence $M \subset \sigma_e(T) \cup \sigma_p(T)$. As the reverse inclusion is obvious, the proof is complete.

Remark 1 The class $\mathscr{S}$ is based on self-adjoint operators in a Hilbert space, which have empty residual spectrum (see [EEv], pp. 5, 414–15; [Tri1], 4.2.3, pp. 219–21). For corresponding assertions concerning normal operators in Hilbert spaces, see [Rud], pp. 312–313.

We recall that the *geometric multiplicity* of an eigenvalue λ of T is $\dim \ker(T - \lambda\mathrm{id})$; its *algebraic multiplicity* is $\dim \bigcup_{n=1}^{\infty} \ker(T - \lambda\mathrm{id})^n$.

Proposition 2 *(i) If $T \in \mathscr{S}_0$, then*

$$\sigma(T) = \{\lambda \in \sigma_p(T) : \lambda \text{ has finite geometric multiplicity}\}.$$

(ii) If $T \in \mathscr{S}_1$, then $(T - \lambda\mathrm{id})^{-1} \in \mathscr{K}$ for all $\lambda \in \rho(T)$.

Proof (i) In view of (2) and Proposition 1, we merely have to exclude eigenvalues of infinite geometric multiplicity. Suppose that $\lambda \in \sigma_p(T)$ is such that $\dim \ker(T - \lambda\mathrm{id}) = \infty$. Then given any $\varepsilon \in (0, 1)$, there is a sequence $\{u_j\}$ in $\ker(T - \lambda\mathrm{id})$ such that $\|u_j\| = 1$ for each j and $\|u_j - u_k\| > 1 - \varepsilon$ for $j \neq k$: this follows by using the same proof as in the Banach space case (see [Tri1], pp. 23, 24). Hence $\lambda \in \sigma_e(T)$ and we have a contradiction.

(ii) That $(T - \lambda\mathrm{id})^{-1} \in \mathscr{K}$ for some $\lambda \in \rho(T)$ if, and only if, $(T - \mu\mathrm{id})^{-1} \in \mathscr{K}$ for all $\mu \in \rho(T)$ follows immediately from the resolvent equation

$$T_\lambda = T_\mu + (\lambda - \mu)T_\lambda T_\mu, \qquad T_\lambda = (T - \lambda\mathrm{id})^{-1}, \; T_\mu = (T - \mu\mathrm{id})^{-1},$$

which holds just as in the Banach space case.

Remark 2 As in the case of self-adjoint operators acting in a Hilbert space (see [Tri1], 4.2.3 and 4.2.4), eigenvalues of infinite geometric multiplicity belong to both $\sigma_p(T)$ and $\sigma_e(T)$.

Remark 3 Proposition 2(ii) simply shows that the definition of $T \in \mathscr{S}_1$ is independent of the particular $\lambda \in \rho(T)$.

Remark 4 If B is a Banach space and $T \in \mathscr{S}_1$, then (see [EEv], Theorem IX.3.1, p. 423)

$$\sigma(T) = \{\lambda \in \sigma_p(T) : \lambda \text{ has finite algebraic multiplicity}\},$$

and for all $\lambda \in \rho(T)$, $(T - \lambda \mathrm{id})^{-1} \in \mathscr{K}$ is a Riesz operator. (For Riesz operators we refer to [Pi1], Chapters 26 & 27 and to [Kön], p. 18.) Hence $\mathscr{S}_1 \subset \mathscr{S}_0$ in the Banach space case.

We now aim to develop a Riesz theory for compact operators, which amounts to extending Remark 4 to quasi-Banach spaces.

Theorem *Let B be a (complex) infinite-dimensional quasi-Banach space and let $K \in \mathscr{K}$.*

  (i) $K \in \mathscr{S}$,
 (ii) $\sigma_e(K) = \{0\}$, $\sigma(K) = \{0\} \cup \sigma_p(K)$,
(iii) *$\sigma(K) \backslash \{0\}$ consists of an at most countably infinite number of eigenvalues of finite algebraic multiplicity which may accumulate only at the origin.*

Proof We proceed in several steps.

Step 1 Let $\lambda \in \sigma_e(K)$, assume that $\lambda \neq 0$, and let $\{u_j\}$ be a related Weyl sequence. Then since K is compact, we may suppose that $\{Ku_j\}$ is convergent, and as

$$\lambda u_j = (\lambda u_j - K u_j) + K u_j,$$

we see that $\{u_j\}$ must be convergent. This contradiction shows that $\sigma_e(K) \subset \{0\}$.

Step 2 To prove that $0 \in \sigma_e(K)$, let $\varepsilon \in (0,1)$. Let $\{u_j\}_{j \in \mathbb{N}}$ be a sequence in B such that for all $j \in \mathbb{N}$, $\|u_j\| = 1$ and

$$\mathrm{dist}\left(u_j, \mathrm{span}(u_1, ..., u_{j-1})\right) > 1 - \varepsilon, \quad j > 1.$$

The existence of such a sequence follows just as in the Banach space case; see [Tri1], pp. 23, 24. Put $v_j = u_{2j} - u_{2j-1}$, $j \in \mathbf{N}$. Then for each $j \in \mathbf{N}$, $1 - \varepsilon \le \|v_j\| \le C$; and if $k < j$, since $u_{2j-1} + u_{2k} - u_{2k-1} \in \text{span}(u_1, ..., u_{2j-1})$, we have

$$\|v_j - v_k\| = \|u_{2j} - u_{2j-1} - u_{2k} + u_{2k-1}\| > 1 - \varepsilon.$$

Since K is compact we may suppose that $\{Ku_j\}$ is convergent; thus $Kv_j \to 0$ as $j \to \infty$. Hence $\{v_j / \|v_j\|\}$ is a Weyl sequence corresponding to 0, and so $0 \in \sigma_e(K)$. This proves the first part of (ii).

Step 3 Let $\lambda \in \sigma_p(K)$, $\lambda \ne 0$. Then λ has finite algebraic multiplicity: this follows from the Banach space arguments given in [EEv], Lemma I.1.18, p. 9 with obvious changes. In the same way, natural adaptations of the proof of [Ka], Theorem III.6.26, p. 185 show that $\sigma_p(K)$ has no accumulation point different from zero and is, of course, countable.

Step 4 First renorm B with an equivalent p-norm. We claim that if $\mu \in \mathbf{C}$, $|\mu| > \|K\| C$ for a suitable positive C, then $\mu \in \rho(K)$. To establish this, the Banach space arguments, involving Neumann series, from [Tri1], pp. 67–70, can be applied, working with the pth power of this p-norm rather than the p-norm itself.

Step 5 Let $\lambda \in \mathbf{C}$, $\lambda \ne 0$, $\lambda \notin \sigma_p(K)$. By Step 1, $\lambda \notin \sigma_e(K)$ and so there exists $c > 0$ such that

$$\|Ku - \lambda u\| \ge c \|u\| \quad \text{for all } u \in B. \tag{5}$$

Now let $\mu \in \mathbf{C}$ be such that $|\mu| > \|K\| C$ in the sense of Step 4, and let γ be a path in $\mathbf{C}$ joining μ to λ which does not pass through any eigenvalues. A simple contradiction argument shows that (5) holds with λ replaced by any κ on γ and with the same constant c for all κ. Let $\kappa \in \rho(K)$. Then

$$\left\|(K - \kappa \text{id})^{-1}\right\| \le c^{-1}.$$

Let v be near κ on γ. Then for all $u \in B$,

$$v = Ku - vu = Ku - \kappa u - (v - \kappa)u \tag{6}$$

and hence

$$(K - \kappa \text{id})^{-1}v = u - (v - \kappa)(K - \kappa \text{id})^{-1}u. \tag{7}$$

We arrange that $|v - \kappa|$ should be so small that $|v - \kappa| \, \|(K - \kappa I)^{-1}\| < 1$. Thus by the Neumann series argument, (7), and so also (6), has a unique solution. Hence by (5) with λ replaced by v, it follows that $v \in \rho(K)$. Iteration of this shows that $\lambda \in \rho(K)$. The proof of the theorem is complete.

Proposition 3 $\mathscr{S}_1 \subset \mathscr{S}_0$.

Proof Let $\lambda = 0$ in (4).The result follows from the theorem and the fact that $0 \neq \mu \in \sigma_e(T)$ if, and only if, $\mu^{-1} \in \sigma_e(T^{-1})$.

1.3 Entropy numbers and approximation numbers

1.3.1 Definitions and elementary properties

We begin with entropy numbers.

Definition 1 *Let A, B be quasi-Banach spaces and let $T \in L(A, B)$; put $U_A = \{a \in A : \|a \,|A\| \leq 1\}$. Then for all $k \in \mathbf{N}$, the kth entropy number $e_k(T)$ of T is defined by*

$$e_k(T) = \inf \left\{ \varepsilon > 0 : T(U_A) \subset \bigcup_{j=1}^{2^{k-1}} (b_j + \varepsilon U_B) \text{ for some } b_1, ..., b_{2^{k-1}} \in B \right\}.$$

Remark 1 This definition has its roots in the notion of the metric entropy of a set which Kolmogorov introduced in the 1930s: given any compact subset K of a metric space and any $\varepsilon > 0$, Kolmogorov denoted by $N_\varepsilon(K)$ the least $N \in \mathbf{N}$ such that K can be covered by N balls of radius ε, and called $H_\varepsilon(K) := \log_2 N_\varepsilon(K)$ the metric entropy of K. If A, B are Banach spaces and $T \in L(A, B)$ is compact, then the entropy numbers of T are obtained, roughly speaking, by solving the equation $H_\varepsilon(T(U_A)) = k - 1$ for $\varepsilon = e_k(T)$.

The following lemma gives some elementary properties of entropy numbers.

Lemma 1 *Let A, B, C be quasi-Banach spaces, let $S, T \in L(A, B)$ and suppose that $R \in L(B, C)$.*

    (i) $\|T\| \geq e_1(T) \geq e_2(T) \geq \cdots \geq 0$; $e_1(T) = \|T\|$ *if B is a Banach space.*

(ii) *For all* $k, \ell \in \mathbf{N}$,

$$e_{k+\ell-1}(R \circ S) \leq e_k(R)e_\ell(S).$$

(iii) *If* B *is a p-Banach space* $(0 < p \leq 1)$, *then for all* $k, \ell \in \mathbf{N}$,

$$e_{k+\ell-1}^p(S + T) \leq e_k^p(S) + e_\ell^p(T).$$

Proof (i) Since $\|T\| = \inf\{\mu \geq 0 : T(U_A) \subset \mu U_B\}$, it is plain that $e_1(T) \leq \|T\|$. If $T(U_A) \subset b_0 + \mu U_B$ for some $b_0 \in B$ and some $\mu \geq 0$, then given any $a \in U_A$, there are elements $b_1, b_2 \in U_B$ such that $Ta = b_0 + \mu b_1$ and $-Ta = b_0 + \mu b_2$. Hence $2Ta = \mu(b_1 - b_2)$ and $\|Ta|B\| \leq \mu 2^{(1/p)-1}$ for a $p \in (0, 1]$ such that B is a p Banach space. It follows immediately that if B is a Banach space, then $e_1(T) = \|T\|$. The rest of (i) is obvious.

(iii) Let $\mu > e_k(S)$ and $v > e_\ell(T)$. Then there are points $b_1, ..., b_M$, $b'_1, ..., b'_N \in B$ such that

$$S(U_A) \subset \bigcup_{i=1}^{M}(b_i + \mu U_B), \quad T(U_A) \subset \bigcup_{j=1}^{N}(b'_j + v U_B),$$

where $M \leq 2^{k-1}$ and $N \leq 2^{\ell-1}$. Given any $a \in U_A$, there are points b_i, b'_j such that $Sa \in b_i + \mu U_B$, $Ta \in b'_j + v U_B$. Hence

$$(S + T)a \in b_i + b'_j + \mu U_B + v U_B \subset b_i + b'_j + (\mu^p + v^p)^{1/p} U_B,$$

and thus

$$(S + T)U_A \subset \bigcup_{i=1}^{M}\bigcup_{j=1}^{N}\left\{b_i + b'_j + (\mu^p + v^p)^{1/p} U_B\right\}.$$

The number of points $b_i + b'_j$ with $i \in \{1, ..., M\}$, $j \in \{1, ..., N\}$ is at most $MN \leq 2^{(k+\ell-1)-1}$. The result follows.

(ii) The proof of this is similar to that of (iii), and is left to the reader.

Remark 2 Since the $e_k(T)$ decrease as k increases, and are non-negative, $\lim_{k \to \infty} e_k(T)$ exists and plainly equals

$\inf\{\varepsilon > 0 : T(U_A)$ can be covered by finitely many B-balls of radius $\varepsilon\}$.

Following the terminology used in the Banach space case (see [EEv], II.1, p. 47), we shall refer to this limit as the *ball measure of non-compactness* of T and denote it by $\tilde{\beta}$ (T); plainly $\tilde{\beta}(T) = 0$ if, and only if, T is compact.

Remark 3 Unless $T \in L(A, B)$ is the zero operator, it is clear that for all $k \in \mathbf{N}$, $e_k(T) \neq 0$. The dependence of $e_k(T)$ upon k is, of course, a function of the particular operator T, and we shall devote considerable space to the determination of this dependence for cases of special importance. For the moment we merely note that if A is a (complex) Banach space of dimension $m < \infty$ and id: $A \to A$ is the identity map, then (see [EEv], II.1, p. 49) for all $k \in \mathbf{N}$,

$$1 \leq 2^{(k-1)/2m} e_k(\text{id}) \leq 4.$$

Moreover, if A and B are (real or complex) Banach spaces and $T \in L(A, B)$, $T \neq 0$ has the property that there are positive numbers c and ρ such that for all $k \in \mathbf{N}$, $e_k(T) \leq c 2^{-\rho k}$, then it turns out that T is of finite rank (see [CaS], 1.3, especially pp. 20–21).

Remark 4 Let A be a quasi-Banach space with a non-convex unit ball U_A (ℓ_p, with $0 < p < 1$, is such a space). Then it may happen that there exists $a \in U_A$ such that

$$a + \{b : b = \lambda b_0, \|b_0\| = 1, \lambda \in [-1, 1]\} \subset \mu U_A$$

for some $\mu \in (0, 1)$. In that case it is easy to find an operator $T \in L(A)$ with $e_1(T) < \|T\|$.

The relationship between the entropy numbers of a map $T \in L(A, B)$, A and B being Banach spaces, and those of its adjoint T^* has attracted a good deal of interest (see Carl [Carl3], Edmunds and Tylli [ETy], Gordon, König and Schütt [GKS] and Bourgain et al [BPST]) and is still not completely resolved. However, when A and B are Hilbert spaces the question is settled by the following result:

Theorem *Let $T \in L(H_1, H_2)$, where H_1 and H_2 are Hilbert spaces. Then for all $k \in \mathbf{N}$, $e_k(T) = e_k(T^*) = e_k(|T|)$, where $|T|$ is the positive square root of T.*

Proof By the Polar Decomposition Theorem (see [EEv], IV.3, p. 180) there is a partial isometry $U \in L(H_1, H_2)$ from $(\ker T)^\perp$ to $\overline{\operatorname{im} T}$ such that $T = U|T|$ and $|T| = U^* T$. It follows that for all $k \in \mathbf{N}$,

$$e_k(T) \leq e_k(|T|) \|U\| = e_k(|T|)$$

and

$$e_k(|T|) \leq e_k(T) \|U^*\| = e_k(T).$$

Hence $e_k(T) = e_k(|T|)$. Use of the facts that $T^* = |T| U^*$ and $|T| = T^* U$ (see [EEv], Theorem IV.3.2) leads to $e_k(T^*) = e_k(|T|)$, and the proof is complete.

Remark 5 Outside of Hilbert spaces the results of Bourgain et al [BPST] will be of particular interest to us. To explain these we need to introduce some terminology. Given any bounded sequence $x = \{x_n\}_{n \in \mathbb{N}}$ of scalars, put

$$x_n^* = \inf \{\sigma \geq 0 : \#\{k \in \mathbb{N} : |x_k| \geq \sigma\} < n\}, \qquad n \in \mathbb{N};$$

the sequence $\{x_n^*\}_{n \in \mathbb{N}}$ is called the *non-increasing rearrangement* of x. A Banach space $(X, \|\cdot |X\|)$ consisting of sequences $\{x_n\}_{n \in \mathbb{N}}$ of scalars is called a *symmetric sequence space* if $\|\{x_n\}|X\| = \|\{x_n^*\}|X\|$ for all $\{x_n\} \in X$; and in that case, $\|\cdot |X\|$ is called a *symmetric norm*. Particular examples of symmetric spaces are the Lorentz spaces: given any $p, q \in (1, \infty]$, the Lorentz space $\ell_{p,q}$ is the linear space of all $x = \{x_n\}_{n \in \mathbb{N}} \in \ell_\infty$ such that

$$\|x |\ell_{p,q}\| = \begin{cases} \left(\sum_{n=1}^{\infty} \left(n^{\frac{1}{p} - \frac{1}{q}} x_n^*\right)^q\right)^{1/q} & \text{if } 1 < q < \infty, \\ \sup_{n \in \mathbb{N}} \left(n^{1/p} x_n^*\right) & \text{if } q = \infty, \end{cases}$$

is finite.

The result of [BPST] is as follows:

Let A be a uniformly convex Banach space, let B be a Banach space and let $T \in L(A, B)$ be compact.

(i) There is a positive constant c, depending only on A, such that for all $m \in \mathbb{N}$ and all $p \in [1, \infty)$,

$$c^{-1} \left(\sum_{k=1}^{m} e_k^p(T^*)\right)^{1/p} \leq \left(\sum_{k=1}^{m} e_k^p(T)\right)^{1/p} \leq c \left(\sum_{k=1}^{m} e_k^p(T^*)\right)^{1/p}.$$

The same holds with the ℓ_p norm replaced by any symmetric norm.

(ii) If, for some $k \in \mathbb{N}$,

$$e_k(T) \leq C e_{2k}(T), \quad e_k(T^*) \leq C e_{2k}(T^*),$$

then

$$C_1^{-1} e_k(T^*) \leq e_k(T) \leq C_1 e_k(T^*),$$

where C_1 depends only on A and C.

Note that (i) implies, by taking the $\ell_{p,\infty}$ norm, that if $e_k(T) = O(k^{-1/p})$ as $k \to \infty$, then $e_k(T^*) = O(k^{-1/p})$.

We now turn to approximation numbers.

Definition 2 *Let A, B be quasi-Banach spaces and let $T \in L(A, B)$. Then given any $k \in \mathbf{N}$, the kth approximation number $a_k(T)$ of T is defined by*

$$a_k(T) = \inf \{ \|T - L\| : L \in L(A, B), \operatorname{rank} L < k \},$$

where $\operatorname{rank} L$ is the dimension of the range of L.

These numbers have various properties similar to those of the entropy numbers and given in the following lemma.

Lemma 2 *Let A, B, C be quasi-Banach spaces, let $S, T \in L(A, B)$ and suppose that $R \in L(B, C)$.*

  (i) $\|T\| = a_1(T) \geq a_2(T) \geq \cdots \geq 0.$
  (ii) *For all $k, \ell \in \mathbf{N}$, $a_{k+\ell-1}(R \circ S) \leq a_k(R) a_\ell(S)$.*
  (iii) *If B is a p-Banach space $(0 < p \leq 1)$, then for all $k, \ell \in \mathbf{N}$,*

$$a_{k+\ell-1}^p(S + T) \leq a_k^p(S) + a_\ell^p(T).$$

Proof It is clear that (i) holds. To prove (ii), note that given any $\varepsilon > 0$, there are maps $R_1 \in L(B, C)$ and $S_1 \in L(A, B)$, with $\operatorname{rank} R_1 < k$ and $\operatorname{rank} S_1 < \ell$, such that $\|R - R_1\| < a_k(R) + \varepsilon$ and $\|S - S_1\| < a_\ell(S) + \varepsilon$. Since $\operatorname{rank} \{ R_1(S - S_1) + RS_1 \} < k + \ell - 1$, it follows that

$$a_{k+\ell-1}(R \circ S) \leq \|RS - R_1(S - S_1) - RS_1\| \leq \|R - R_1\| \|S - S_1\|$$
$$\leq \{a_k(R) + \varepsilon\} \{a_\ell(S) + \varepsilon\},$$

and we obtain (ii).

As for (iii), we observe that given any $\varepsilon > 0$, there are maps $S', T' \in L(A, B)$, with $\operatorname{rank} S' < k$ and $\operatorname{rank} T' < \ell$, such that $\|S - S'\| < a_k(S) + \varepsilon$ and $\|T - T'\| < a_\ell(T) + \varepsilon$. Since $\operatorname{rank}(S' + T') < k + \ell - 1$, we have

$$a_{k+\ell-1}^p(S + T) \leq \|S + T - S' - T'\|^p \leq \|S - S'\|^p + \|T - T'\|^p$$
$$< \{a_k(S) + \varepsilon\}^p + \{a_\ell(T) + \varepsilon\}^p,$$

from which (iii) follows.

Remark 6 (i) While the approximation numbers are similar to the entropy numbers in so far as the above properties are concerned, there are nevertheless radical differences between them. For example, if A and B are Banach spaces it can be shown (see [EEv], Prop.II.2.3, p. 54) that :

  (a) if $T \in L(A, B)$, then $a_k(T) = 0$ if, and only if, $\operatorname{rank} T < k$,

(b) if dim$A \geq n$ and id: $A \to A$ is the identity map, then $a_k(\text{id}) = 1$ for $k = 1, ..., n$.

(ii) The approximation numbers behave quite well with respect to the taking of adjoints. Thus if A and B are Banach spaces and $T \in L(A, B)$ is compact, then $a_k(T) = a_k(T^*)$ for all $k \in \mathbf{N}$, while if the compactness requirement is dropped, it can be shown that for all $k \in \mathbf{N}$, $a_k(T^*) \leq a_k(T) \leq 5a_k(T^*)$; see [EEv], II.2, pp. 55–6 for these results. Comparable results are not known, in general, for entropy numbers, but see Remark 5.

(iii) Let $T \in L(A, B)$, where A and B are Banach spaces. Plainly $\alpha(T) = \lim\limits_{k \to \infty} a_k(T)$ exists. If we put

$$\|T\|_{\mathscr{K}} = \text{dist}\,[T, \mathscr{K}(A, B)] = \inf\{\|T - K\| : K \in \mathscr{K}(A, B)\},$$

where $\mathscr{K}(A, B)$ is the space of all compact linear maps from A to B, then (see [EEv], Prop. II.2.7, p. 57),

$$\tilde{\beta}(T) \leq \|T\|_{\mathscr{K}} \leq \alpha(T),$$

where $\tilde{\beta}(T)$ is given by Remark 2 above. Since there are Banach spaces A and B, and a map $T \in \mathscr{K}(A, B)$, such that T cannot be approximated arbitrarily closely by elements of $L(A, B)$ with finite rank, it may happen that $\|T\|_{\mathscr{K}} < \alpha(T)$ (observe that $\alpha(T) = \text{dist}[T, \mathscr{E}(X, Y)]$, where $\mathscr{E}(X, Y) = \{T \in L(X, Y) : \dim T(A) < \infty\}$). This cannot happen if A and B are Hilbert spaces when, as is well known, $\alpha(T) = \|T\|_{\mathscr{K}}$.

(iv) Let A, B be quasi-Banach spaces and let $T \in L(A, B)$ be compact. It is natural to ask whether $e_k(T)$ and $a_k(T)$ can be compared: in particular, is it the case that there is a constant $C > 0$ such that for all $k \in \mathbf{N}$,

$$e_k(T) \leq Ca_k(T)? \tag{1}$$

Sometimes this holds in a trivial way: as indicated in (iii), if B does not have the approximation property, then there exists T such that $a_k(T) \to \alpha(T) > 0$ while $e_k(T) \to 0$ as $k \to \infty$. However, in the absence of some additional restrictions, (1) is false. For example, we have already seen in Remark 3 that if $T \neq 0$ and for some positive c and ρ, $e_k(T) \leq c2^{-\rho k}$ for all $k \in \mathbf{N}$, then T is of finite rank and so $a_k(T) = 0$ for all large enough k. Since $e_k(T)$ is never zero, (1) cannot hold. Moreover, in [CaS], p. 106 we find an example of a diagonal map $D : \ell_p \to \ell_p$ for which $a_k(D) = 2^{-k}$ while

$$\frac{1}{2}2^{-\sqrt{2k}} \leq e_{k+1}(D) \leq 3\sqrt{2}\,2^{-\sqrt{2k}}$$

for all $k \in \mathbf{N}$: again (1) cannot hold. We shall return later, in 1.3.3, to the question of conditions sufficient to ensure that (1) holds.

1.3.2 Interpolation properties of entropy numbers

Here we establish the interpolation properties of the entropy numbers of a bounded linear map acting between quasi-Banach spaces, one end point being fixed. In a Banach space setting these results, expressed in terms of entropy functions and entropy ideals, were developed by Peetre and Triebel (see [Triα], 1.16.2); their reformulation in terms of entropy numbers may be found in [Pi1], 12.1. The proof which we give is that presented in [HaT1]: as pointed out in that paper, it is essentially the same as that already available for the Banach space case, but it must be given in full because accurate knowledge of the constants involved will be important later (see 4.3). We take for granted standard ideas and notation from real interpolation theory (see [Triα], Ch.1). In particular, if $\{A_0, A_1\}$ is an interpolation couple of quasi-Banach spaces (that is, both A_0 and A_1 are linearly and continuously embedded in a common quasi-Banach space A) then Peetre's K-functional is defined for all $a \in A_0 + A_1$ and all $t \in (0, \infty)$ by

$$K(t, a) = K(t, a, A_0, A_1)$$
$$= \inf \{ \|a_0 | A_0 \| + t \|a_1 | A_1 \| : a = a_0 + a_1, \; a_0 \in A_0, \; a_1 \in A_1 \}. \quad (1)$$

Given any quasi-Banach spaces A, B and any $T \in L(A, B)$, we shall for extra clarity in the theorem below denote the kth entropy number of T by $e_k(T : A \to B)$.

Theorem *(i) Let A be a quasi-Banach space and let $\{B_0, B_1\}$ be an interpolation couple of p-Banach spaces. Let $0 < \theta < 1$ and let B_θ be a quasi-Banach space such that $B_0 \cap B_1 \subset B_\theta \subset B_0 + B_1$ and*

$$\|b | B_\theta \| \leq \|b | B_0 \|^{1-\theta} \|b | B_1 \|^{\theta} \text{ for all } b \in B_0 \cap B_1. \quad (2)$$

Let $T \in L(A, B_0 \cap B_1)$, where $B_0 \cap B_1$ is given the quasi-norm $\max\left(\|b | B_0 \|, \|b | B_1 \|\right)$. Then for all $k_0, k_1 \in \mathbf{N}$,

$$e_{k_0+k_1-1}(T : A \to B_\theta) \leq 2^{1/p} e_{k_0}^{1-\theta}(T : A \to B_0) e_{k_1}^{\theta}(T : A \to B_1). \quad (3)$$

(ii) Let $\{A_0, A_1\}$ be an interpolation couple of quasi-Banach spaces and let B be a p-Banach space. Let $0 < \theta < 1$ and let A be a quasi-Banach

space such that $A \subset A_0 + A_1$ and

$$t^{-\theta}K(t,a) \leq \|a\,|A\,\|\,for\ all\ a \in A\ and\ all\ t \in (0,\infty). \tag{4}$$

Let $T : A_0 + A_1 \to B$ be linear and such that its restrictions to A_0 and A_1 are continuous. Then its restriction to A is also continuous and for all $k_0, k_1 \in \mathbf{N}$,

$$e_{k_0+k_1-1}(T : A \to B) \leq 2^{1/p}e_{k_0}^{1-\theta}(T : A_0 \to B)e_{k_1}^{\theta}(T : A_1 \to B). \tag{5}$$

Proof (i) Given any $\varepsilon > 0$, $T(U_A)$ can be covered by 2^{k_0-1} B_0-balls K_j of radius $(1 + \varepsilon)e_{k_0}(T : A \to B_0)$. Each of the sets $K_j \cap T(U_A)$ can be covered in B_1 by 2^{k_1-1} balls of radius $(1 + \varepsilon)2^{1/p}e_{k_1}(T : B \to B_1)$ with centres in $K_j \cap T(U_A)$: the factor $2^{1/p}$ which appears here arises from the requirement that the centres should lie in $K_j \cap T(U_A)$ and from the form of the triangle inequality in p-Banach spaces. There are $2^{k_0+k_1-2}$ such centres b_ℓ. Given any $a \in U_A$, there exists one such centre b_ℓ with

$$\|Ta - b_\ell\,|B_i\,\|^p \leq 2(1 + \varepsilon)^p e_{k_i}^p(T : A \to B_i), \qquad i = 0, 1.$$

Together with (2), this establishes (i).

(ii) Let $a \in U_A$ and $\varepsilon, t > 0$; choose $a_0 \in A_0$, $a_1 \in A_1$ such that $a = a_0 + a_1$ and

$$\|a_0\,|A_0\,\| + t\,\|a_1\,|A_1\,\| \leq (1 + \varepsilon)K(t,a) \leq (1 + \varepsilon)t^{\theta};$$

the final inequality follows from (4). Hence

$$\|a_0\,|A_0\,\| \leq (1 + \varepsilon)t^{\theta}, \quad \|a_1\,|A_1\,\| \leq (1 + \varepsilon)t^{\theta-1}.$$

Put $t = e_{k_1}(T : A_1 \to B)e_{k_0}^{-1}(T : A_0 \to B)$ (there is no loss in generality in supposing that $e_{k_0}(T : A_0 \to B) \neq 0$) and choose 2^{k_0-1} elements $b_\ell \in B$ and 2^{k_1-1} elements $b^m \in B$ such that $(1 + \varepsilon)t^{\theta}T(U_{A_0})$ can be covered by 2^{k_0-1} balls of radius

$$(1+\varepsilon)^2 t^{\theta} e_{k_0}(T : A_0 \to B) = (1+\varepsilon)^2 e_{k_0}^{1-\theta}(T : A_0 \to B)e_{k_1}^{\theta}(T : A_1 \to B) \tag{6}$$

centred at b_ℓ, and $(1 + \varepsilon)t^{\theta-1}T(U_{A_1})$ can be covered by 2^{k_1-1} balls of radius

$$(1 + \varepsilon)^2 t^{\theta-1}e_{k_1}(T : A_1 \to B) = (1 + \varepsilon)^2 e_{k_0}^{1-\theta}(T : A_0 \to B)e_{k_1}^{\theta}(T : A_1 \to B) \tag{7}$$

centred at b^m. Given any $a \in A$, the result follows from

$$\|Ta - b_\ell - b^m\,|B\,\|^p \leq \|Ta_0 - b_\ell\,|B\,\|^p + \|Ta_1 - b^m\,|B\,\|^p,$$

for an optimal choice of $a = a_0 + a_1$, b_ℓ and b^m, together with (6) and (7).

1.3.3 Relationships between entropy and approximation numbers

We have already seen, in Remark 1.3.1/6(iv), that if T is a compact linear map from a quasi-Banach space A to another such space B, then in general no inequality of the form $e_k(T) \le Ca_k(T)$, $k \in \mathbf{N}$, can hold. Our concern here, motivated by concrete examples and the wish to shed fresh light on the abstract background, is to find conditions sufficient to ensure that such an inequality holds, or some 'global' version of it. The following result, due to Triebel [Tri5], does just this.

Theorem *Let A and B be quasi-Banach spaces and let $T \in L(A, B)$ be compact.*

(i) Suppose that for some $c > 0$,

$$a_{2^{j-1}}(T) \le ca_{2^j}(T) \text{ for all } j \in \mathbf{N}. \tag{1}$$

Then there is a positive constant C such that for all $j \in \mathbf{N}$,

$$e_j(T) \le Ca_j(T). \tag{2}$$

(ii) Let $f : \mathbf{N} \to \mathbf{R}$ be positive and increasing, and suppose that for some $c > 0$,

$$f(2^j) \le cf(2^{j-1}) \text{ for all } j \in \mathbf{N}. \tag{3}$$

Then there is a positive constant C such that for all $n \in \mathbf{N}$,

$$\sup_{1 \le j \le n} f(j)e_j(T) \le C \sup_{1 \le j \le n} f(j)a_j(T). \tag{4}$$

Proof Since B is a quasi-Banach space, there is a constant $K \ge 1$ such that for all $u, v \in B$,

$$\|u + v \,|B\, \| \le K \left(\|u \,|B\, \| + \|v \,|B\, \| \right). \tag{5}$$

Iteration of (5) shows that there exists $\lambda \ge 0$ such that for all $k \in \mathbf{N}$; and all $y_1, ..., y_k \subset B$,

$$\left\| \sum_{j=1}^{k} y_j \,|B \right\| \le \sum_{j=1}^{k} 2^{\lambda(k+1-j)} \|y_j \,|B\, \|. \tag{6}$$

For each $j \in \mathbb{N}$ let $L_j \in L(A, B)$ be such that

$$\text{rank} L_j < 2^{j-1} \text{ and } \|T - L_j\| \le 2a_{2^{j-1}}(T), \quad j \ge 2, \quad L_1 = 0. \quad (7)$$

Thus for some $c > 0$,

$$\text{rank}\left(L_{j+1} - L_j\right) < 2^{j+1}, \quad \|L_{j+1} - L_j\| \le c_1 a_{2^{j-1}}(T). \quad (8)$$

Put $m_j = \mu 2^j (k - j)$ for some $\mu > 0$ to be chosen later, and note that

$$\sum_{j=1}^{k-1} m_j = \mu 2^k \sum_{j=1}^{k-1} 2^{-(k-j)}(k - j) \le \mu c_2 2^k \quad (9)$$

for some $c_2 > 0$. The representation of T as

$$T = T - L_k + \sum_{j=1}^{k-1}\left(L_{j+1} - L_j\right)$$

now shows, together with (6) and (7), that for all $x \in U_A$ and $y_j \in B$, $j = 1, ..., k - 1$,

$$\left\| Tx - \sum_{j=1}^{k-1} y_j \,|B \right\| \le 2^\lambda \,\|(T - L_k)x\,|B\| + \sum_{j=1}^{k-1} 2^{\lambda(k+1-j)} \,\|(L_{j+1} - L_j)x - y_j \,|B\,\|$$

$$\le 2^{\lambda+1} a_{2^{k-1}}(T) + \sum_{j=1}^{k-1} 2^{\lambda(k+1-j)} \,\|(L_{j+1} - L_j)x - y_j \,|B\,\|. \quad (10)$$

For each $j \in \{1, ..., k - 1\}$ let ε_{m_j} be the infimum of all $\varepsilon > 0$ such that there are 2^{m_j-1} balls in B of radius ε which cover the unit ball in the image of $L_{j+1} - L_j$. It follows that for some positive constants c_3, c_4,

$$\varepsilon_{m_j} \le c_3 2^{-c_4 m_j/2^j} = c_3 2^{-c_4 \mu(k-j)}; \quad (11)$$

see [CaS], p. 21 and [EEv], Proposition II.1.3, p. 48 for details related to the Banach space case, there being an immediate counterpart for quasi-Banach spaces. Now (10), together with (8) and (11), gives

$$e_{\mu c_2 2^k}(T) \le 2^{\lambda+1} a_{2^{k-1}}(T) + c_3 c_1 \sum_{j=1}^{k} 2^{\lambda(k+1-j)} a_{2^{j-1}}(T) 2^{-c_4 \mu(k-j)}, \quad (12)$$

where $\mu > 0$ is at our disposal; c_1, c_3, c_4 are independent of μ.

We can now prove part (i) of the theorem. By (1),

$$a_{2^{j-1}}(T) \le 2^{\kappa(k-j)} a_{2^{k-1}}(T), \quad j = 1, ..., k, \quad (13)$$

for some $\kappa > 0$. Insertion of (13) in (12) and choice of a large enough μ now give

$$e_{C2^k}(T) \leq c' a_{2^{k-1}}(T), \qquad k \in \mathbf{N}, \tag{14}$$

for some $c' > 0$. We may assume that $C = 2^n$ for some $n \in \mathbf{N}$; thus by (1) and (14),

$$e_{2^{k+n}}(T) \leq c'' a_{2^{k+n}}(T), \tag{15}$$

from which the result follows, using (1) and the monotonicity properties of the a_j and e_j.

To prove (ii), let $b = \sup \left\{ f(j) a_j(T) : j = 1, ..., 2^{k+n} \right\}$. Then

$$a_j(T) \leq b/f(j), \; j = 1, ..., 2^{k+n}. \tag{16}$$

We may now replace the $a_{2^{j-1}}(T)$ in the right-hand side of (12) by $b/f(2^{j-1})$; (3) is now the direct analogue of (1), and so by the respective counterparts of (13)–(15) we obtain

$$e_{2^{k+n}}(T) \leq Cbf(2^{k+n})^{-1} \text{ for all } k \in \mathbf{N}; \tag{17}$$

where $C(> 0)$ is independent of b. Together with the monotonicity properties of f and $e_j(T)$, plus the fact that $e_1(T) \leq \|T\| = a_1(T)$, this gives (4).

Remark 1 Arguments similar to those used in the proof will be found in [CaS], pp. 96–100, for the case of Banach spaces.

Remark 2 Given any $\rho > 0$, the functions f_1 and f_2 given by

$$f_1(j) = j^\rho, \; f_2(j) = (\log(j+1))^\rho, \quad j \in \mathbf{N},$$

both satisfy (3). Hence we have for all $n \in \mathbf{N}$,

$$\sup_{1 \leq j \leq n} j^\rho e_j(T) \leq C \sup_{1 \leq j \leq n} j^\rho a_j(T) \tag{18}$$

and

$$\sup_{1 \leq j \leq n} (\log(j+1))^\rho e_j(T) \leq C \sup_{1 \leq j \leq n} (\log(j+1))^\rho a_j(T). \tag{19}$$

In Banach spaces, (18) coincides with Carl's inequality: see [Carl1], p. 294 and [CaS], p. 96. Our interest in the generalisation (4), and in particular the case (19), stems from the fact that, as we shall see in Chapter 3, the approximation numbers of embeddings between function spaces typically have power-type or logarithmic-type decay.

Remark 3 Even if (1) holds it may happen that the behaviour of a_j and e_j is completely different. Examples may be found in 3.4 and 4.3; see especially Remark 4.3.3/2.

1.3.4 Connections with eigenvalues

We begin with a (complex) quasi-Banach space A and a compact map $T \in L(A)$. We know from Theorem 1.2 that the spectrum of T, apart from the point 0, consists solely of eigenvalues of finite algebraic multiplicity: let $\{\lambda_k(T)\}_{k \in \mathbf{N}}$ be the sequence of all non-zero eigenvalues of T, repeated according to algebraic multiplicity and ordered so that

$$|\lambda_1(T)| \geq |\lambda_2(T)| \geq \cdots \geq 0.$$

If T has only m ($< \infty$) distinct eigenvalues and M is the sum of their algebraic multiplicities, we put $\lambda_n(T) = 0$ for all $n > M$.

Perhaps the most useful connection for our purposes between the eigenvalues of T and its entropy numbers is the following extension to quasi-Banach spaces of a result established by Carl and Triebel [CaT] when A is a Banach space.

Theorem *Let T and $\{\lambda_k(T)\}_{k \in \mathbf{N}}$ be as above. Then*

$$\left(\prod_{m=1}^{k} |\lambda_m(T)| \right)^{1/k} \leq \inf_{n \in \mathbf{N}} 2^{n/(2k)} e_n(T), \quad k \in \mathbf{N}. \tag{1}$$

Proof Let λ be a non-zero eigenvalue of T and let $a \in A$ be an (associated) eigenvector with

$$(T - \lambda \mathrm{id})^{k-1} a \neq 0, \; (T - \lambda \mathrm{id})^k a = 0 \text{ for some } k \in \mathbf{N}. \tag{2}$$

Then the elements $a_1, ..., a_k$ given by

$$a_j = (T - \lambda \mathrm{id})^{j-1} a, \quad j = 1, ..., k, \tag{3}$$

are linearly independent and $\Lambda = \mathrm{span}\,\{a_1, ..., a_k\}$ is an invariant subspace of T with $T(\Lambda) = \Lambda$ since $\lambda \neq 0$. We have

$$a_{j+1} = (T - \lambda \mathrm{id}) a_j, \quad j = 1, ..., k - 1,$$

and hence

$$Ta_j = a_{j+1} + \lambda a_j, \; j = 1, ..., k - 1, \quad Ta_k = \lambda a_k. \tag{4}$$

Let $a_0 \in \Lambda$, with $a_0 = \sum\limits_{j=1}^{k} \rho_j a_j$ for some $\rho_j \in \mathbf{C}$. Then

$$
\begin{aligned}
T a_0 &= \sum_{j=1}^{k} \rho_j T a_j = \sum_{j=1}^{k-1} \rho_j \left(a_{j+1} + \lambda a_j \right) + \lambda \rho_k a_k \\
&= \rho_1 \lambda a_1 + \sum_{j=2}^{k} \left(\rho_{j-1} + \rho_j \lambda \right) a_j.
\end{aligned}
$$

Let Π be the related matrix, so that

$$
\Pi \begin{pmatrix} \rho_1 \\ \cdot \\ \cdot \\ \cdot \\ \rho_k \end{pmatrix} = \begin{pmatrix} \lambda & & & \\ 1 & \lambda & & \\ & 1 & \cdot & \\ & & \cdot & \cdot \\ 0 & & & \cdot \\ & & & 1 & \lambda \end{pmatrix} \begin{pmatrix} \rho_1 \\ \cdot \\ \cdot \\ \cdot \\ \rho_k \end{pmatrix}. \tag{5}
$$

We interpret the k-dimensional complex subspace Λ as $\mathbf{R}^{2k}$, equipped with Lebesgue measure. Decompose λ and the ρ_j into their real and imaginary parts and let Π' be the corresponding matrix. Then

$$
\Pi' \begin{pmatrix} \mathrm{Re}\rho_1 \\ \mathrm{Im}\rho_1 \\ \mathrm{Re}\rho_2 \\ \mathrm{Im}\rho_2 \\ \cdot \\ \cdot \\ \cdot \\ \cdot \\ \mathrm{Im}\rho_k \end{pmatrix} = \begin{pmatrix} \boxed{\begin{array}{cc} \mathrm{Re}\lambda & -\mathrm{Im}\lambda \\ \mathrm{Im}\lambda & \mathrm{Re}\lambda \end{array}} & & & \\ \begin{array}{cc} 1 & 0 \\ 0 & 1 \end{array} & \Box & 0 & \\ & & & \\ 0 & & & \end{pmatrix} \begin{pmatrix} \mathrm{Re}\rho_1 \\ \mathrm{Im}\rho_1 \\ \mathrm{Re}\rho_2 \\ \mathrm{Im}\rho_2 \\ \cdot \\ \cdot \\ \cdot \\ \cdot \\ \mathrm{Im}\rho_k \end{pmatrix}. \tag{6}
$$

Let $M \subset \Lambda$, with Λ interpreted as $\mathbf{R}^{2k}$. Then

$$
\mathrm{vol}\, T(M) = |\lambda|^{2k}\, \mathrm{vol}\, M. \tag{7}
$$

Now let $|\lambda_1(T)| \geq |\lambda_2(T)| \geq \cdots \geq |\lambda_k(T)| > 0$. Then the counterparts of (5) and (6) have a block structure with (5) and (6) as blocks. By the same argument as above we have

$$
\mathrm{vol}\, T(M) = \left(\prod_{j=1}^{k} |\lambda_j(T)| \right)^2 \mathrm{vol}\, M \tag{8}
$$

for any M contained in the span of the (associated) eigenvectors corresponding to $\lambda_1, ..., \lambda_k$: denote this span again by Λ, for simplicity, and note that it is an invariant subspace of T.

Suppose that $T(U_A)$ is covered by 2^{n-1} balls of radius $(1 + \varepsilon)e_n(T)$, for some $\varepsilon > 0$. Let B be one of these balls with $B \cap \Lambda \neq \emptyset$ and let $a \in B \cap \Lambda$. Then B is contained in a ball centred at a and with radius $ce_n(T)$, where $c > 0$ depends upon the geometry of A, that is, upon the constant appearing in the triangle inequality for A. Hence all these balls cover $T(U_A) \cap \Lambda$. By (8),

$$\prod_{j=1}^{k} |\lambda_j(T)|^2 \operatorname{vol}(U_A \cap \Lambda) = \operatorname{vol}(T(U_A) \cap \Lambda)$$
$$\leq 2^{n-1}e_n^{2k}(T)c^{2k}\operatorname{vol}(U_A \cap \Lambda). \tag{9}$$

Thus

$$\left(\prod_{j=1}^{k} |\lambda_j(T)|\right)^{1/k} \leq c2^{(n-1)/(2k)}e_n(T) \leq c2^{n/(2k)}e_n(T). \tag{10}$$

It remains to prove that (10) holds with c replaced by 1. To do this, apply (10) to T^m instead of T for each $m \in \mathbf{N}$. Note that $\lambda_j(T^m) = \lambda_j^m(T)$, and that (by induction) $e_{mn}(T^m) \leq e_n^m(T)$. Now (10), applied to T^m and with $e_{mn}(T^m)$ in place of $e_n(T)$, gives

$$\left(\prod_{j=1}^{k} |\lambda_j(T)|\right)^{m/k} \leq c2^{mn/(2k)}e_n^m(T),$$

and hence

$$\left(\prod_{j=1}^{k} |\lambda_j(T)|\right)^{1/k} \leq c^{1/m}2^{n/(2k)}e_n(T).$$

As this is true for all $m \in \mathbf{N}$, we may let $m \to \infty$ and obtain the desired estimate.

Corollary *For all $k \in \mathbf{N}$,*

$$|\lambda_k(T)| \leq \sqrt{2}e_k(T). \tag{11}$$

Proof This follows immediately from (1) on setting $k = n$.

Remark 1 The inequality (11) was originally proved by Carl [Carl1] in the context of Banach spaces. Using it, Zemánek [Zem] (see also [EEv], Corollary II.1.7, p. 52) was able to prove that when A is a Banach space,

$$\lim_{m \to \infty} (e_k(T^m))^{1/m} = r(T), \qquad k \in \mathbf{N},$$

where $r(T)$ is the spectral radius of T.

The approximation numbers also have important connections with eigenvalues, the picture being clearest in a Hilbert space setting. Thus if H is a complex Hilbert space and $T \in L(H)$ is compact, with non-zero eigenvalues denoted as above by $\{\lambda_k(T)\}$, then T^*T has a non-negative, self-adjoint, compact square root $|T|$, and for all $k \in \mathbf{N}$,

$$a_k(T) = \lambda_k (|T|)$$

(see [EEv], Theorem II.5.10, p. 91). (Hence if in addition T is non-negative and self-adjoint, then the approximation numbers of T coincide with its eigenvalues.) Moreover, a famous inequality of Weyl (see [Kön], Theorem 1.b.5, p. 31) states that for all $n \in \mathbf{N}$,

$$\prod_{j=1}^{n} |\lambda_j(T)| \le \prod_{j=1}^{n} a_j(T),$$

from which it follows that for all $n \in \mathbf{N}$ and all $p \in (0, \infty)$,

$$\sum_{j=1}^{n} |\lambda_j(T)|^p \le \sum_{j=1}^{n} a_j^p(T).$$

Outside Hilbert spaces the results are naturally less good but are nevertheless extremely interesting. Thus let A be a complex Banach space and let $T \in L(A)$ be compact, with eigenvalues $\lambda_k(T)$ as usual. Then König has shown (see [Kön], Proposition 2.d.6, p. 134) that for all $n \in \mathbf{N}$,

$$|\lambda_n(T)| = \lim_{k \to \infty} a_n^{1/k}(T^k).$$

He has also proved that for all $N \in \mathbf{N}$ and all $p \in (0, \infty)$,

$$\left(\sum_{k=1}^{N} |\lambda_k(T)|^p \right)^{1/p} \le K_p \left(\sum_{k=1}^{N} a_k^p(T) \right)^{1/p},$$

where

$$K_p = \begin{cases} 2e / \sqrt{p} & \text{if } 0 < p < 1, \\ 2^{1/p} \sqrt{2e} & \text{if } 1 \le p < \infty. \end{cases}$$

Moreover, given any $p, q \in (0, \infty)$, there is a constant c, depending only on p and q, such that

$$\|(\lambda_n(T)) \,|\ell_{p,q}\| \leq c \,\|(a_n(T)) \,|\ell_{p,q}\|$$

where $\|\cdot\,|\ell_{p,q}\|$ is the usual quasi-norm on the Lorentz sequence space $\ell_{p,q}$, defined for $p, q \in (0, \infty)$ just as in Remark 1.3.1/5 for $p, q \geq 1$. Thus, in particular, if for some $\rho > 0$, $a_n(T) = O(n^{-\rho})$ (or $o(n^{-\rho})$) as $n \to \infty$, then $|\lambda_n(T)| = O(n^{-\rho})$ (or $o(n^{-\rho})$) as $n \to \infty$. For details of these results, and many others, we refer to [CaS], [EEv], [Kön], [Pi2].

2

Function Spaces

2.1 Introduction

This chapter deals mainly with the function spaces B_{pq}^s and F_{pq}^s on $\mathbf{R}^n$ and on domains. These spaces have been treated in detail in [Triβ] and [Triγ] and this chapter might be considered as a continuation of these two books. On the other hand, with our later applications in mind, we are now especially interested in limiting situations. For that purpose we complement the above-mentioned spaces by spaces of Orlicz type, in particular $L_p(\log L)_a$, and logarithmic Sobolev spaces. Our approach is based on a new characterisation of $L_p(\log L)_a$ in terms of L_p spaces. This both is of intrinsic interest and paves the way for the natural introduction of logarithmic Sobolev spaces.

Our aim is threefold. Firstly, to make this and the following chapters self-contained we give all the necessary definitions, but hardly mention those properties which are proved in [Triβ] or [Triγ]. If we need assertions of that type we give precise references. On the other hand, we rely in what follows on some very recent results about the function spaces above which we developed, at least partly, in close connection with the subjects treated in the following chapters. Hence, secondly, we discuss and establish these properties in detail; this forms the main bulk of the chapter. Thirdly, we have occasionally yielded to the temptation to report on selected recent results which provide a better understanding of what is going on. Of course, these are related to our subject, but do not enter the main stream of our arguments. In these cases we do not give proofs, but simply refer to the original papers.

2.2 The spaces B_{pq}^s and F_{pq}^s on $\mathbf{R}^n$

2.2.1 Definitions

Let $\mathbf{R}^n$ be Euclidean n-space and let $\mathbf{N}$ be the collection of all natural numbers. Let $\mathscr{S}(\mathbf{R}^n)$ be the Schwartz space of all complex-valued, rapidly decreasing, infinitely differentiable functions on $\mathbf{R}^n$. By $\mathscr{S}'(\mathbf{R}^n)$ we denote its topological dual, the space of all tempered distributions on $\mathbf{R}^n$. Furthermore, $L_p(\mathbf{R}^n)$, with $0 < p \leq \infty$, is the usual quasi-Banach space with respect to Lebesgue measure, quasi-normed by

$$\|f \mid L_p(\mathbf{R}^n)\| = \left(\int_{\mathbf{R}^n} |f(x)|^p \, dx \right)^{1/p} \tag{1}$$

with the natural modification if $p = \infty$. If $\varphi \in \mathscr{S}(\mathbf{R}^n)$ then

$$\overset{\wedge}{\varphi}(\xi) = (F\varphi)(\xi) = (2\pi)^{-\frac{n}{2}} \int_{\mathbf{R}^n} e^{-i\xi \cdot x} \varphi(x) \, dx, \ \xi \in \mathbf{R}^n, \tag{2}$$

denotes the Fourier transform of φ. As usual, $F^{-1}\varphi$ or $\overset{\vee}{\varphi}$ stands for the inverse Fourier transform, given by the right-hand side of (2) with i instead of $-i$. Whether we use $Ff, F^{-1}f$ or $\hat{f}, \overset{\vee}{f}$ will be a matter of convenience. Of course, $\xi \cdot x$ or ξx denotes the scalar product in $\mathbf{R}^n$. Both F and F^{-1} are extended to $\mathscr{S}'(\mathbf{R}^n)$ in the standard way. Let $\varphi \in \mathscr{S}(\mathbf{R}^n)$ with

$$\varphi(x) = 1 \text{ if } |x| \leq 1 \text{ and } \varphi(x) = 0 \text{ if } |x| \geq \frac{3}{2}. \tag{3}$$

We put $\varphi_{0=}\varphi$, $\varphi_1(x) = \varphi(x/2) - \varphi(x)$ and

$$\varphi_k(x) = \varphi_1\left(2^{-k+1}x\right), \ x \in \mathbf{R}^n, k \in \mathbf{N}. \tag{4}$$

Then since

$$1 = \sum_{k=0}^{\infty} \varphi_k(x) \text{ for all } x \in \mathbf{R}^n, \tag{5}$$

the φ_k form a dyadic resolution of unity. Recall that $\left(\varphi_k \hat{f}\right)^{\vee}$ is an entire analytic function on $\mathbf{R}^n$ for any $f \in \mathscr{S}'(\mathbf{R}^n)$. In particular, $\left(\varphi_k \hat{f}\right)^{\vee}(x)$ makes sense pointwise. Put $\mathbf{R} = \mathbf{R}^1$.

Definition (i) *Let $s \in \mathbf{R}$; $p, q \in (0, \infty]$. Then $B_{pq}^s(\mathbf{R}^n)$ is the collection of* *all $f \in \mathcal{S}'(\mathbf{R}^n)$ such that*

$$\|f \mid B_{pq}^s(\mathbf{R}^n)\|_\varphi = (\sum_{j=0}^{\infty} 2^{jsq} \left\| (\varphi_j f^\wedge)^\vee \mid L_p(\mathbf{R}^n) \right\|^q)^{\frac{1}{q}} \tag{6}$$

(with the usual modification if $q = \infty$) is finite.

(ii) Let $s \in \mathbf{R}, 0 < p < \infty$ and $0 < q \le \infty$. Then $F_{pq}^s(\mathbf{R}^n)$ is the collection of all $f \in \mathcal{S}'(\mathbf{R}^n)$ such that

$$\|f \mid F_{pq}^s(\mathbf{R}^n)\|_\varphi = \left\| \left(\sum_{j=0}^{\infty} 2^{jsq} \mid (\varphi_j f^\wedge)^\vee (\cdot) \mid^q \right)^{\frac{1}{q}} \mid L_p(\mathbf{R}^n) \right\| \tag{7}$$

(with the usual modification if $q = \infty$) is finite.

Remark These spaces, including their forerunners and special cases, have a long history, see also the following subsection. The theory of these spaces in its full extent has been developed in [Triβ] and [Triγ]. In particular, both $B_{pq}^s(\mathbf{R}^n)$ and $F_{pq}^s(\mathbf{R}^n)$ are quasi-Banach spaces which are independent of the function φ chosen according to (3), in the sense of equivalent quasi-norms. This justifies our omission of the subscript φ in (6) and (7) in what follows. If $p \ge 1$ and $q \ge 1$ then both $B_{pq}^s(\mathbf{R}^n)$ and $F_{pq}^s(\mathbf{R}^n)$ are Banach spaces. For sake of clarity we shall occasionally write $B_{p,q}^s(\mathbf{R}^n)$ or $F_{p,q}^s(\mathbf{R}^n)$ instead of $B_{pq}^s(\mathbf{R}^n)$ or $F_{pq}^s(\mathbf{R}^n)$, especially when p and q are concrete numbers; for example we prefer $B_{2,1}^s(\mathbf{R}^n)$ compared with $B_{21}^s(\mathbf{R}^n)$ which looks somewhat ambiguous.

2.2.2 Concrete spaces

We list some special cases of the above spaces $B_{pq}^s(\mathbf{R}^n)$ and $F_{pq}^s(\mathbf{R}^n)$ without further comments. Details may be found in [Triβ], especially 2.2.2, p. 35, and [Triγ], especially Ch. 1.

(i) Let $1 < p < \infty$; then

$$F_{p,2}^0(\mathbf{R}^n) = L_p(\mathbf{R}^n). \tag{1}$$

This is a Paley–Littlewood theorem, see [Triβ], 2.5.6, p. 87.

(ii) Let $1 < p < \infty$ and $s \in \mathbf{N}_0 = \mathbf{N} \cup \{0\}$; then

$$F_{p,2}^s(\mathbf{R}^n) = W_p^s(\mathbf{R}^n) \tag{2}$$

are the classical Sobolev spaces usually normed by

$$\|f \mid W_p^s(\mathbf{R}^n)\| = \sum_{|\alpha| \le s} \|D^\alpha f \mid L_p(\mathbf{R}^n)\|. \tag{3}$$

This generalises assertion (i), see again [Triβ], 2.5.6, p. 87.

(iii) Let $\sigma \in \mathbf{R}$; then

$$I_\sigma : f \mapsto \left(\langle \xi \rangle^\sigma \hat{f} \right)^\vee , \tag{4}$$

with $\langle \xi \rangle = \left(1 + |\xi|^2 \right)^{\frac{1}{2}}$, is a one-to-one map of $\mathscr{S}(\mathbf{R}^n)$ onto itself and from $\mathscr{S}'(\mathbf{R}^n)$ onto itself. As for the spaces $B^s_{pq}(\mathbf{R}^n)$ and $F^s_{pq}(\mathbf{R}^n)$ with $s \in \mathbf{R}, 0 < p \le \infty$ ($p < \infty$ for the F scale), $0 < q \le \infty$, I_σ acts as a lift (equivalent quasi-norms):

$$I_\sigma B^s_{pq}(\mathbf{R}^n) = B^{s-\sigma}_{pq}(\mathbf{R}^n) \text{ and } I_\sigma F^s_{pq}(\mathbf{R}^n) = F^{s-\sigma}_{pq}(\mathbf{R}^n) ; \tag{5}$$

see [Triβ], 2.3.8, p. 58. In particular, let

$$H^s_p(\mathbf{R}^n) = I_{-s} L_p(\mathbf{R}^n), \, s \in \mathbf{R}, \, 1 < p < \infty, \tag{6}$$

be the (fractional) Sobolev spaces. Then (1), (2) and (5) yield

$$H^s_p(\mathbf{R}^n) = F^s_{p,2}(\mathbf{R}^n), \, s \in \mathbf{R}, \, 1 < p < \infty, \tag{7}$$

and

$$H^s_p(\mathbf{R}^n) = W^s_p(\mathbf{R}^n) \text{ if } s \in \mathbf{N}_0 \text{ and } 1 < p < \infty. \tag{8}$$

As for notation we follow the recent custom and refer to the members of the full scale $H^s_p(\mathbf{R}^n)$ given by (6) as Sobolev spaces (Sobolev in person would presumably disagree, but the other candidates for names of these spaces, such as Lebesgue spaces, Liouville spaces, Bessel-potential spaces or fractional Sobolev spaces, all have their own shortcomings); see also (2.3.3/11) for a further extension of (7).

(iv) Denote

$$\mathscr{C}^s(\mathbf{R}^n) = B^s_{\infty,\infty}(\mathbf{R}^n), \, s \in \mathbf{R}, \tag{9}$$

as Hölder–Zygmund spaces. Let

$$\left(\Delta^1_h f \right)(x) = f(x+h) - f(x), \quad \left(\Delta^{l+1}_h f \right)(x) = \Delta^1_h (\Delta^l_h f)(x),$$
$$x \in \mathbf{R}^n, \, h \in \mathbf{R}^n, \, l \in \mathbf{N}, \tag{10}$$

be the iterated differences in $\mathbf{R}^n$. Let $0 < s < m \in \mathbf{N}$; then

$$\| f \mid \mathscr{C}^s(\mathbf{R}^n) \|_m = \sup_{x \in \mathbf{R}^n} |f(x)| + \sup |h|^{-s} |\Delta^m_h f(x)|, \tag{11}$$

where the second supremum is taken over all $x \in \mathbf{R}^n$, $h \in \mathbf{R}^n$ with $0 < |h| \le 1$, are equivalent norms on $\mathscr{C}^s(\mathbf{R}^n)$. Some differences in (11) can be replaced by some derivatives. We refer to [Triβ], 2.2.2, 2.5.12. Again we have extended the Hölder–Zygmund spaces to $s \le 0$.

(v) Assertion (iv) can be generalised as follows. Once more let $0 < s < m \in \mathbf{N}$ and $p, q \in [1, \infty]$. Then

$$\|f \mid L_p(\mathbf{R}^n)\| + \left(\int_{|h| \le 1} |h|^{-sq} \, \|\Delta_h^m f \mid L_p(\mathbf{R}^n)\|^q \, \frac{dh}{|h|^n} \right)^{\frac{1}{q}} \tag{12}$$

(with the usual modification if $q = \infty$) are equivalent norms on $B_{pq}^s(\mathbf{R}^n)$; see again [Triβ], 2.2.2, 2.5.12. These are the classical Besov spaces (with the notationally incorporated limiting cases $p = 1$ and $p = \infty$). Again some differences in (12) can be replaced by some derivatives.

(vi) Let $0 < p < \infty$ and let φ be the same function as in (2.2.1/3). Then the (local or inhomogeneous) Hardy space $h_p(\mathbf{R}^n)$ consists of all $f \in \mathscr{S}'(\mathbf{R}^n)$ such that

$$\|f \mid h_p(\mathbf{R}^n)\|_\varphi = \left\| \sup_{0 < t < 1} \left| \left(\varphi(t \cdot) \hat{f} \right)^\vee (\cdot) \right| \mid L_p(\mathbf{R}^n) \right\| \tag{13}$$

is finite. All these (quasi-)norms are mutually equivalent. In extension of the Paley–Littlewood assertion (1) we have

$$F_{p,2}^0(\mathbf{R}^n) = h_p(\mathbf{R}^n), \ 0 < p < \infty; \tag{14}$$

see [Triβ], 2.2.2, 2.5.8, especially p. 92.

(vii) Let $f \in L_1^{\mathrm{loc}}(\mathbf{R}^n)$; then

$$f_Q = |Q|^{-1} \int_Q f(x) dx, \text{ for any cube } Q \text{ in } \mathbf{R}^n, \tag{15}$$

where $|Q|$ is the volume of Q, is the mean value of f with respect to Q. The (local, inhomogeneous) space $\mathrm{bmo}(\mathbf{R}^n)$ of all functions with bounded mean oscillation consists of all $f \in L_1^{\mathrm{loc}}(\mathbf{R}^n)$ such that

$$\|f \mid \mathrm{bmo}(\mathbf{R}^n)\| = \sup_{|Q|>1} |Q|^{-1} \int_Q |f(x)| \, dx + \sup_{|Q| \le 1} |Q|^{-1} \int_Q |f(x) - f_Q| \, dx \tag{16}$$

is finite. There is a modification of Definition 2.2.1(ii) which includes $p = \infty$, $1 < q \le \infty$. Although we shall never use this possibility we mention for the sake of completeness that then (14) can be extended to $p = \infty$,

$$F_{\infty,2}^0(\mathbf{R}^n) = \mathrm{bmo}(\mathbf{R}^n); \tag{17}$$

see [Triβ], 2.2.2, 2.5.8, especially p. 93.

Remark The list above makes the point that many more or less classical function spaces, with their own historical roots, are special cases of the two scales B_{pq}^s and F_{pq}^s. It also enabled us to introduce some notation. However, in the later parts of this book we shall often avoid dealing with the full scales of spaces B_{pq}^s and F_{pq}^s, choosing rather to restrict ourselves to some classes. We have a special predisposition towards H_p^s, occasionally also with $p \leq 1$, and $\mathscr{C}^s$.

2.2.3 Atomic representations

Atomic representations nowadays play a decisive role in the theory of function spaces. In [Triγ], 1.9 we gave a description from an historical point of view, which will not be repeated here. The corresponding theory for the spaces B_{pq}^s and F_{pq}^s is mainly due to M. Frazier and B. Jawerth, see [FrJ1], [FrJ2], [FrJW] 5, [Torr2] and [Triγ], 1.9, 3.2. We present here two modifications without proofs since they are more or less covered by the work of Frazier and Jawerth. Especially the second version is somewhat more general than needed in this book but it paves the way to the brief description of atoms in domains and on closed sets given in 2.5. We closely follow [TrW1], see also [Tri4].

Let $\mathbf{Z}^n$ be the lattice of all points in $\mathbf{R}^n$ with integer-valued components. Let $b > 0$ be given, $v \in \mathbf{N}_0$ and $k \in \mathbf{Z}^n$. Then Q_{vk} denotes a closed cube in $\mathbf{R}^n$ with sides parallel to the axes, centred at $x^{v,k} \in \mathbf{R}^n$, with

$$\left| x^{v,k} - 2^{-v}k \right| \leq b2^{-v} \tag{1}$$

and with side length 2^{-v}. Let Q be a cube in $\mathbf{R}^n$ and $r > 0$; then rQ is the cube in $\mathbf{R}^n$ concentric with Q and with side length r times that of Q. We always tacitly assume in the sequel that $d > 0$ is chosen in dependence on b so that for all choices $v \in \mathbf{N}_0$ and all possible choices of $x^{v,k}$ according to (1),

$$\bigcup_{k \in \mathbf{Z}^n} dQ_{vk} = \mathbf{R}^n. \tag{2}$$

Let $K \in \mathbf{N}_0$. Then $C^K(\mathbf{R}^n)$ is the usual space of all complex-valued functions f having classical bounded continuous derivatives $D^\alpha f$ with $|\alpha| \leq K$ on $\mathbf{R}^n$.

Definition 1$_K$ *(i) Let $K \in \mathbf{N}_0$. Then a function a is called a 1-atom (more precisely, a 1_K-atom) if*

$$\text{supp } a \subset dQ_{0k} \text{ for some } k \in \mathbf{Z}^n \tag{3}$$

and

$$a \in C^K(\mathbf{R}^n) \text{ with } |D^\alpha a(x)| \le 1 \text{ if } |\alpha| \le K. \tag{4_K}$$

(ii) Let $s \in \mathbf{R}$, $0 < p \le \infty$, $K \in \mathbf{N}_0$ and $L + 1 \in \mathbf{N}_0$. Then a function a is called an (s,p)-atom (more precisely an $(s,p)_{K,L}$-atom) if

$$\text{supp } a \subset dQ_{vk} \text{ for some } v \in \mathbf{N} \text{ and some } k \in \mathbf{Z}^n, \tag{5}$$

$$a \in C^K(\mathbf{R}^n) \text{ with } |D^\alpha a(x)| \le 2^{-v(s-\frac{n}{p})+v|\alpha|} \text{ if } |\alpha| \le K, \tag{6_K}$$

and

$$\int_{\mathbf{R}^n} x^\beta a(x) dx = 0 \text{ if } |\beta| \le L. \tag{7}$$

Remark 1 We begin with some technical explanations. Of course, Q_{0k} is a cube with side length l. Assume that b in (1) is given. Then (2) and all the assertions made below are true if d is chosen sufficiently large. We shall not stress this point in the sequel. Recall that $x^\beta = x_1^{\beta_1} ... x_n^{\beta_n}$ if $x = (x_1, ..., x_n) \in \mathbf{R}^n$ and β is a multi-index. The moment conditions (7) are equivalent to

$$\left(D^\beta \hat{a}\right)(0) = 0 \text{ for } |\beta| \le L. \tag{8}$$

If $L = -1$, then (7) simply means that there are no moment conditions. The reason for the normalising factor in (6_K) and (4_K) is that there exists a constant c such that for all these atoms,

$$\left\|a \mid B_{pq}^s(\mathbf{R}^n)\right\| \le c, \quad \left\|a \mid F_{pq}^s(\mathbf{R}^n)\right\| \le c. \tag{9}$$

Hence, atoms are normalised building blocks, satisfying some moment conditions. The preferred case in $\mathbf{R}^n$ is, of course, given by $x^{v,k} = 2^{-v}k$ in (1). Then we have essentially the atoms introduced by Frazier and Jawerth. The above slight generalisation is of interest if $\mathbf{R}^n$ is replaced by domains or closed sets. For the same reason it is desirable to replace $C^K(\mathbf{R}^n)$ in the above definition by the Hölder–Zygmund spaces $\mathscr{C}^\sigma(\mathbf{R}^n)$ as defined by (2.2.2/9) and (2.2.2/11) with $\sigma > 0$.

Definition 1_σ *(i) Let $0 < \sigma \notin \mathbf{N}$. Then a function a is called a 1-atom (more precisely, a 1_σ-atom) if (3) holds and*

$$a \in \mathscr{C}^\sigma(\mathbf{R}^n) \text{ with } \|a \mid \mathscr{C}^\sigma(\mathbf{R}^n)\| \le 1. \tag{4_σ}$$

(ii) Let $s \in \mathbf{R}$, $0 < p \le \infty$, $0 < \sigma \notin \mathbf{N}$ and $L + 1 \in \mathbf{N}_0$. Then a

function a is called an (s, p)-atom (more precisely an $(s,p)_{\sigma,L}$-atom) if (5) and (7) hold and

$$a \in \mathscr{C}^\sigma(\mathbf{R}^n) \text{ with } \|a(2^{-\nu}\cdot) \mid \mathscr{C}^\sigma(\mathbf{R}^n)\| \le 2^{-\nu(s-\frac{n}{p})}. \qquad (6_\sigma)$$

Remark 2 Of course Definitions 1_K and 1_σ complement each other. If $0 < \sigma \notin \mathbf{N}$ then in addition to (2.2.2/11),

$$\|f \mid \mathscr{C}^\sigma(\mathbf{R}^n)\| = \sum_{|\alpha| \le [\sigma]} \|D^\alpha f \mid L_\infty(\mathbf{R}^n)\| + \sum_{|\alpha|=[\sigma]} \sup_{x \ne y} \frac{|D^\alpha f(x) - D^\alpha f(y)|}{|x-y|^{\{\sigma\}}}$$

$$(10)$$

with $\sigma = [\sigma] + \{\sigma\}, [\sigma] \in \mathbf{N}_0, 0 < \{\sigma\} < 1$, is also an equivalent norm which can be used in (4_σ) and (6_σ). Naturally, (6_K) can be rephrased in the sense of (6_σ).

Next we introduce two sequence spaces. Let

$$\lambda = \{\lambda_{\nu k} : \lambda_{\nu k} \in \mathbf{C}, \nu \in \mathbf{N}_0 \text{ and } k \in \mathbf{Z}^n\} \qquad (11)$$

and let $\chi_{\nu k}^{(p)}(x)$ be the *p*-normalised characteristic function with respect to the above cube $Q_{\nu k}$; this means that

$$\chi_{\nu k}^{(p)}(x) = 2^{\frac{\nu n}{p}} \text{ if } x \in Q_{\nu k} \text{ and } \chi_{\nu k}^{(p)}(x) = 0 \text{ if } x \in \mathbf{R}^n \backslash Q_{\nu k}, \qquad (12)$$

where $0 < p \le \infty$.

Definition 2 *Let $p, q \in (0, \infty]$.*

(i) Then b_{pq} is the collection of all sequences λ given by (11) such that

$$\|\lambda \mid b_{pq}\| = \left(\sum_{\nu=0}^{\infty} \left(\sum_{k \in \mathbf{Z}^n} |\lambda_{\nu k}|^p \right)^{\frac{q}{p}} \right)^{\frac{1}{q}} \qquad (13)$$

(with the usual modification if p and/or q is infinite) is finite.

(ii) Then f_{pq} is the collection of all sequences λ given by (11) such that

$$\|\lambda \mid f_{pq}\| = \left\| \left(\sum_{\nu=0}^{\infty} \sum_{k \in \mathbf{Z}^n} \left| \lambda_{\nu k} \chi_{\nu k}^{(p)}(\cdot) \right|^q \right)^{\frac{1}{q}} \mid L_p(\mathbf{R}^n) \right\| \qquad (14)$$

(with the usual modification if $q = \infty$) is finite.

Remark 3 We have

$$\left\| \chi_{\nu k}^{(p)} \mid L_p(\mathbf{R}^n) \right\| = 1, \quad \|\lambda \mid b_{pp}\| = \|\lambda \mid f_{pp}\| \qquad (15)$$

and

$$b_{p,\min(p,q)} \subset f_{pq} \subset b_{p,\max(p,q)}. \tag{16}$$

The last assertion follows from (13) and (14) and the triangle inequality, distinguishing between the cases $p \geq q$ and $p \leq q$.

If $c \in \mathbf{R}$, then we put $c_+ = \max(c,0)$ and as above denote by $[c]$ the largest integer less than or equal to c. Furthermore, if $0 < p \leq \infty$ and $0 < q \leq \infty$ then we use throughout this book the standard abbreviations

$$\sigma_p = n\left(\frac{1}{p} - 1\right)_+ \quad \text{and} \quad \sigma_{pq} = n\left(\frac{1}{\min(p,q)} - 1\right)_+. \tag{17}$$

If the atom a is supported by dQ_{vk} according to (3) and (5), then we write it as a_{vk}, so that

$$\operatorname{supp} a_{vk} \subset dQ_{vk}, v \in \mathbf{N}_0 \text{ and } k \in \mathbf{Z}^n. \tag{18}$$

In the proposition and the theorem below we assume $\sigma \geq 0$. Then the corresponding assertions refer to Definition 1_K or 1_σ depending on whether $\sigma = K \in \mathbf{N}_0$ or $\sigma \notin \mathbf{N}_0$.

Proposition *Let $s \in \mathbf{R}$, $0 < p \leq \infty$ and $0 < q \leq \infty$. Let $\sigma \geq 0$, $\sigma > s$ and $L + 1 \in \mathbf{N}_0$ with $L \geq \max([\sigma_p - s], -1)$. Suppose that either $\lambda \in b_{pq}$ or $\lambda \in f_{pq}$ in the sense of Definition 2. Finally assume that the a_{vk} with $v \in \mathbf{N}_0$ and $k \in \mathbf{Z}^n$ are either 1_σ-atoms ($v = 0$) or $(s,p)_{\sigma,L}$-atoms ($v \in \mathbf{N}$) in the sense of Definition 1_K or 1_σ, respectively, with (18). Then*

$$\sum_{v=0}^{\infty} \sum_{k \in \mathbf{Z}^n} \lambda_{vk} a_{vk}(x) \tag{19}$$

converges in $\mathscr{S}'(\mathbf{R}^n)$.

Remark 4 The proof is essentially covered by what has been done by Frazier and Jawerth; see in particular [FrJ2]. Our modifications are immaterial in this context.

Theorem *(i) Let $s \in \mathbf{R}$, $0 < p \leq \infty$ and $0 < q \leq \infty$. Let $\sigma \geq 0$, $\sigma > s$ and $L + 1 \in \mathbf{N}_0$ with $L \geq \max([\sigma_p - s], -1)$. Then $f \in \mathscr{S}'(\mathbf{R}^n)$ belongs to $B_{pq}^s(\mathbf{R}^n)$ if, and only if, it can be represented as*

$$f = \sum_{v=0}^{\infty} \sum_{k \in \mathbf{Z}^n} \lambda_{vk} a_{vk}, \quad \text{convergence being in } \mathscr{S}'(\mathbf{R}^n), \tag{20}$$

where the a_{vk} are 1_σ-atoms ($v = 0$) or $(s,p)_{\sigma,L}$-atoms ($v \in \mathbf{N}$) in the sense of Definition 1_K or 1_σ, respectively, with (18), and $\lambda \in b_{pq}$. Furthermore,

$$\inf \|\lambda \mid b_{pq}\|, \tag{21}$$

where the infimum is taken over all admissible representations (20), is an equivalent quasi-norm on $B_{pq}^s(\mathbf{R}^n)$.

(ii) Let $s \in \mathbf{R}$, $0 < p < \infty$, $0 < q \le \infty$. Let $\sigma \ge 0$, $\sigma > s$ and $L+1 \in \mathbf{N}_0$ with $L \ge \max([\sigma_{pq} - s], -1)$. Then $f \in \mathscr{S}'(\mathbf{R}^n)$ belongs to $F_{pq}^s(\mathbf{R}^n)$ if, and only if, it can be represented by (20) where the atoms a_{vk} have the same meaning as in part (i) (now, perhaps, with a different value of L) and $\lambda \in f_{pq}$. Furthermore

$$\inf \|\lambda \mid f_{pq}\|, \tag{22}$$

where the infimum is taken over all admissible representations (20), is an equivalent quasi-norm on $F_{pq}^s(\mathbf{R}^n)$.

Remark 5 This theorem coincides essentially with the theorem in [TrW1], 2.3. But as we emphasised above, apart from some technical modifications characterised, for example, by (1) or some generalisations such as in Definition 1_σ, it is covered by the work of Frazier and Jawerth. We use it almost exclusively in $\mathbf{R}^n$. Then it is completely sufficient to take $\sigma = K \in \mathbf{N}_0$ and $x^{v,k} = 2^{-v}k$ in (1). However, for atomic representations of function spaces on domains and closed sets these generalisations and modifications are of great help; see [TrW1] and [TrW2]. We shall return briefly to this subject in 2.5.

2.3 Special properties

2.3.1 Dilations

We use properties of the spaces $B_{pq}^s(\mathbf{R}^n)$ and $F_{pq}^s(\mathbf{R}^n)$ available in the literature, especially in [Triβ] and [Triγ], without describing them again. The aim of section 2.3 is to complement this theory by some very recent results, mostly obtained in close connection with the main themes of this book. Recall that σ_p is given by (2.2.3/17).

Proposition 1 *Let $0 < p \le \infty$ ($p < \infty$ in the F case) and $0 < q \le \infty$.*

(i) Let $s > \sigma_p$. There exists a constant $c > 0$ such that for all $R \ge 1$,

$$\left\| f(R\cdot) \mid B_{pq}^s(\mathbf{R}^n) \right\| \le cR^{s-\frac{n}{p}} \left\| f \mid B_{pq}^s(\mathbf{R}^n) \right\| \text{ for all } f \in B_{pq}^s(\mathbf{R}^n) \tag{1}$$

and

$$\|f(R\cdot) \mid F_{pq}^s(\mathbf{R}^n)\| \le cR^{s-\frac{n}{p}} \|f \mid F_{pq}^s(\mathbf{R}^n)\| \quad \text{for all } f \in F_{pq}^s(\mathbf{R}^n). \quad (2)$$

(ii) If for some $s \in \mathbf{R}$ there exists a constant c such that either (1) or (2) holds for all $R \ge 1$, then $s \ge \sigma_p$.

Proof Step 1 We prove (i) for the F case; the proof for the B case is the same. Let $\psi = \varphi_1$ be the same function as in (2.2.1/3) and (2.2.1/4). We omit "$\mathbf{R}^n$" in this proof. Since $s > \sigma_p$,

$$\|f \mid L_p\| + \left\| \left(\int_0^\infty t^{-sq} \mid \psi(t\cdot) \hat{f})^\vee(\cdot) \mid^q \frac{dt}{t} \right)^{\frac{1}{q}} \mid L_p \right\| \quad (3)$$

is an equivalent quasi-norm on F_{pq}^s; see [Triγ], 2.3.3, p. 99. Elementary calculations show that

$$(\psi(t\cdot)f(R\cdot)^\wedge(\cdot))^\vee(x) = (\psi(t\cdot)\hat{f}(R^{-1}\cdot))^\vee(x)R^{-n}$$
$$= (\psi(Rt\cdot)\hat{f}(\cdot))^\vee(Rx). \quad (4)$$

From (3) with $f(Rx)$ in place of $f(x)$ and (4) we obtain

$$\|f(R\cdot) \mid F_{pq}^s\| \le cR^{-\frac{n}{p}}\|f \mid L_p\| + cR^{s-\frac{n}{p}} \left\| \left(\int_0^\infty t^{-sq} \left| (\psi(t\cdot)\hat{f})^\vee(\cdot) \right|^q \frac{dt}{t} \right)^{1/q} \mid L_p \right\|$$
$$\quad (5)$$

Then (2) follows from (5), $R \ge 1, s > 0$ and (3).

Step 2 We prove (ii). Assume that we have (1) or (2) for all $R \ge 1$. We choose non-trivial functions f_R such that both $\hat{f}_R$ and $f_R(R\cdot)^\wedge$ are supported near the origin. Then by Definition 2.2.1 both (1) and (2) reduce to

$$\|f_R(R\cdot) \mid L_p\| \le R^{s-\frac{n}{p}} \|f_R \mid L_p\|. \quad (6)$$

If $s < 0$ and if R tends to infinity then we obtain a contradiction. To complete the proof of (ii) it remains to disprove that there exists a constant $c > 0$ such that

$$\|f(R\cdot) \mid B_{pq}^s\| \le cR^{s-\frac{n}{p}} \|f \mid B_{pq}^s\| \quad \text{for all } R \ge 1 \text{ and all } f \in B_{pq}^s \quad (7)$$

(and its F-counterpart) if $0 < p < 1$ and $s < \sigma_p$. But this is a by-product of Step 9 of the proof of Theorem 2.3.2; see Remark 2.3.2/2.

Proposition 2 *Let $0 < p \leq \infty$ ($p < \infty$ in the F case) and $0 < q \leq \infty$.*

(i) Let $s < 0$. There exists a constant $c > 0$ such that for all $R \in (0, 1]$,

$$\left\| f(R\cdot) \mid B_{pq}^s(\mathbf{R}^n) \right\| \leq cR^{s-\frac{n}{p}} \left\| f \mid B_{pq}^s(\mathbf{R}^n) \right\| \quad \text{for all } f \in B_{pq}^s(\mathbf{R}^n) \tag{8}$$

and

$$\left\| f(R\cdot) \mid F_{pq}^s(\mathbf{R}^n) \right\| \leq cR^{s-\frac{n}{p}} \left\| f \mid F_{pq}^s(\mathbf{R}^n) \right\| \quad \text{for all } f \in F_{pq}^s(\mathbf{R}^n). \tag{9}$$

(ii) If for some $s \in \mathbf{R}$ there exists a constant c such that either (8) or (9) holds whenever $0 < R \leq 1$, then $s \leq 0$.

Proof Step 1 We prove (i) for the F case, the proof for the B case being the same. Again we omit "$\mathbf{R}^n$". Let φ and $\psi = \varphi_1$ be the same functions as in (2.2.1/3) and (2.2.1/4). Then

$$\left\| \left(\varphi \hat{f} \right)^\vee \mid L_p \right\| + \left\| \left(\int_0^1 \left(t^{-sq} \mid (\psi(t\cdot) \hat{f})^\vee(\cdot) \mid^q \frac{dt}{t} \right)^{\frac{1}{q}} \mid L_p \right\| \tag{10}$$

is an equivalent quasi-norm on F_{pq}^s; see [Triγ], 2.4.1, p. 100. By (4) the second term in (10) with $f(R\cdot)$ in place of f equals

$$R^{s-\frac{n}{p}} \left\| \left(\int_0^R t^{-sq} \mid (\psi(t\cdot) \hat{f})^\vee(\cdot) \mid^q \frac{dt}{t} \right)^{\frac{1}{q}} \mid L_p \right\|. \tag{11}$$

Since $R \leq 1$ it follows from the equivalent quasi-norm (10) that this term can be estimated from above by the right-hand side of (9). We have to consider the first term in (10) with $f(R\cdot)$ in place of f. Assume $2^{k-1} \leq R^{-1} \leq 2^k$ for some $k \in \mathbf{N}$. Then we have $\varphi(Rx) = \sum_{j=0}^{k+2} \varphi(Rx)\varphi_j(x)$ for the function φ_j given by (2.2.1/3) and (2.2.1/4). From (4) with φ in place of $\psi(t\cdot)$ it follows that

$$\left\| (\varphi f(R\cdot)^\wedge)^\vee \mid L_p \right\| \leq R^{-\frac{n}{p}} \left\| \sum_{j=0}^{k+2} \mid (\varphi(R\cdot)\varphi_j \hat{f})^\vee(\cdot) \mid \mid L_p \right\|. \tag{12}$$

Using the Fourier multiplier theorem in [Triβ], 1.6.3, p. 31, the right-hand side of (12) can be estimated from above by

$$cR^{-n/p} \left\| \sum_{j=0}^{k+2} \left| (\varphi_j \hat{f})^\vee(\cdot) \right| \mid L_p \right\|, \tag{13}$$

which in turn, since $s < 0$, can be estimated from above by

$$cR^{-\frac{n}{p}} 2^{-ks} \left\| \sup_{0 \leq j \leq k+2} 2^{js} \left| (\varphi_j \hat{f})^\vee(\cdot) \right| \mid L_p \right\| \leq c' R^{s-\frac{n}{p}} \left\| f \mid F_{pq}^s \right\|. \tag{14}$$

But this is the desired estimate and (9) is proved.

Step 2 We prove (ii). Assume that we have (8) or (9) for all R with $0 < R \leq 1$. We argue in the same way as in Step 2 of the proof of Proposition 1. If, as there, f_R is appropriately chosen then we have (6). If $s > 0$ and if R tends to zero then we obtain a contradiction. The proof of (ii) is complete.

Remark We followed essentially [Tri4]; see also [Triβ], 3.4.1, p. 206 as far as part (i) of Proposition 1 is concerned. The parts (ii) of the two propositions show that the restrictions for s in the respective parts (i) are natural. Only the limiting cases $s = \sigma_p$ and $s = 0$, respectively, remain open.

2.3.2 Localisation

Again let $\mathbf{Z}^n$ be the lattice of all points in $\mathbf{R}^n$ having integer-valued components. Let $x^{j,k} = 2^{-j}k$ with $k \in \mathbf{Z}^n$ and $j \in \mathbf{N}$. Let $f \in S'(\mathbf{R}^n)$ with

$$\text{supp } f \subset Q_d = \{x \in \mathbf{R}^n : x = (x_1,...,x_n), |x_l| < d \text{ if } l = 1,...,n\} \quad (1)$$

where $d > 0$ is assumed to be small, at least $d \leq \frac{1}{2}$. Let

$$f^j(x) = \sum_{k \in \mathbf{Z}^n} a_k f(2^{j+1}(x - x^{j,k})) , \ a_k \in \mathbf{C}, j \in \mathbf{N}: \quad (2)$$

Of course, the terms in (2) have mutually disjoint supports. So far as notation is concerned, we remind the reader that σ_p is defined by (2.2.3/17); if $b \in \mathbf{R}$ then $[b]$ is the largest integer less than or equal to b, whereas $[b]^-$ is the largest integer strictly less than b. All function spaces and corresponding quasi-norms in this subsection are taken with respect to $\mathbf{R}^n$. This justifies our temporary omission of "$\mathbf{R}^n$". Let A_{pq}^s be either B_{pq}^s or F_{pq}^s. Of course, A_{pq}^s in a given equality or inequality refers always to the same space; thus it is always B_{pq}^s or always F_{pq}^s. Also all integrals are taken over $\mathbf{R}^n$.

Theorem *Let $s \in \mathbf{R}$, $0 < p \leq \infty$ ($p < \infty$ for the F scale), $0 < q \leq \infty$. Let $0 < d \leq \frac{1}{4}$.*
(i) There exist two constants $c_1 > 0$ and $c_2 > 0$ such that for all $f \in A_{pq}^s$ with $\text{supp} f \subset Q_d$ and

$$\int x^{\beta} f(x) dx = 0 \text{ for } |\beta| \leq L = \max([\sigma_p - s], -1), \quad (3)$$

all $j \in \mathbf{N}$ *and all* f^j *given by* (2),

$$c_1 \left\| f^j \mid A^s_{pq} \right\| \leq 2^{j(s-\frac{n}{p})} (\sum_k |a_k|^p)^{\frac{1}{p}} \left\| f \mid A^s_{pq} \right\| \leq c_2 \left\| f^j \mid A^s_{pq} \right\|. \qquad (4)$$

(ii) *Let* $c > 0$ *and* $M + 1 \in \mathbf{N}_0$ $(= \mathbf{N} \cup \{0\})$ *be such that for all* $f \in A^s_{pq}$ *with* supp$f \subset Q_d$ *and*

$$\int x^\beta f(x)dx = 0 \ for \ |\beta| \leq M, \qquad (5)$$

all $j \in \mathbf{N}$ *and all* f^j *given by* (2),

$$c \left\| f^j \mid A^s_{pq} \right\| \leq 2^{j(s-\frac{n}{p})} (\sum_k |a_k|^p)^{\frac{1}{p}} \left\| f \mid A^s_{pq} \right\|. \qquad (6)$$

Then

$$M \geq L^- = \max([\sigma_p - s]^-, -1). \qquad (7)$$

Proof Step 1 We begin with a few technical explanations. Of course, (3) with $L = -1$ or (5) with $M = -1$ simply means that there are no moment conditions. Furthermore, (3), (5) and also (2) must be understood in the context of distributions. Since $\hat{f}$ is analytic, (3) can easily be reformulated as

$$(D^\beta \hat{f})(0) = 0 \ for \ |\beta| \leq L. \qquad (8)$$

Step 2 In the following step we prove the right-hand side of (4). For that purpose we need a little preparation. By (2) we have

$$f^j(2^{-j-1}x) = \sum_{k \in \mathbf{Z}^n} a_k f(x - 2k), \ a_k \in \mathbf{C}, \ j \in \mathbf{N}, \qquad (9)$$

with $f \in A^s_{pq}$ and (1). We claim that

$$\left\| \sum_{k \in \mathbf{Z}^n} a_k f(\cdot - 2k) \mid A^s_{pq} \right\| \sim \left(\sum_{k \in \mathbf{Z}^n} |a_k|^p \right)^{\frac{1}{p}} \left\| f \mid A^s_{pq} \right\|, \qquad (10)$$

where the equivalence relation "$\sim$" means that each side of (10) can be estimated (from above) by the other side times a constant which is independent of f and $\{a_k\}$. Since the elements $f(x - 2k)$ have disjoint supports in unit cubes centred at $2k$, the relation (10) with $A^s_{pq} = F^s_{pq}$ follows immediately from the localisation property of the spaces F^s_{pq}; see [Triγ], 2.4.7, p. 124. To prove (10) for $A^s_{pq} = B^s_{pq}$ we use the characterisation of B_{pq}^s via local means; see [Triγ], 2.5.3, p. 138. Let K_0 be a C^∞ function in $\mathbf{R}^n$ with

$$\operatorname{supp}K_0 \subset \{y \in \mathbf{R}^n : |y| < c\}, \ \overset{\wedge}{K_0}(0) \neq 0, \tag{11}$$

for some $c > 0$, and $K(x) = \Delta^N K_0(x)$ for some $N \in \mathbf{N}$, where Δ is the Laplacian in $\mathbf{R}^n$. The local means are given by

$$K(t,g)(x) = \int_{\mathbf{R}^n} K(y)g(x+ty)dy, \ 0 < t < 1, \ g \in \mathscr{S}'(\mathbf{R}^n), \tag{12}$$

with its obvious counterpart $K_0(1,g)$. If $2N > \max(s, \sigma_p)$ then

$$\|g \mid B_{pq}^s\| \sim \|K_0(1,g) \mid L_p\| + \left(\int_0^1 t^{-sq} \|K(t,g) \mid L_p\|^q \frac{dt}{t}\right)^{\frac{1}{q}} \tag{13}$$

as equivalent quasi-norms. The proof in [Triγ] shows that $c > 0$ in (11) may assumed to be small. We insert (9) in (12) and obtain, by the support properties of $f(x - 2k)$,

$$K\left(t, \sum_k a_k f(\cdot - 2k)\right)(x) = \sum_k a_k K(t,f)(x - 2k), \tag{14}$$

$$\left\|K\left(t, \sum_k a_k f(\cdot - 2k)\right) \mid L_p\right\| = \left(\sum_k |a_k|^p\right)^{\frac{1}{p}} \|K(t,f) \mid L_p\| \tag{15}$$

and, by (13), finally (10) with $A_{pq}^s = B_{pq}^s$.

Step 3 We prove the right-hand inequality of (4). Let $s < 0$. Then the desired result follows from (9), (10) and Proposition 2.3.1/2:

$$\left(\sum_k |a_k|^p\right)^{1/p} \|f \mid A_{pq}^s\| \leq c \|f^j (2^{-j-1} \cdot) \mid A_{pq}^s\|$$

$$\leq c' 2^{-j(s-n/p)} \|f^j \mid A_{pq}^s\|. \tag{16}$$

Let $s \geq 0$ and $s - m < 0$ for some $m \in \mathbf{N}$. Recall that

$$\|f^j \mid A_{pq}^s\| \sim \|f^j \mid A_{pq}^{s-m}\| + \sum_{|\alpha|=m} \|D^\alpha f^j \mid A_{pq}^{s-m}\|, \tag{17}$$

see [Triβ], 2.3.8, p. 59. From (16) with $s - m$ in place of s and since

$(D^\alpha f^j)(x) = 2^{(j+1)m}(D^\alpha f)^j(x)$, we have

$$\|f^j \mid A^s_{pq}\|$$

$$\geq c \left(\sum_k |a_k|^p\right)^{1/p} 2^{j(s-m-n/p)}\left[\|f \mid A^{s-m}_{pq}\| + \sum_{|\alpha|=m} 2^{jm}\|D^\alpha f \mid A^{s-m}_{pq}\|\right]$$

$$\geq c \left(\sum_k |a_k|^p\right)^{1/p} 2^{j(s-n/p)}\sum_{|\alpha|=m}\|D^\alpha f \mid A^{s-m}_{pq}\|. \tag{18}$$

In Remark 1 below we give a short proof of the following fact, special cases of which are certainly known: there is a constant $c > 0$ such that for all $g \in A^s_{pq}$ with support in the unit cube,

$$\|g \mid A^{s-m}_{pq}\| \leq c \sum_{|\alpha|=m}\|D^\alpha g \mid A^{s-m}_{pq}\|. \tag{19}$$

Hence the right-hand side of (18) is equivalent to the middle term of (4). Then (18) coincides with right-hand side of (4).

Step 4 When $s > \sigma_p$ the left-hand inequality of (4) is an easy consequence of Proposition 2.3.1/1 and (10):

$$\|f^j \mid A^s_{pq}\| \leq c2^{j(s-n/p)}\left\|\sum_k a_k f(\cdot - 2k) \mid A^s_{pq}\right\|$$

$$\leq c'2^{j(s-n/p)}\left(\sum_k |a_k|^p\right)^{1/p}\|f \mid A^s_{pq}\|. \tag{20}$$

Step 5 We prove the left-hand inequality of (4) for B^s_{pq} with $s \leq \sigma_p$. Consider an optimal atomic decomposition of f given by Theorem 2.2.3(i), where now we denote the atoms a_{vk} in (2.2.3/20) by b_{vk} to avoid confusion with the coefficients a_k. Since f is supported near the origin we assume that (2.2.3/20) has the form

$$f = \lambda_0 b_0 + \sum_{v=1}^\infty \sum_{k \in Z^n} \lambda_{vk} b_{vk}, \tag{21}$$

where b_0 is the only 1-atom, located near the origin. We may also assume that the moment conditions (3) and those ones needed in Theorem 2.2.3 are the same. Then b_0 satisfies these moment conditions and we may incorporate this term in the sum in (21). Hence without loss of generality we may take $\lambda_0 = 0$ in (21) and

$$\|\lambda \mid b_{pq}^s\| \sim \|f \mid B_{pq}^s\|, \lambda = \{\lambda_{vk} : v \in \mathbf{N}, k \in \mathbf{Z}^n\}. \tag{22}$$

By (2) we then have

$$f^j(x) = 2^{(j+1)(s-n/p)} \sum_{l \in \mathbf{Z}^n} a_l \sum_{v,k} \lambda_{vk} 2^{-(j+1)(s-n/p)} b_{vk} \left(2^{j+1} \left(x - x^{j,l}\right)\right). \tag{23}$$

In view of (1), with $d > 0$ and small, we may assume that all non-vanishing terms in (21) have supports near the origin, say, in $Q_{1/2}$. Hence the atoms $2^{-(j+1)(s-n/p)} b_{vk} \left(2^{j+1} \left(x - x^{j,l}\right)\right)$ belonging to different l-terms have disjoint supports. This means that (23) is an atomic representation of f^j and we have

$$\|f^j \mid B_{pq}^s\| \le c 2^{j(s-\frac{n}{p})} \left(\sum_{v=1}^{\infty} \left(\sum_{l,k} |a_l|^p |\lambda_{vk}|^p\right)^{\frac{q}{p}}\right)^{\frac{1}{q}}$$

$$= c 2^{j(s-\frac{n}{p})} \left(\sum_l |a_l|^p\right)^{\frac{1}{p}} \|\lambda \mid b_{pq}^s\|. \tag{24}$$

Now the left-hand inequality in (4) with $A_{pq}^s = B_{pq}^s$ follows from (24) and (22).

Step 6 To prove the left-hand inequality of (4) for F_{pq}^s with $s \le \sigma_p$ in the same way as in Step 5 we are faced with the difficulty that the number L in (3) and the number L needed to apply Theorem 2.2.3(ii) may be different if $q < p$. We assume temporarily that we have (3) with $L = \left[\sigma_{pq} - s\right]$ (and $s \le \sigma_p$). Then we can argue as in the previous step: f is given by (21) with $\lambda_0 = 0$,

$$\|\lambda \mid f_{pq}^s\| \sim \|f \mid F_{pq}^s\|, \tag{25}$$

and (23) is a corresponding atomic decomposition of f^j. By Theorem 2.2.3(ii), (2.2.3/14) and the fact that different l-terms in (23) have disjoint supports, we have

$$\|f^j \mid F_{pq}^s\| \le c 2^{j(s-n/p)} \left(\sum_l |a_l|^p\right)^{1/p} \|\lambda \mid f_{pq}^s\|. \tag{26}$$

Now the left-hand inequality in (4) with $A_{pq}^s = F_{pq}^s$ follows from (26) and (25).

Step 7 Here we let $s \leq \sigma_p$ and $L = [\sigma_{pq} - s]$ as assumed in (3). We wish to prove the left-hand inequality in (4) for F^s_{pq}. By Step 6 we may suppose that $q < p$ and temporarily put $\tilde{L} = [\sigma_{pq} - s]$. We have (21), where the b_{vk} are now $(s, p)_{K, \tilde{L}}$-atoms according to Definition 2.2.3/1_K, and b_0 may be considered as an $(s, p)_{K,L}$-atom located at $Q_{0,0}$. Again we assume that (21) is an optimal atomic decomposition,

$$|\lambda_0| + \|\lambda \mid f_{pq}\| \sim \|f \mid F^s_{pq}\|, \lambda = \{\lambda_{vk} : v \in \mathbf{N}, k \in \mathbf{Z}^n\}. \quad (27)$$

We put

$$f = f_1 + f_2 \text{ with } f_1(x) = \lambda_0 b_0(x), \quad (28)$$

and apply Step 6 to f_2:

$$\left\| f_2^j \mid F^s_{pq} \right\| \leq c 2^{j(s - n/p)} \left(\sum_l |a_l|^p \right)^{1/p} \|f \mid F^s_{pq}\| \quad (29)$$

in analogy to (26). As for f_1^j we may use Step 5 and obtain

$$\left\| f_1^j \mid B^s_{pq} \right\| \leq c 2^{j(s - n/p)} \left(\sum_l |a_l|^p \right)^{1/p} |\lambda_0| \|b_0 \mid B^s_{pq}\|. \quad (30)$$

Since $\|b_0 \mid B^s_{pq}\| \leq c$ uniformly for all admissible atoms b_0, the inclusion $B^s_{pq} \subset F^s_{pq}$, (27) and (30) yield

$$\left\| f_1^j \mid F^s_{pq} \right\| \leq c 2^{j(s - n/p)} \left(\sum_l |a_l|^p \right)^{1/p} \|f \mid F^s_{pq}\|. \quad (31)$$

The left-hand inequality in (4) now follows from (28), (29) and (31). The proof of part (i) of the theorem is complete.

Step 8 We prove part (ii) of the theorem and assume first that $1 \leq p \leq \infty$; hence $\sigma_p = 0$. There is nothing to prove if $s \geq 0$. Let $-m - 1 \leq s < -m$ for some $m \in \mathbf{N}_0$. Then $L^- = m$ in (7). Hence we have to find a function $f \in A^s_{pq}$, where $0 < q \leq \infty$, with

$$\int x^\beta f(x) dx = 0 \text{ if } |\beta| \leq m - 1, \quad (32)$$

for which there is no $c > 0$ such that (6) holds for all $j \in \mathbf{N}$ and all a_k. Let $n = 1$ and let χ be the characteristic function of the interval $[0, 1]$.

Then $\chi \in A_{pq}^{s+m}$, since $L_p \subset A_{pq}^{s+m}$, and $f = \frac{d^m}{dx^m}\chi \in A_{pq}^s$. We thus have (32) interpreted as in (8). Let $x^{j,k} = 2^{-j}k$; then

$$f^j(x) = \sum_{k=0}^{2^j-1} f\left(2^j\left(x - x^{j,k}\right)\right) = 2^{-jm}f(x) \tag{33}$$

essentially coincides with (2) if $a_k = 1$. Assume that (6) holds for some $c > 0$; then

$$
\begin{aligned}
c2^{-jm}\left\|f \mid A_{pq}^s\right\| &= c\left\|f^j \mid A_{pq}^s\right\| \\
&\leq 2^{j(s-\frac{1}{p})}\left(\sum_{k=0}^{2^j-1} 1\right)^{\frac{1}{p}}\left\|f \mid A_{pq}^s\right\| \\
&= 2^{js}\left\|f \mid A_{pq}^s\right\|
\end{aligned}
\tag{34}
$$

for all $j \in \mathbf{N}$. But this contradicts $s < -m$. The number d in (1), (2) and the theorem may be so small that the above argument cannot be applied immediately. But this is immaterial. If (6) or (2) holds for all admissible $f(x)$, then it also holds for $f(bx)$ for any given fixed $b > 0$. Upon replacing the interval $[0,1]$ by the unit cube in $\mathbf{R}^n$ the above arguments can be extended from $n = 1$ to $n > 1$; this completes the proof of part (ii) of the theorem if $1 \leq p \leq \infty$.

Step 9 Finally we prove (ii) of the theorem for $0 < p < 1$, in which case $\sigma_p = n\left(\frac{1}{p} - 1\right)$. If $s \geq \sigma_p$ there is nothing to prove. Let $-m - 1 \leq s - \sigma_p < -m$ for some $m \in \mathbf{N}_0$ and let $0 < q \leq \infty$. Then $L^- = m$ in (7). Let $f \in A_{pq}^s$, with supp $f \subset Q_d$, be such that (32) is satisfied, which can be reformulated as (8). Hence $\hat{f}$ is an analytic function and

$$\hat{f}(x) = \sum_{|\alpha|\geq m} b_\alpha x^\alpha, \tag{35}$$

with convergence in $\mathbf{R}^n$. Assume that (6) holds at least for one summand, so that

$$\left\|f\left(2^j\cdot\right) \mid A_{pq}^s\right\| \leq c2^{j(s-n/p)}\left\|f \mid A_{pq}^s\right\|. \tag{36}$$

Then we have, by 2.2.1/3 and Definition 2.2.1,

$$\left\|\left(\varphi f\left(2^j\cdot\right)^\wedge\right)^\vee \mid L_p\right\| \leq c2^{j(s-n/p)}\left\|f \mid A_{pq}^s\right\|. \tag{37}$$

By (35) and standard arguments we obtain

$$\varphi(x) f \left(2^j \cdot\right)^\wedge (x) = 2^{-jn} \varphi(x) \hat{f} \left(2^{-j}x\right)$$
$$= \varphi(x) 2^{-jn-jm} \sum_{|\alpha| \geq m} b_\alpha 2^{-j(|\alpha|-m)} x^\alpha, \tag{38}$$

$$\varphi(x) \sum_{|\alpha| \geq m} b_\alpha 2^{-j(|\alpha|-m)} x^\alpha \rightarrow \varphi(x) \sum_{|\alpha|=m} b_\alpha x^\alpha \text{ in } \mathscr{S} \tag{39}$$

as $j \rightarrow \infty$, and hence

$$\left(\varphi(x) \sum_{|\alpha| \geq m} b_\alpha 2^{-j(|\alpha|-m)} x^\alpha\right)^\vee (\xi) \rightarrow (-i)^m \sum_{|\alpha|=m} b_\alpha D^\alpha \overset{\vee}{\varphi} (\xi) \text{ in } \mathscr{S} \tag{40}$$

as $j \rightarrow \infty$. By (37), (38) and (40) we have

$$\left\| \sum_{|\alpha|=m} b_\alpha D^\alpha \overset{\vee}{\varphi} \mid L_p \right\| \leq c \left\| f \mid A_{pq}^s \right\| \lim_{j \rightarrow \infty} 2^{j(s-\sigma_p+m)} = 0, \tag{41}$$

since $s - \sigma_p + m < 0$. But there are functions f (and φ) such that the left-hand side of (41) is positive. These functions disprove (36). The proof of part (ii) of the theorem is complete.

Remark 1 We prove that (19) holds for an appropriate constant $c > 0$ for all $g \in A_{pq}^s$ with support in the unit cube, and $m \in \mathbf{N}$. Assume that there is no such constant. Then we find a sequence $\{g_l\}_{l=1}^\infty \subset A_{pq}^s$, with supports in the unit cube, such that

$$\left\| g_l \mid A_{pq}^{s-m} \right\| = 1 \text{ and } \sum_{|\alpha|=m} \left\| D^\alpha g_l \mid A_{pq}^{s-m} \right\| \rightarrow 0 \text{ as } l \rightarrow \infty. \tag{42}$$

Thus $\{g_l\}$ is bounded in A_{pq}^s (see (17)) and hence is pre-compact in A_{pq}^{s-m}. By (42), $\{g_l\}$ is also pre-compact in A_{pq}^s. Assume that $g_l \rightarrow g$ in A_{pq}^s as $l \rightarrow \infty$. Then g has support in the unit cube,

$$\left\| g \mid A_{pq}^{s-m} \right\| = 1 \text{ and } D^\alpha g = 0 \text{ if } |\alpha| = m. \tag{43}$$

By the last part of (43), g is a compactly supported polynomial. Hence $g = 0$, which contradicts the first part of (43).

Remark 2 The disproof of (36) fills the gap left in Step 2 of the proof of Proposition 2.3.1/1.

Remark 3 As for the numbers L and L^- in (3) and (7) respectively, we have $L = L^- + 1$ if $\sigma_p - s \in \mathbf{N}_0$ and $L = L^-$ otherwise. Hence with the possible exception of the cases $\sigma_p - s \in \mathbf{N}_0$ the moment conditions (3) are necessary and sufficient for (4). If $\sigma_p - s \in \mathbf{N}_0$, then the question as to whether L or L^- is the correct number may depend on q.

Remark 4 We followed essentially [Tri4], a forerunner of which may be found in [Tri3]. Constructions of type (2) are now rather fashionable: they involve a generating function which is dyadically dilated and translated. This is a typical procedure in connection with wavelets, spline bases and, in a more qualitative version, atomic representations, see 2.2.3. We use assertions of type (4) later on to obtain estimates from *below* for entropy numbers of compact embeddings between function spaces of the above type, defined on bounded domains in $\mathbf{R}^n$, by reducing these problems to l_p (or more precisely to their finite-dimensional counterparts l_p^N).

2.3.3 Embeddings

Embedding theorems for the spaces $B_{pq}^s(\mathbf{R}^n)$ and $F_{pq}^s(\mathbf{R}^n)$ introduced in 2.2.1 and their special cases described in 2.2.2 have a long history. They are of crucial importance, not only for the theory itself, but even more for applications to general pseudodifferential operators. The "if"-parts of these theorems, that is, the conditions under which the respective embeddings hold, have been known for some 10 or 15 years and are more or less covered by [Triβ]. The "only if"-parts, that is, the assertions that these conditions are sharp, are more recent (although there was never any doubt about the outcome). These sharp embeddings and a detailed discussion of some related delicate limiting problems may be found in [SicT]. We restrict ourselves here to a discussion of only a very few of these results. We emphasise that we shall not need the "only if"-parts in what follows in this book, whereas the "if"-parts, which will be used in the sequel, are well covered by the literature. This justifies our omission of proofs.

Recall that "$\subset$" in connection with two quasi-Banach spaces always means that there is a continuous linear embedding. Furthermore, all spaces in this subsection are spaces on $\mathbf{R}^n$, introduced in Definition 2.2.1. This justifies our omission of "$\mathbf{R}^n$" in this subsection.

(i) Embeddings with constant smoothness

Let $s \in \mathbf{R}$ be the smoothness parameter, let $p \in (0, \infty)$ and let $q, u, v \in (0, \infty]$. Then

$$B_{pu}^s \subset F_{pq}^s \subset B_{pv}^s \tag{1}$$

if, and only if, $0 < u \leq \min (p, q)$ and $\max (p, q) \leq v \leq \infty$;

$$B_{1,u}^0 \subset L_1 \subset B_{1,v}^0 \text{ if, and only if, } 0 < u \leq 1 \text{ and } v = \infty; \tag{2}$$

$$B_{\infty,u}^0 \subset L_\infty \subset B_{\infty,v}^0 \text{ if, and only if, } 0 < u \leq 1 \text{ and } v = \infty, \tag{3}$$

where L_∞ in (3) can be replaced by C, the space of all complex-valued bounded and uniformly continuous functions on $\mathbf{R}^n$. The "if"-parts of (1) on the one hand and of both (2) and (3) on the other hand are covered by [Triβ], Proposition 2.3.2, p. 47, and Proposition 2.5.7, p. 89, respectively. The more complicated "only if"-parts are proved in [SicT]. A completely different proof of the "only if"-part of (1), based on entropy numbers in weighted function spaces on $\mathbf{R}^n$, may be found in [HaT1], Corollary 4.3.

(ii) Embeddings with constant differential dimension

We recall that $s - n/p$ is called the *differential dimension* of both B_{pq}^s and F_{pq}^s. It is a characteristic number which plays a crucial role in the theory of these spaces. Let

$$s \in \mathbf{R}, 0 < p_0 < p < p_1 \leq \infty, s_0 - \frac{n}{p_0} = s - \frac{n}{p} = s_1 - \frac{n}{p_1}, \tag{4}$$

and suppose that $q, u, v \in (0, \infty]$. Then

$$B_{p_0u}^{s_0} \subset F_{pq}^s \subset B_{p_1v}^{s_1} \tag{5}$$

if, and only if, $0 < u \leq p \leq v \leq \infty$. The "if"-part is due to Jawerth and Franke, see [Jaw], [Fra2], [Triβ], p. 131, and the Russian translation of [Triβ] (Moskva, 1986), p. 191. As for the "only if"-part we refer again to [SicT]. For our purpose the following observation will be of great service in the sequel. Let

$$s_0 \in \mathbf{R}, 0 < p_0 < p_1 < \infty, s_0 - \frac{n}{p_0} = s_1 - \frac{n}{p_1}, u, v \in (0, \infty]; \tag{6}$$

then

$$F_{p_0u}^{s_0} \subset F_{p_1v}^{s_1}; \tag{7}$$

see [Triβ], 2.7.1, p. 129. The crucial point about (7) is that u and v are unrelated, in contrast to (5). In particular, by formula (11) below, we have

$$H_{p_0}^{s_0} \subset H_{p_1}^{s_1}, \; s_0 \in \mathbf{R}, \; 0 < p_0 < p_1 < \infty, \; s_0 - \frac{n}{p_0} = s_1 - \frac{n}{p_1}. \qquad (8)$$

Remark In a $(\frac{1}{p}, s)$-diagram, see Fig. 2.4.3, the assertions (i) correspond to horizontal lines, whereas the embeddings in part (ii) correspond to lines of slope n. Later on there will be special interest in the case $s = 0$ and $p = \infty$, i.e. the origin in the diagram. Then one can ask for embeddings along the ray $s = n/p$ with $(0,0)$ as an end point. In addition to $B_{\infty,q}^0$, L_∞, C and bmo are also possible target spaces for limiting embeddings along the ray $s = n/p$. We shall return to this subject later on in great detail, but for the moment content ourselves by complementing the assertions in (ii).

(iii) Limiting embeddings

The general assumptions are again $0 < p \le \infty$ (with $p < \infty$ for F spaces) and $0 < q \le \infty$. On the ray of interest we have $s = n/p$. Then

$$F_{pq}^{n/p} \subset L_\infty \text{ if, and only if, } 0 < p \le 1 \text{ and } 0 < q \le \infty, \qquad (9)$$

$$B_{pq}^{n/p} \subset L_\infty \text{ if, and only if, } 0 < p \le \infty \text{ and } 0 < q \le 1. \qquad (10)$$

In both assertions L_∞ can be replaced by C. For proofs we refer to [Fra2] and [Triβ], 2.8.3, p. 146. The space bmo is larger than L_∞; see 2.2.2 for the definition. Furthermore we extend (2.2.2/7) by

$$H_p^s(\mathbf{R}^n) = F_{p,2}^s(\mathbf{R}^n), s \in \mathbf{R}, 0 < p < \infty. \qquad (11)$$

Then we have

$$H_p^{n/p} \subset \text{bmo and } B_{pp}^{n/p} \subset \text{bmo if } 0 < p < \infty; \qquad (12)$$

see [Tri3], p. 595. We give a short proof. By (6), (7) and the fact that $B_{pp}^{n/p} = F_{pp}^{n/p}$ it is sufficient to prove the first part of (12) when $1 < p < \infty$. Then we have by (7), $H_1^0 \subset H_p^{-n+\frac{n}{p}}$. The rest is a matter of duality; see [Triβ], pp. 93 and 178. The second part of (12) cannot be extended to $p = \infty$; see [SicT], Theorem 3.3/2 and Corollary 3.3.

2.4 Hölder inequalities

2.4.1 Preliminaries

The (Hardy–)Sobolev spaces $H_p^s(\mathbf{R}^n)$ are defined in (2.3.3/11). Let the smoothness $s \in \mathbf{R}$ be given. Then we ask for which p_1, p_2, p the pointwise multiplier property

$$H^s_{p_1}(\mathbf{R}^n) H^s_{p_2}(\mathbf{R}^n) \subset H^s_p(\mathbf{R}^n) \tag{1}$$

makes sense. Here (1) must be understood to mean that there exists a constant $c > 0$ with

$$\|f_1 f_2 \mid H^s_p(\mathbf{R}^n)\| \le c \|f_1 \mid H^s_{p_1}(\mathbf{R}^n)\| \, \|f_2 \mid H^s_{p_2}(\mathbf{R}^n)\| \tag{2}$$

for all $f_j \in H^s_{p_j}(\mathbf{R}^n)$, $j = 1, 2$. If $s = 0$ then one has as a special case the classical Hölder inequality

$$L_{r_1}(\mathbf{R}^n) L_{r_2}(\mathbf{R}^n) \subset L_r(\mathbf{R}^n), \tag{3}$$

with $r_1, r_2 \in [1, \infty]$, $\frac{1}{r_1} + \frac{1}{r_2} = \frac{1}{r} \le 1$. Estimates for entropy numbers and inequalities of type (1) are the cornerstones of the spectral theory of degenerate elliptic operators which we shall present in Chapter 5. No additional effort is required to deal with the slightly more general problems

$$B^s_{p_1 q_1}(\mathbf{R}^n) B^s_{p_2 q_2}(\mathbf{R}^n) \subset B^s_{pq}(\mathbf{R}^n) \tag{4}$$

and

$$F^s_{p_1 q_1}(\mathbf{R}^n) F^s_{p_2 q_2}(\mathbf{R}^n) \subset F^s_{pq}(\mathbf{R}^n). \tag{5}$$

We call these assertions Hölder inequalities. Roughly speaking, it turns out that the desired relations (1), (4) and (5) can be obtained by shifting (3) along the lines of slope n in the $(\frac{1}{p}, s)$-diagram to the level s; see Fig. 2.4.3.

We are not interested in *multiplication algebras*, that is, (1) with $p_1 = p_2 = p$, or (4), (5) with $p_1 = p_2 = p$, $q_1 = q_2 = q$. In that case there is a final answer; see [SicT], 3.3 for a description with references to [Tri2], 2.6.2, [Fra2]; see also [Triβ], 2.8.3 and [Sic], pp. 56–7, where there are also further references about the history of this problem. It emerges that for a space $B^s_{pq}(\mathbf{R}^n)$ or $F^s_{pq}(\mathbf{R}^n)$ to be a multiplication algebra it must be continuously embedded in C (this is almost sufficient). Hence $s \ge n/p$ is necessary; $s > n/p$ is sufficient; as for $s = n/p$ we refer to the literature cited above; see also the description of limiting embeddings in 2.3.3(iii). We wish to apply the results of this section to degenerate elliptic operators with coefficients, roughly speaking, in H^s_p on $\mathbf{R}^n$ or on domains. The coefficients and their inverses may be essentially unbounded. This makes it clear that we are mainly interested in spaces of type H^s_p with $s < n/p$, or, in limiting situations, $s = n/p$, which is just the complement of the situation described above in connection with multiplication algebras. Together with the indicated shifting procedure of (3) to obtain assertions of type (1), or (4), (5), this shows plainly that we should confine ourselves to spaces such that $(\frac{1}{p}, s)$ belongs to the strips G_1 or G_2 in connection with Fig. 2.4.3.

One might try to generalise (1), and also (4), (5), by replacing $H_{p_1}^s$ and $H_{p_2}^s$ by $H_{p_1}^{s_1}$ and $H_{p_2}^{s_2}$ respectively with possibly different s_1 and s_2. These more general multiplication properties are not of interest for us here. We refer to the extensive survey by W. Sickel [Sic]; see also [SicT] for some more recent references. The only exception treated in 2.4.5 in connection with $s < 0$ fits completely into our general intention to shift (3) along lines of slope n in the $(\frac{1}{p}, s)$-diagram in Fig. 2.4.3.

2.4.2 Paramultiplication

To prove assertions of type (2.4.1/1), (2.4.1/4) or (2.4.1/5) we use the method of paramultiplication. This method had been used by J. Peetre and H. Triebel around 1976/77; see [Peet], Ch. 7 and [Tri2], 2.6 and the references given there. Shortly afterwards, around 1980, J.M. Bony and Y. Meyer introduced similar techniques in connection with microlocal analysis, non-linear problems, Calderón–Zygmund operators, etc. They also coined the word "paramultiplication". In what follows this method will prove to be a crucial instrument, also in connection with problems of type (2.4.1/1) and more general pointwise multiplier assertions. Detailed references to the original papers and historical comments may be found in [Sic]; see also [SicT], Remark 2.2/2. Here we need a somewhat special but rather sharp assertion which is the basis of the subsequent considerations. We follow [SicT].

Paramultiplication All spaces are defined on $\mathbf{R}^n$, which again justifies our omission of "$\mathbf{R}^n$" from now on. Let φ and φ_j be the functions defined by (2.2.1/3) and (2.2.1/4). Let $f \in \mathscr{S}'$ and put

$$f_j(x) = \left(\varphi_j \hat{f}\right)^{\vee}(x) \text{ and } f^j(x) = \left(\varphi\left(2^{-j}\cdot\right)\hat{f}\right)^{\vee}(x), \quad j \in \mathbf{N}_0. \quad (1)$$

Of course, $f^j \to f$ in $\mathscr{S}'$, endowed with the weak topology. Both f_j and f^j are analytic functions of exponential type. Hence the product $f^j g^j$ makes sense for any $j \in \mathbf{N}_0$ and any $f, g \in \mathscr{S}'$. We define

$$fg = \lim_{j \to \infty} f^j g^j = \lim_{j \to \infty} \left(\sum_{k=0}^{j} f_k\right) \left(\sum_{k=0}^{j} g_k\right) \text{ in } \mathscr{S}' \quad (2)$$

if this limit exists. Here we used $f^j = \sum_{k=0}^{j} f_k$. We obtain

$$
\begin{aligned}
fg &= \sum_{j=2}^{\infty} f^{j-2} g_j + \sum_{k=2}^{\infty} f_k g^{k-2} + \sum_{|j-k| \le 1} f_j g_k \\
&= {\sum}'(f,g) + {\sum}''(f,g) + {\sum}'''(f,g)
\end{aligned}
\tag{3}
$$

with

$$
\operatorname{supp} \left(f^{k-2} g_k\right)^{\wedge} \subset \left\{\xi : 2^{k-3} \le |\xi| \le 15 \cdot 2^{k-3}\right\}, k-2 \in \mathbf{N}_0,
\tag{4}
$$

and

$$
\operatorname{supp} \left(f_k g_j\right)^{\wedge} \subset \left\{\xi : |\xi| \le 9 \cdot 2^{k-1}\right\}, |j-k| \le 1.
\tag{5}
$$

Here (4) and (5) follow from the support conditions on φ_j and $\varphi\left(2^{-j}\cdot\right)$, and, for example, $\left(f_k g_j\right)^{\wedge} = \left(\widehat{\varphi_k \hat{f}}\right) * \left(\widehat{\varphi_j \hat{g}}\right)$, where $*$ stands for convolution; see (1). This crosswise multiplication is called paramultiplication. The advantage is the control of the supports in (4) and (5).

The Fatou property We discuss the meaning of the limit in (2). Assume that

$$
f^k g^k = h^k \to h \text{ in } \mathscr{S}' \text{ and, say, } \sup_k \left\| h^k \mid F_{pq}^s \right\| = A < \infty.
\tag{6}
$$

Then it follows from Definition 2.2.1(ii) and Fatou's lemma that $h \in F_{pq}^s$ and $\left\| h \mid F_{pq}^s \right\| \le cA$, where c depends only on F_{pq}^s and the chosen quasi-norm. Of course, one can replace F_{pq}^s by B_{pq}^s. In other words, when asking whether or not the limit in (2) belongs to a space B_{pq}^s or F_{pq}^s one has only to deal with the quasi-norms according to (6). In general, subspaces of $\mathscr{S}'$ do not have the Fatou property. The simplest counterexamples are C and L_1: the characteristic function of a cube can be approximated in $\mathscr{S}'$ by a bounded sequence of continuous functions, and the δ-distribution can be approximated in $\mathscr{S}'$ by a bounded sequence of L_1 functions. Furthermore, any $h \in B_{pq}^s$ (or F_{pq}^s) can be approximated in $\mathscr{S}'$ by functions $h^k \in \mathscr{S}$ with property (6). However, if $p = \infty$ and/or $q = \infty$, then the completion of $\mathscr{S}$ in B_{pq}^s, denoted by $\overset{0}{B}{}_{pq}^s$, does not coincide with B_{pq}^s. Then by the above argument, B_{pq}^s does not have the Fatou property. See also [Fra2], 2.6 for further details.

Recall that the h_p are the Hardy spaces given by (2.2.2/14) and that $\sigma_p = n\left(\frac{1}{p} - 1\right)_+$; we put $h_\infty = L_\infty$. Let the functions φ_k, f_k, g_k have the same meaning as above. We put $f_j = g_j = 0$ if $j < 0$. Of course, ${\sum}'(f,g)^{\wedge}$ denotes the Fourier transform of ${\sum}'(f,g)$, etc.

Proposition *Let $p_1, p_2, q \in (0, \infty]$ and $\frac{1}{p} = \frac{1}{p_1} + \frac{1}{p_2}$.*
(i) Let $0 < p < \infty$. Then

$$\left\| \left(\sum_{k=0}^{\infty} \left| \left(\varphi_k \sum{}' (f, g)^\wedge \right)^\vee (\cdot) \right|^q \right)^{1/q} \mid L_p \right\|$$

$$\leq c \, \|f \mid h_{p_2}\| \left\| \left(\sum_{k=0}^{\infty} |g_k(\cdot)|^q \right)^{1/q} \mid L_{p_1} \right\| \tag{7}$$

(and the usual modification if $q = \infty$) where c is independent of f and g.
(ii) Let $0 < p \leq \infty$ and $\bar{p} = \min(1, p)$. Then

$$\left\| \left(\varphi_k \sum{}''' (f, g)^\wedge \right)^\vee \mid L_p \right\|^{\bar{p}}$$

$$\leq c \sum_{l=0}^{\infty} \sum_{|l-u|+|l-v| \leq 2} 2^{nl\left(1-\bar{p}\right)} \|f_{k+u} \mid L_{p_1}\|^{\bar{p}} \|g_{k+v} \mid L_{p_2}\|^{\bar{p}} \tag{8}$$

where c is independent of f, g and $k \in \mathbf{N}_0$.
(iii) Let $0 < p < \infty$ and $s > \sigma_p$. Then

$$\left\| \sup_k 2^{ks} \left| \left(\varphi_k \sum{}''' (f, g)^\wedge \right)^\vee (\cdot) \right| \mid L_p \right\|$$

$$\leq c \left\| \sup_k 2^{ks/2} |f_k| \mid L_{p_1} \right\| \left\| \sup_k 2^{ks/2} |g_k| \mid L_{p_2} \right\| \tag{9}$$

where c is independent of f and g.

Proof Step 1 We prove (i). By (2.2.1/3) and (2.2.1/4) we have

$$\text{supp } \varphi_k \subset \left\{ \xi : 2^{k-1} \leq |\xi| \leq 3 \cdot 2^{k-1} \right\}, k \in \mathbf{N}. \tag{10}$$

Hence by (3) and (4), $\varphi_k \sum' (f, g)^\wedge = \varphi_k \sum_{|j| \leq J} \left(f^{k+j-2} g_{k+j} \right)^\wedge$, where J is independent of k. Then we find by the Fourier multiplier theorem in [Triβ], 1.6.3, p. 31, that the left-hand side of (7) can be estimated from above by

$$c \left\| \left(\sum_{k=2}^{\infty} |f^{k-2} g_k|^q \right)^{1/q} \mid L_p \right\| \leq c \left\| \sup_l |f^l| \left(\sum_{k=2}^{\infty} |g_k|^q \right)^{1/q} \mid L_p \right\|$$

$$\leq c \left\| \sup_l |f^l| \mid L_{p_2} \right\| \cdot \left\| \left(\sum_{k=0}^{\infty} |g_k|^q \right)^{1/q} \mid L_{p_1} \right\| \tag{11}$$

where we used Hölder's inequality in the final stage. However, by (1) and
[Triβ], p. 37, formula (12), the first factor on the right-hand side can be
estimated from above by $\|f \mid h_{p_2}\|$. The proof of (i) is complete.

Step 2 We prove (ii). By (10) and (5), apart from immaterial index-
shiftings, we must estimate

$$\left\| \left(\varphi_k \sum_{l=0}^{\infty} (f_{k+l}g_{k+l})^\wedge \right)^\vee \mid L_p \right\|^{\bar{p}} \le \sum_{l=0}^{\infty} \left\| \left(\varphi_k (f_{k+l}g_{k+l})^\wedge \right)^\vee \mid L_p \right\|^{\bar{p}}. \quad (12)$$

By [Triβ], Proposition 1.5.1, p. 25, and formula (13) on p. 28 with $b = 2^{k+l}$
we have

$$\begin{aligned}
\left\| \left(\varphi_k (f_{k+l}g_{k+l})^\wedge \right)^\vee \mid L_p \right\| &\le c 2^{-\frac{n}{p}(k+l)} \left\| \varphi_k \left(2^{k+l} \cdot \right)^\vee \mid L_{\bar{p}} \right\| \\
&\quad \times \left\| (f_{k+l}g_{k+l}) \left(2^{-k-l} \cdot \right) \mid L_p \right\| \\
&\le c' \left\| \varphi_1 \left(2^{l+1} \cdot \right)^\vee \mid L_{\bar{p}} \right\| \|f_{k+l} \mid L_{p_1}\| \|g_{k+l} \mid L_{p_2}\|
\end{aligned}$$
$$(13)$$

where we used (2.2.1/4) and Hölder's inequality. The first factor on the
right-hand side can now be estimated in the desired way. The proof of
(8) is complete.

Step 3 We prove (iii) and again let $\bar{p} = \min(1,p)$. As in Step 2 it is
sufficient to estimate

$$\left\| \sup_k 2^{ks} \left| \left(\varphi_k \sum_{l=0}^{\infty} (f_{k+l}g_{k+l})^\wedge \right)^\vee \right| \mid L_p \right\|^{\bar{p}}$$
$$\le \sum_{l=0}^{\infty} 2^{-ls\bar{p}} \left\| \sup_k 2^{(k+l)s} \left| \left(\varphi_k (f_{k+l}g_{k+l})^\wedge \right)^\vee \right| \mid L_p \right\|^{\bar{p}}. \quad (14)$$

We apply the Fourier multiplier theorem in [Triβ], 1.6.3, 1.6.4, pp. 31–2,
with $L_p(l_\infty)$. For fixed $l \in N_0$ we have the estimate, in analogy to (13),

$$\left\| \varphi_1 \left(2^{l+1} \cdot \right) \mid H_2^\kappa \right\| \le c_\varepsilon 2^{l(\kappa - \frac{n}{2})}, \quad \kappa = n\left(\frac{1}{p} - \frac{1}{2} \right) + \varepsilon, \; \varepsilon > 0. \quad (15)$$

This follows from Proposition 2.3.1/1. With the help of (15) and the
multiplier theorem mentioned above, the right-hand side of (14) can be

estimated from above by

$$c_\varepsilon \sum_{l=0}^{\infty} 2^{-l\bar{p}(s-\sigma_p-\varepsilon)} \left\| \sup_k 2^{(k+l)s} |f_{k+l} g_{k+l}| \mid L_p \right\|^{\bar{p}}. \tag{16}$$

Now choose $\varepsilon > 0$ so small that $s > \sigma_p + \varepsilon$ and apply Hölder's inequality to the right-hand side of (16). This gives the right-hand side of (9). The proof is complete.

Remark By Step 1 it follows immediately that the scalar case of (7),

$$\left\| \left(\varphi_k \sum{}' (f,g)^\wedge \right)^\vee \mid L_p \right\| \le c \|f \mid h_{p_2}\| \max_{-1 \le j \le 1} \|g_{k+j} \mid L_{p_1}\|, \tag{17}$$

also holds for $p = \infty$. Let $0 < p < \infty$, $0 < q \le \infty$ and $s > \sigma_{pq}$ (see (2.2.3/17)); then one can replace $L_p(l_\infty)$, $L_{p_1}(l_\infty)$, $L_{p_2}(l_\infty)$ in (9) and in Step 3 by $L_p(l_q)$, $L_{p_1}(l_q)$, $L_{p_2}(l_q)$, respectively. In this formulation, part (iii) is essentially covered by [Yam], Theorem 3.7.

2.4.3 The main theorem

In the $(\frac{1}{p}, s)$-diagram in Fig. 2.1 we introduce the strip

$$G_1 = \left\{ (\frac{1}{p}, s) : 0 < p < \infty, n \left(\frac{1}{p} - 1 \right)_+ < s < \frac{n}{p} \right\} \tag{1}$$

and the extended strip

$$G_2 = \left\{ (\frac{1}{p}, s) : 0 < p < \infty, n \left(\frac{1}{p} - 1 \right) < s < \frac{n}{p} \right\}. \tag{2}$$

Any line of slope n in these strips is characterised by the point at which it meets the axis $s = 0$: we shall refer to this point, $(\frac{1}{r}, 0)$ say, as the "foot point" of the line. Thus any point on G_2 has coordinates

$$\left(\frac{1}{r} + \frac{s}{n}, s \right) = \left(\frac{1}{r^s}, s \right), \text{ say, with } \frac{1}{r} + \frac{s}{n} > 0, 1 < r < \infty. \tag{3}$$

According to 2.3.3(ii), lines of slope n are characterised by constant differential dimensions of the spaces B_{pq}^s and F_{pq}^s with the sharp embeddings (2.3.3/4)–(2.3.3/8) which will be of some service to us later on.

Starting with (2.4.1/3) in non-limiting cases we try to shift the corresponding assertion, motivated by the embeddings with constant differential dimension, along the lines of slope n to the situation indicated in the figure by the large dots. That means we ask for assertions of type (2.4.1/1), (2.4.1/4) and (2.4.1/5) with $p_j = r_j^s$ ($j = 1, 2$) and $p = r^s$. In

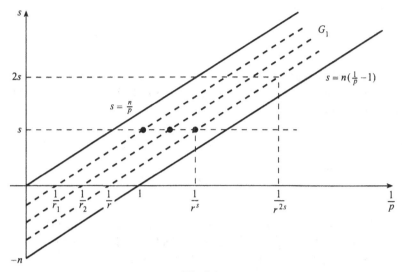

Fig. 2.1.

[SicT], Theorem 4.2, we obtained necessary and sufficient conditions for this problem as formulated in the theorem below. We restrict ourselves here to the "if"-part (sufficient conditions). As the proof of the "only if"-part (necessary conditions) is rather sophisticated and this part will not be subsequently needed here, we refer to [SicT] for details of it. All spaces are defined on $\mathbf{R}^n$, which again justifies our omission of "$\mathbf{R}^n$" in what follows.

Theorem *Let $s > 0$; q_1, q_2, $q \in (0, \infty]$;*

$$r_1, r_2 \in (1, \infty), \quad \frac{1}{r} = \frac{1}{r_1} + \frac{1}{r_2} < 1. \tag{4}$$

(i) Then

$$B^s_{r_1 q_1} B^s_{r_2 q_2} \subset B^s_{r^s q} \tag{5}$$

if, and only if,

$$0 < q_1 \le r_1, \ 0 < q_2 \le r_2, \ \max(q_1, q_2) \le q \le \infty. \tag{6}$$

(ii) Also

$$F^s_{r_1 q_1} F^s_{r_2 q_2} \subset F^s_{r^s q} \tag{7}$$

if, and only if,

$$\max(q_1, q_2) \leq q \leq \infty. \tag{8}$$

Proof (if-part) Step 1 We prove (5). By the Fatou property described in 2.4.2 we have only to establish the corresponding estimates of the quasi-norms in analogy to (2.4.1/2). Assume that $f \in B^s_{r_1^s q_1}$ and $g \in B^s_{r_2^s q_2}$ with $q_j \leq r_j$ $(j = 1, 2)$. Recall that $L_{r_j} = h_{r_j} = F^0_{r_j, 2}$. By (2.3.3/4), (2.3.3/5) and (3) we have

$$B^s_{r_j^s q_j} \subset L_{r_j} (j = 1, 2). \tag{9}$$

We use the decomposition (2.4.2/3). By (9), $\frac{1}{r^s} = \frac{1}{r_1} + \frac{1}{r_2}$ and the respective scalar case of Proposition 2.4.2(i) we have

$$2^{ks} \left\| \left(\varphi_k \sum' (f, g)^\wedge \right)^\vee \mid L_{r^s} \right\| \leq c \|f \mid L_{r_1}\| \, 2^{ks} \sup_{|j| \leq J} \left\| g_{k+j} \mid L_{r_2^s} \right\|$$

$$\leq c' \left\| f \mid B^s_{r_1^s q_1} \right\| \, 2^{ks} \sup_{|j| \leq J} \left\| g_{k+j} \mid L_{r_2^s} \right\| \tag{10}$$

and a corresponding assertion for $\sum'' (f, g)$ in place of $\sum' (f, g)$. Taking the qth power and summing, one obtains the desired estimates as far as $\sum' (f, g)$ and $\sum'' (f, g)$ are concerned, where the fact that $q \geq \max(q_1, q_2)$ must be used. It remains to estimate $\sum''' (f, g)$. We introduce the auxiliary space $B^{2s}_{r^{2s} q}$ as indicated in the figure. In addition to (2.3.3/5) we have $B^{2s}_{r^{2s} q} \subset B^s_{r^s q}$; see [Triβ], 2.7.1, p. 129. Note that

$$\frac{1}{r^{2s}} = \frac{1}{r} + \frac{2s}{n} = \frac{1}{r_1^s} + \frac{1}{r_2^s}. \tag{11}$$

We put $\lambda = \overline{r^{2s}} = \min(1, r^{2s})$, apply Proposition 2.4.2(ii) with respect to (11) and obtain (if $q < \infty$)

$$\left\| \sum''' (f, g) \mid B^s_{r^s q} \right\|^q \leq c \left\| \sum''' (f, g) \mid B^{2s}_{r^{2s} q} \right\|^q$$

$$\leq c' \sum_{k=0}^{\infty} 2^{2ksq} \left(\sum_{l=0}^{\infty} 2^{nl(1-\lambda)} \left\| f_{l+k} \mid L_{r_1^s} \right\|^\lambda \left\| g_{l+k} \mid L_{r_2^s} \right\|^\lambda \right)^{q/\lambda} + + \tag{12}$$

where $++$ indicates similar terms caused by the index-shifting in (2.4.2/8).

The right-hand side of (12) can be estimated from above by

$$c \sum_{k=0}^{\infty} \left(\sum_{l=0}^{\infty} 2^{nl(1-\lambda)-2sl\lambda} 2^{2s(l+k)\lambda} \left\| f_{l+k} \mid L_{r_1^s} \right\|^{\lambda} \left\| g_{l+k} \mid L_{r_2^s} \right\|^{\lambda} \right)^{q/\lambda} + +$$

$$\leq c_{\varepsilon} \sum_{k=0}^{\infty} \sum_{l=0}^{\infty} 2^{-l\left[2s-n\left(\frac{1}{\lambda}-1\right)-\varepsilon\right]q} 2^{2s(l+k)q} \left\| f_{l+k} \mid L_{r_1^s} \right\|^{q} \left\| g_{l+k} \mid L_{r_2^s} \right\|^{q} + +.$$
$$(13)$$

We have in any case $2s > n\left(\frac{1}{\lambda} - 1\right)$ (see the figure) and hence $2s > n\left(\frac{1}{\lambda} - 1\right) + \varepsilon$ if $\varepsilon > 0$ is small. But then it follows that the right-hand side of (13) can be estimated from above by

$$c \left\| f \mid B_{r_1^s q_1}^s \right\|^{q} \left\| g \mid B_{r_2^s q_2}^s \right\|^{q} \tag{14}$$

as required. The modifications necessary when $q = \infty$ are obvious. Now (2.4.2/3) and the above estimates for $\sum'$ and $\sum''$ prove (5).

Step 2 We prove (7). The counterpart of (9) with F in place of B holds without any restrictions on q_j; see (2.3.3/6) and (2.3.3/7). Then Proposition 2.4.2(i) (now the vector-valued version) yields

$$\begin{aligned} \left\| \sum{}' (f,g) \mid F_{r^s q}^s \right\| &\leq c \left\| f \mid L_{r_1} \right\| \left\| g \mid F_{r_2 q}^s \right\| \\ &\leq c' \left\| f \mid F_{r_1^s q_1}^s \right\| \left\| g \mid F_{r_2^s q_2}^s \right\|; \end{aligned} \tag{15}$$

here we used $q_2 \leq q$. Similarly $\sum'' (f,g)$ may be estimated. As for $\sum''' (f,g)$, we apply Proposition 2.4.2(iii) with $2s$ in place of s and with respect to (11). Using (2.3.3/7) in addition we have

$$\begin{aligned} \left\| \sum{}''' (f,g) \mid F_{r^s q}^s \right\| &\leq c \left\| \sum{}''' (f,g) \mid F_{r^{2s} \infty}^{2s} \right\| \\ &\leq c' \left\| f \mid F_{r_1^s \infty}^s \right\| \left\| g \mid F_{r_2^s \infty}^s \right\| \\ &\leq c'' \left\| f \mid F_{r_1^s q_1}^s \right\| \left\| g \mid F_{r_2^s q_2}^s \right\|. \end{aligned} \tag{16}$$

The proof (of the "if"-part) is complete.

Remark We essentially followed [SicT] where one also finds a proof of the "only if"-part. Recall that $H_p^s = F_{p,2}^s$ are the (Hardy–)Sobolev spaces if $s \in \mathbf{R}, 0 < p < \infty$; see (2.3.3/11). Let $s > 0$ and let r_1, r_2, r be given by (4). Then

$$H_{r_1^s}^s H_{r_2^s}^s \subset H_{r^s}^s \tag{17}$$

is a special case of (7) which is of crucial importance for our later applications.

2.4.4 Limiting cases

Limiting embeddings play a crucial role in our subsequent considerations. The related results in 2.3.3(iii) correspond to the line $s = n/p$ in Fig. 2.4.3. The foot point of this ray is $(0,0)$, corresponding to $r = \infty$ in (2.4.3/3). In this sense we have $\left(\frac{s}{n}, s\right) = \left(\frac{1}{\infty^s}, s\right)$ with $s > 0$. We ask for the counterpart of Theorem 2.4.3, now with, say, $r_2 = \infty$ and $r = r_1 \in (1, \infty)$. It turns out that the counterparts of (2.4.3/6) and (2.4.3/8) must be modified. In [SicT], 4.3 necessary and sufficient conditions were obtained; these will be formulated in the theorem below. As in 2.4.3 we prove the "if"-part, which we shall use later on, and refer to [SicT] for the the "only if"-part (which, by the way, is now much easier to prove than the "only if"-part of Theorem 2.4.3). Again we omit "$\mathbf{R}^n$" in what follows since all spaces in this subsection are defined on $\mathbf{R}^n$. Although $\infty^s = n/s$ with $s > 0$ looks somewhat curious we use this notation to make the assertion below comparable with Theorem 2.4.3.

Theorem *Let $s > 0$; q_1, q_2, $q \in (0, \infty]$ and $r \in (1, \infty)$.*
 (i) Then

$$B^s_{r^s q_1} B^s_{\infty^s q_2} \subset B^s_{r^s q} \tag{1}$$

if, and only if,

$$0 < q_1 \leq r, \ \ 0 < q_2 \leq 1, \ \ \infty \geq q \geq \max(q_1, q_2). \tag{2}$$

 (ii) Also

$$F^s_{r^s q_1} F^s_{\infty^s q_2} \subset F^s_{r^s q} \tag{3}$$

if, and only if,

$$s \geq n, \ \ \infty \geq q \geq \max(q_1, q_2). \tag{4}$$

Proof Compared with the proof of Theorem 2.4.3 we now have $r_2 = \infty$. Here the related counterparts of (2.4.3/9) are

$$B^s_{\infty^s q_2} \subset L_\infty \ \text{and} \ F^s_{\infty^s q_2} \subset L_\infty. \tag{5}$$

But these follow from (2.3.3/10) and (2.3.3/9), since $q_2 \leq 1$ and $s \geq n$, respectively. With these modifications in mind one can follow the proof of Theorem 2.4.3.

Remark 1 The counterpart of (2.4.3/17) is given by

$$H^s_{r^s} H^s_{\infty^s} \subset H^s_{r^s}, r \in (1, \infty), s \geq n. \tag{6}$$

If $s < n$ then (6) is false. Recall that $H_\infty^s = H_p^{n/p}$ with $s = n/p > 0$.

Remark 2 The limiting case above corresponds to the "upper" boundary of G_1 in (2.4.3/1). Another interesting limiting case is the "bottom" of G_1, by which we mean $s = 0$ and $1 \le p \le \infty$. In that case we have the classical Hölder inequality (2.4.1/3). It is rather surprising that any assertion of type (2.4.1/4) or (2.4.1/5) follows from (2.4.1/3) by elementary embeddings. In [SicT], 4.3 the following assertion was proved: Let

$$q_1, q_2, q \in (0, \infty], 1 \le r_1 \le \infty, 1 \le r_2 < \infty, \frac{1}{r} = \frac{1}{r_1} + \frac{1}{r_2} \le 1. \qquad (7)$$

Let $A^0_{r_j q_j}$ be either $B^0_{r_j q_j}$ or $F^0_{r_j q_j}$, $j = 1, 2$; and let A^0_{rq} be either B^0_{rq} or F^0_{rq} (with $A^0_{r_1 q_1} = B^0_{r_1 q_1}$ if $r_1 = \infty$). Then

$$A^0_{r_1 q_1} A^0_{r_2 q_2} \subset A^0_{rq} \qquad (8)$$

if, and only if,

$$A^0_{r_j q_j} \subset L_{r_j} \text{ and } L_r \subset A^0_{rq}. \qquad (9)$$

The "if"-part is obvious, but not the "only if"-part. A reformulation of (9) in terms of conditions for q_j and q can be given by using 2.3.3(i). The situation is different if Lorentz spaces, say $L_{r_1 \infty}$, are admitted as factors in (8); see [Triβ], 2.8.6, formula (13) on p. 156.

2.4.5 Hölder inequalities for H_p^s

Recall that the $H_p^s = F_{p,2}^s$ (with $s \in \mathbf{R}, 0 < p < \infty$) are the (Hardy–)-Sobolev spaces; see (2.3.3/11). Again we omit "$\mathbf{R}^n$" in what follows since all spaces in this subsection are defined on $\mathbf{R}^n$. The spaces in Theorem 2.4.3 are characterised by points within the strip G_1; see (2.4.3/1). We wish to extend these assertions to G_2, given by (2.4.3/2), where we restrict ourselves to the spaces H_p^s. Now (2.4.3/3) is applied to G_2.

Theorem *Let $r_1, r_2 \in (1, \infty)$ and suppose $\frac{1}{r} = \frac{1}{r_1} + \frac{1}{r_2} < 1$. Suppose also that $s \in \mathbf{R}$ and $\frac{1}{r_1^s} = \frac{1}{r_1} + \frac{s}{n} > 0$. Then*

$$H_{r_1^s}^s H_{r_2^{|s|}}^{|s|} \subset H_{r^s}^s. \qquad (1)$$

Proof If $s > 0$ then (1) coincides with (2.4.3/17). If $s = 0$, then (1) is the classical Hölder inequality (2.4.1/3). Suppose that $s < 0$ and let $g \in H_{r_2^{|s|}}^{|s|}$.

Then by (1) with $|s|$ instead of s we have a continuous map $f \longmapsto fg$ from $H_{(r')^{|s|}}^{|s|}$ into $H_{(r_1')^{|s|}}^{|s|}$ since $\frac{1}{r_1} + \frac{1}{r'} = \frac{1}{r} + \frac{1}{r'} = 1$ and $\frac{1}{r'} + \frac{1}{r_2} = \frac{1}{r_1'}$. Also

$$\frac{1}{(r')^{-s}} + \frac{1}{r^s} = \frac{1}{(r_1')^{-s}} + \frac{1}{r_1^s} = 1, \quad |s| = -s. \tag{2}$$

We use the duality

$$\left(H_p^s\right)' = H_{p'}^{-s}, s \in \mathbf{R}, \frac{1}{p} + \frac{1}{p'} = 1, 1 < p < \infty, \tag{3}$$

see [Triβ], 2.11.2, p. 178. Since $f \longmapsto fg$ is formally self-adjoint, the above assertion and (2) yield (1).

Remark We followed essentially [ET3], 2.4.

2.5 The spaces B_{pq}^s and F_{pq}^s on domains

2.5.1 Definitions

Let Ω be a domain (an open connected set) in $\mathbf{R}^n$. As usual $\mathscr{D}'(\Omega)$ stands for all complex distributions on Ω. The restriction of $g \in \mathscr{S}'(\mathbf{R}^n)$ to Ω is denoted by $g \mid \Omega$ and is considered as an element of $\mathscr{D}'(\Omega)$.

Definition 1 *Let Ω be a domain in $\mathbf{R}^n$, let $s \in \mathbf{R}$ and $0 < q \leq \infty$.*

(i) Let $0 < p \leq \infty$. Then $B_{pq}^s(\Omega)$ is the restriction of $B_{pq}^s(\mathbf{R}^n)$ to Ω, quasi-normed by

$$\left\| f \mid B_{pq}^s(\Omega) \right\| = \inf \left\| g \mid B_{pq}^s(\mathbf{R}^n) \right\| \tag{1}$$

where the infimum is taken over all $g \in B_{pq}^s(\mathbf{R}^n)$ with $g \mid \Omega = f$.

(ii) Let $0 < p < \infty$. Then $F_{pq}^s(\Omega)$ is the restriction of $F_{pq}^s(\mathbf{R}^n)$ to Ω, quasi-normed by

$$\left\| f \mid F_{pq}^s(\Omega) \right\| = \inf \left\| g \mid F_{pq}^s(\mathbf{R}^n) \right\| \tag{2}$$

where the infimum is taken over all $g \in F_{pq}^s(\mathbf{R}^n)$ with $g \mid \Omega = f$.

Remark 1 When introducing spaces on domains one is faced with a dilemma. There are two possibilities: spaces on Ω might be defined

(i) by restriction as above, or
(ii) intrinsically.

We prefer (i). But then one has to ask, for which domains the above definition is reasonable, if there exist linear and bounded extension operators into the corresponding spaces on $\mathbf{R}^n$, and whether convincing intrinsic characterisations are available. If Ω is a bounded C^∞ domain, then the problem about a linear and bounded extension operator is settled, see [Triγ], 5.1.3; and also, if in addition $s > \sigma_p$, some intrinsic characterisations are known, see [Triγ], Ch.5. Up to fairly recent times not very much was known if Ω is non-smooth or if $s \leq \sigma_p$. In [TrW1] we studied intrinsic characterisations of all the above spaces $B_{pq}^s(\Omega)$ and $F_{pq}^s(\Omega)$ under rather mild restrictions for the (non-smooth) domains using atoms. We give here a rather brief description of this, without proofs as we shall not need these results in what follows. But we have the feeling that these characterisations shed light upon spaces on domains beyond the points of view usually adopted. Further comments and, in particular, references to the existing literature, especially when starting with intrinsic definitions, may be found in [TrW1].

Remark 2 The above definition of $B_{pq}^s(\Omega)$ and $F_{pq}^s(\Omega)$ applies to any domain Ω in $\mathbf{R}^n$. But in this generality it is not very reasonable. For example, let g be a continuous function on $\mathbf{R}^n$. Then, of course, $f = g \mid \Omega$ can be extended continuously to $\overline{\Omega}$. Coming that way, isolated points, or surfaces of lower dimension etc. within the interior of $\overline{\Omega}$ are completely ignored. The atomic decomposition given in 2.2.3 shows that this argument applies also to more general functions and distributions, and related spaces. In other words, it seems to be reasonable to restrict consideration to those domains Ω which coincide with the interior of their closure, $\Omega = \text{int}(\overline{\Omega})$. Furthermore, it is convenient for us, although not necessary, to assume that Ω is bounded. We summarise this discussion as follows.

Definition 2 *Let* $\text{MR}(n)$ *(minimally regular) be the collection of all bounded domains* Ω *in* $\mathbf{R}^n$ *with* $\Omega = \text{int}(\overline{\Omega})$.

Remark 3 We refer to [TrW1], 3.1 for further details and discussions, and especially for what this means in connection with intrinsically defined Sobolev spaces. Our aim is to give intrinsic characterisations of the above spaces by using atoms in domains. For that purpose we need a few further mild restrictions on the underlying domains. In these, cubes will always be cubes in $\mathbf{R}^n$ with sides parallel to the axes.

Definition 3 *(i) Let* IR(n) *(interior regular) be the collection of all domains* $\Omega \in$ MR(n) *for which there is a positive number c such that for any cube Q centred on $\partial\Omega$ with sidelength less than or equal to 1,*

$$|\Omega \cap Q| \geq c\,|Q|.\tag{3}$$

(ii) Let ER(n) *(exterior regular) be the collection of all domains* $\Omega \in$ MR(n) *for which there is a positive number c such that for any cube Q centred on $\partial\Omega$ with sidelength l less than or equal to 1 there exists a subcube Q^e with sidelength cl and*

$$Q^e \subset Q \cap \left(\mathbf{R}^n \backslash \overline{\Omega}\right).\tag{4}$$

(iii) Let

$$R(n) = \mathrm{IR}(n) \cap \mathrm{ER}(n) \ (regular).\tag{5}$$

Remark 4 We again followed [TrW1], 3.2, where not only was the geometrical meaning of these classes of domains discussed, but also the necessary references to the existing literature for non-smooth domains and related function spaces were given; see also 4.3 of that paper. In what follows we shall not rely on these considerations. This may justify our giving merely a brief report here, referring the reader to [TrW1] and [TrW2] for further details, proofs and, in particular, to the respective literature. Note, however, that condition IR(n) is related to the condition of α-plumpness of Martio and Väisälä [MaVä], 2.1 and to the corkscrew condition of Jerison and Kenig [JK], (3.1).

Embeddings Again let $\Omega \in$ MR(n). Some assertions for the spaces $B^s_{pq}(\mathbf{R}^n)$ and $F^s_{pq}(\mathbf{R}^n)$ can be extended immediately to the corresponding spaces on Ω. This applies especially to embeddings. Let

$$s \in \mathbf{R},\ 0 < p_0 < p < p_1 \leq \infty,\ s_0 - \frac{n}{p_0} = s - \frac{n}{p} = s_1 - \frac{n}{p_1}\tag{6}$$

and let $q, u, v \in (0, \infty]$. Then

$$B^{s_0}_{p_0 u}(\Omega) \subset F^s_{pq}(\Omega) \subset B^{s_1}_{p,v}(\Omega)\tag{7}$$

if, and only if, $0 < u \leq p \leq v \leq \infty$. This follows from assertion 2.3.3(ii) and the above Definition 1. Also the other embeddings in 2.3.3 have immediate counterparts. Since Ω is bounded we are now even in a better position than on $\mathbf{R}^n$. Let

$$-\infty < s_2 < s_1 < \infty,\ p_1, p_2, q_1, q_2 \in (0, \infty]\tag{8}$$

and

$$\delta_+ = s_1 - s_2 - n \max \left(\frac{1}{p_1} - \frac{1}{p_2}, 0 \right) > 0. \qquad (9)$$

Then the embedding

$$\text{id} : B^{s_1}_{p_1 q_1}(\Omega) \to B^{s_2}_{p_2 q_2}(\Omega) \qquad (10)$$

is compact. Of course, the B spaces in (10) can be replaced by the F spaces. If Ω is a bounded smooth domain then the entropy numbers of the compact embedding (10) will be studied in great detail in 3.3. If Ω is a non-smooth domain then the compactness of the embedding (10) follows from Definition 1 and the corresponding assertion for bounded smooth domains just mentioned. We return to this problem briefly in 3.5.

2.5.2 Atoms and atomic domains

We follow [TrW1], 3.3 and 3.4. As in the previous subsection we restrict ourselves to a brief report. We are looking for the relevant counterparts on domains of atoms, sequence spaces, and atomic representations on $\mathbf{R}^n$ as described in 2.2.3. It emerges that Definition 2.2.3/1_σ, but not Definition 2.2.3/1_K, can be taken as a starting point for the introduction of atoms on domains. For that purpose we need first the counterparts of the spaces $\mathscr{C}^\sigma(\mathbf{R}^n)$, which can also be normed by (2.2.3/10).

The spaces $\mathscr{C}^\sigma(\overline{\Omega})$ We always assume that $\Omega \in MR(n)$, see Definition 2.5.1/2. Let $0 < \sigma = [\sigma] + \{\sigma\}$ with $[\sigma] \in \mathbf{N}_0$ and $0 < \{\sigma\} < 1$. Then $\mathscr{C}^\sigma(\overline{\Omega})$ is the collection of all complex continuous functions on $\overline{\Omega}$ having classical derivatives $D^\alpha f$ for $|\alpha| \le [\sigma]$ in Ω, which can be extended continuously to $\overline{\Omega}$, and such that

$$\left\| f \mid \mathscr{C}^\sigma(\overline{\Omega}) \right\| = \sum_{|\alpha| \le [\sigma]} \| D^\alpha f \mid L_\infty(\Omega) \| + \sum_{|\alpha| = [\sigma]} \sup \frac{|D^\alpha f(x) - D^\alpha f(y)|}{|x - y|^{\{\sigma\}}} < \infty,$$

$$(1)$$

where the supremum is taken over all $x \in \overline{\Omega}$ and $y \in \overline{\Omega}$ with $x \ne y$. This is the direct counterpart of (2.2.3/10) with respect to the closed set $\overline{\Omega}$.

The Whitney extension Since in our case the compact set $\overline{\Omega}$ is the closure of a domain, the above Banach space $\mathscr{C}^\sigma(\overline{\Omega}), 0 < \sigma \notin \mathbf{N}$; is the Lipschitz space $\text{Lip}(\sigma, \overline{\Omega})$. By Whitney's extension method there exists a linear and bounded extension operator from $\mathscr{C}^\sigma(\overline{\Omega})$ into $\mathscr{C}^\sigma(\mathbf{R}^n) = B^\sigma_{\infty\infty}(\mathbf{R}^n)$. In particular, $\mathscr{C}^\sigma(\overline{\Omega})$ is the restriction of $\mathscr{C}^\sigma(\mathbf{R}^n)$ on $\overline{\Omega}$. Details may be found in [JoW], pp. 22, 44–5 and [Ste], pp. 170–180. There is no counterpart of this assertion for the spaces C^K introduced just before Definition 2.2.3/1_K. Together with the generalisation (2.2.3/1), compared with $x^{v,k} = 2^{-v}k$, this gives just the reason for our complementation of Definition 2.2.3/1_K by Definition 2.2.3/1_σ.

Atoms Let $\Omega \in MR(n)$. Let Q_{vk} with $v \in \mathbf{N}_0$ and $k \in \mathbf{Z}^n$ be the same cubes in $\mathbf{R}^n$ as in 2.2.3, including (2.2.3/1) with $b > 0$ given. As in 2.2.3, just before Definition 2.2.3/1_K and in Remark 2.2.3/1 it is always assumed that d is sufficiently large; in particular we have (2.2.3/2). Now we assume in addition that $x^{v,k} \in \partial\Omega$ for the centres $x^{v,k}$ of those cubes Q_{vk} with $dQ_{vk} \cap \partial\Omega \neq \emptyset$. Accordingly we call Q_{vk}

$$\text{an interior cube if } dQ_{vk} \subset \Omega, v \in \mathbf{N}_0, k \in \mathbf{Z}^n, \tag{2}$$

and

$$\text{a boundary cube if } x^{v,k} \in \partial\Omega, v \in \mathbf{N}_0, k \in \mathbf{Z}^n. \tag{3}$$

Cubes in $\mathbf{R}^n \setminus \overline{\Omega}$ are no longer of interest for us. Let

$$\Omega^v = \{x \in \mathbf{R}^n : 2^{-v}x \in \Omega\}, v \in \mathbf{N}_0, \tag{4}$$

and let $\mathscr{C}^0(\overline{\Omega})$ be the space of all complex continuous functions on $\overline{\Omega}$.

Definition 1 *Let $\Omega \in MR(n)$, $0 \le \sigma \notin \mathbf{N}$, $s \in \mathbf{R}$, $0 < p \le \infty$.*
(i) A function a is called a 1-atom (more precisely, a 1_σ-atom) in Ω if

$$\text{supp } a \subset \overline{\Omega} \cap dQ_{0k} \tag{5}$$

for some interior or boundary cube Q_{0k} with $k \in \mathbf{Z}^n$ and

$$a \in \mathscr{C}^\sigma\left(\overline{\Omega}\right) \text{ with } \left\| a \mid \mathscr{C}^\sigma\left(\overline{\Omega}\right) \right\| \le 1. \tag{6}$$

(ii) In addition suppose that $L + 1 \in \mathbf{N}_0$. Then a function a is called an (s, p)-atom (more precisely, an $(s, p)_{\sigma,L}$-interior-atom) in Ω if

$$\text{supp } a \subset dQ_{vk} \tag{7}$$

for some interior cube Q_{vk} with $v \in \mathbf{N}$ and $k \in \mathbf{Z}^n$,

$$a \in \mathscr{C}^\sigma\left(\overline{\Omega}\right) \text{ with } \left\| a\left(2^{-v}\cdot\right) \mid \mathscr{C}^\sigma\left(\overline{\Omega^v}\right) \right\| < 2^{-v(s-\frac{n}{p})} \tag{8}$$

and

$$\int_\Omega x^\beta a(x)dx = 0 \text{ if } |\beta| \le L. \tag{9}$$

(iii) A function a is called an (s,p)-atom (more precisely, an $(s,p)_\sigma$-boundary-atom) in Ω if

$$\text{supp } a \subset \overline{\Omega} \cap dQ_{vk} \tag{10}$$

and (8) holds for some boundary cube Q_{vk} with $v \in \mathbf{N}$ and $k \in \mathbf{Z}^n$.

Remark 1 This definition is the direct counterpart of Definition $2.2.3/1_\sigma$ (now also with $\sigma = 0$); see also Remark 2.2.3/1 for technical explanations. The most remarkable new aspect is that no moment conditions of type (9) are required for the boundary atoms.

We need the counterparts of the sequence spaces b_{pq} and f_{pq} introduced in Definition 2.2.3/2. In modification of (2.2.3/11) we put

$$\lambda = \{\lambda_{vk} : \lambda_{vk} \in \mathbf{C}, v \in \mathbf{N}_0, \ k \in \mathbf{Z}^n, \ Q_{vk} \text{ interior or boundary cube.}\} \tag{11}$$

Furthermore, $\sum_{k\in\mathbf{Z}^n}^{v,\Omega}$ means that for fixed $v \in \mathbf{N}_0$ the sum is taken over those $k \in \mathbf{Z}^n$ for which Q_{vk} is an interior or boundary cube. Let $\chi_{vk}^{(p)}$ be the same normalised characteristic function of Q_{vk} as in (2.2.3/12).

Definition 2 *Let $\Omega \in \mathrm{MR}(n)$, and let $p, q \in (0, \infty]$.*

(i) Then $b_{pq}(\Omega)$ is the collection of all sequences λ given by (11) such that

$$\|\lambda \mid b_{pq}(\Omega)\| = \left(\sum_{v=0}^{\infty} \left(\sum_{k\in\mathbf{Z}^n}^{v,\Omega} |\lambda_{vk}|^p \right)^{\frac{q}{p}} \right)^{\frac{1}{q}} \tag{12}$$

(with the usual modification if p and/or q is infinite) is finite.

(ii) Also $f_{pq}(\Omega)$ is the collection of all sequences λ given by (11) such that

$$\|\lambda \mid f_{pq}(\Omega)\| = \left\| \left(\sum_{v=0}^{\infty} \sum_{k\in\mathbf{Z}^n}^{v,\Omega} |\lambda_{vk}\chi_{vk}^{(p)}(\cdot)|^q \right)^{\frac{1}{q}} \mid L_p(\Omega) \right\| \tag{13}$$

(with the usual modification if q is infinite) is finite.

Remark 2 This definition coincides with [TrW1], Definition 3.4/1, and it is the obvious counterpart of Definition 2.2.3/2. In [TrW1], Remark 3.4/1, there was a discussion of the differences between these two definitions and how they were related to the types of domains introduced in Definition 2.5.1/3. In particular, an immediate counterpart of Proposition 2.2.3 cannot be expected. So we convert this into a definition of domains having this property. Apart from this we use the same notation as in Proposition 2.2.3; in particular, σ_p and σ_{pq} are given by (2.2.3/17).

Definition 3 *Let $s \in \mathbf{R}$ and $0 < q \leq \infty$.*

(i) Let $0 < p \leq \infty$, $0 \leq \sigma \notin \mathbf{N}$, $\sigma > s$, and $L + 1 \in \mathbf{N}_0$ with $L \geq \max([\sigma_p - s], -1)$. Then $\mathrm{Atom}(B_{pq}^s)^n$ (atomic B_{pq}^s-domain) denotes the collection of all domains $\Omega \in \mathrm{MR}(n)$ such that for any fixed choices of σ and L,

$$\sum_{v=0}^{\infty} \sum_{k \in \mathbf{Z}^n} {}^{v,\Omega} \lambda_{vk} a_{vk}(x), \; x \in \Omega, \; \lambda \in b_{pq}(\Omega), \tag{14}$$

converges in $\mathscr{D}'(\Omega)$ to an element of $B_{pq}^s(\Omega)$, where the a_{vk} are 1_σ-atoms $(v = 0)$, $(s, p)_{\sigma,L}$- interior atoms $(v \in \mathbf{N})$ or $(s, p)_\sigma$-boundary atoms $(v \in \mathbf{N})$ according to Definition 1 with

$$\mathrm{supp} \; a_{vk} \subset \overline{\Omega} \cap dQ_{vk}, \; v \in \mathbf{N}_0. \tag{15}$$

(ii) Let $0 < p < \infty$, $0 \leq \sigma \notin \mathbf{N}$, $\sigma > s$ and $L + 1 \in \mathbf{N}_0$ with $L \geq \max([\sigma_{pq} - s], -1)$. Then $\mathrm{Atom}(F_{pq}^s)^n$ (atomic F_{pq}^s-domain) denotes the collection of all domains $\Omega \in \mathrm{MR}(n)$ such that for all fixed choices of σ and L,

$$\sum_{v=0}^{\infty} \sum_{k \in \mathbf{Z}^n} {}^{v,\Omega} \lambda_{vk} a_{vk}(x), \; x \in \Omega, \; \lambda \in f_{pq}(\Omega), \tag{16}$$

converges in $\mathscr{D}'(\Omega)$ to an element of $F_{pq}^s(\Omega)$, where the $a_{vk}(x)$ are the same atoms as in part (i).

Remark 3 This definition coincides with [TrW1], Definition 3.4/2. Of course, this definition not only is related to Proposition 2.2.3 but also uses the natural restrictions obtained in Theorem 2.2.3. In (14) and (16) we ask only for convergence in $\mathscr{D}'(\Omega)$. But it turns out that these series, under the conditions of the theorem in the next subsection, also converge in $B_{pq}^s(\Omega)$ and $F_{pq}^s(\Omega)$, respectively, if $p < \infty$ and $q < \infty$. If $p = \infty$ and/or $q = \infty$ the situation is different. This is the counterpart of the Fatou property discussed in 2.4.2.

2.5.3 Atomic representations

We are now in a position to formulate the counterpart of 2.2.3 which dealt with the atomic representations of B_{pq}^s and F_{pq}^s on $\mathbf{R}^n$. The conditions in Theorem 2.2.3 are now incorporated in Definition 2.5.2/3. Atomic representations on $\mathbf{R}^n$ are used in this book quite often. This is not the case for the corresponding representations on domains, at least as far as this book is concerned (but we are convinced that the results obtained will be of some use for the further study of general pseudodifferential operators, integral operators etc. in non-smooth domains). We restrict ourselves to formulations. Further details, explanations and, in particular, proofs may be found in [TrW1]. The notation has the same meaning as in the two preceding subsections.

Theorem *(i) Let p, $q \in (0, \infty]$ and $s \in \mathbf{R}$; then*

$$\text{Atom}(B_{pq}^s)^n \supset \text{ER}(n). \tag{1}$$

Suppose, in addition, that $s > \sigma_p$. Then

$$\text{Atom}(B_{pq}^s)^n = \text{MR}(n). \tag{2}$$

(ii) Let $0 < p < \infty$, $0 < q \leq \infty$ and $s \in \mathbf{R}$. Then

$$\text{Atom}(F_{pq}^s)^n \supset \text{R}(n). \tag{3}$$

If, in addition, $s > \sigma_{pq}$, then

$$\text{Atom}(F_{pq}^s)^n \supset \text{IR}(n). \tag{4}$$

Remark 1 This theorem is covered by [TrW1], 3.5, 3.6. To provide some perspective we add a few comments. Let $s > \sigma_p$ in the B case. Then by Definition 2.5.2/3 no moment conditions for the Ω-atoms are required. Now the Whitney extension theorem (see 2.5.2) allows us to transfer this problem from Ω to $\mathbf{R}^n$, which in turn is covered by Theorem 2.2.3. The position is similar if $s > \sigma_{pq}$ in the F case, but here one must try to replace $L_p(\Omega)$ in (2.5.2/13) by $L_p(\mathbf{R}^n)$. This explains the difference between (4) and (2). For general $s \in \mathbf{R}$ one has in addition to take account of the required moment conditions. Then one needs $\Omega \in \text{ER}(n)$, as in (1), which results in the F case in (3).

Representations and equivalent quasi-norms If $f = g \mid \Omega$ with, say $g \in F_{pq}^s(\mathbf{R}^n)$, then it follows immediately from Theorem 2.2.3 that $f \in F_{pq}^s(\Omega)$ can be represented by (2.5.2/16), which means that

$$f = \sum_{v=0}^{\infty} \sum_{k \in \mathbf{Z}^n} {}^{v,\Omega} \lambda_{vk} a_{vk}(x), \; x \in \Omega, \; \lambda \in f_{pq}(\Omega), \tag{5}$$

convergence being in $\mathscr{D}'(\Omega)$. Hence the above theorem covers the harder converse of this assertion. This includes the related statement about equivalent quasi-norms. For example, let $\Omega \in R(n)$, $0 < p < \infty$, $0 < q \leq \infty$ and $s \in \mathbf{R}$, and let σ and L be fixed, satisfying the required conditions according to Definitions 2.5.2/3(ii). Then $f \in \mathscr{D}'(\Omega)$ belongs to $F_{pq}^s(\Omega)$ if, and only if, it can be represented by (5) (with convergence in $\mathscr{D}'(\Omega)$); and

$$\inf \| \lambda \mid f_{pq}(\Omega) \|, \tag{6}$$

where the infimum is taken over all admissible representations (5), is an equivalent quasi-norm on $F_{pq}^s(\Omega)$. This assertion corresponds to (3). There are obvious counterparts related to (1), (2) and (4).

Remark 2 The boundary atoms introduced in Definition 2.5.2/1(iii) are atoms on $\overline{\Omega}$; here no use was made of the additional information that $\overline{\Omega}$ is the closure of a (bounded) domain. On that basis [TrW2] introduced atoms on arbitrary closed (fractal) sets in $\mathbf{R}^n$ with empty interior and related Besov spaces. We do not go into further details here.

2.6 The spaces $L_p(\log L)_a$ and logarithmic Sobolev spaces

2.6.1 Definitions and preliminaries

The last chapter of this book mainly deals with the distribution of eigenvalues of degenerate elliptic operators in bounded domains Ω in $\mathbf{R}^n$. The (singular) coefficients of these operators might belong to some Sobolev spaces $H_p^s(\Omega)$, preferably to $L_p(\Omega)$. However, to handle certain sophisticated limiting situations a more refined tuning is desirable. For that purpose we introduce in this section 2.6 logarithmic Sobolev spaces $H_p^s(\log H)_a$ with a special emphasis on $s = 0$; in other words, on $L_p(\log L)_a$. For these latter spaces we rely on the careful presentation given in [BeS] and the underlying references. However, we have a fresh look at these spaces and offer in 2.6.2 new representation theorems which might be also of intrinsic interest. For us these theorems provide the basis for our later work on entropy numbers of compact embeddings. However, when no extra effort is needed we do not necessarily assume in 2.6.1 and 2.6.2 that the spaces $L_p(\log L)_a$ are distributional spaces, that is, subspaces of L_1^{loc}.

We always suppose that Ω is a domain in $\mathbf{R}^n$ with finite Lebesgue measure $|\Omega|$. For any p with $0 < p \leq \infty$ we let $L_p(\Omega)$ be the usual complex quasi-Banach space with respect to Lebesgue measure and equipped with the quasi-norm $\| f \mid L_p(\Omega) \| = \left(\int_\Omega |f(x)|^p \, dx \right)^{\frac{1}{p}}$, with the obvious modification if $p = \infty$. As usual, functions equal almost everywhere will be identified. To which base log is taken is rather immaterial but one may think of 2 or e.

Definition *(i) Let $0 < p < \infty$ and $a \in \mathbf{R}$. Then $L_p(\log L)_a(\Omega)$ is the set of all measurable functions $f : \Omega \to \mathbf{C}$ such that*

$$\int_\Omega |f(x)|^p \log^{ap}(2 + |f(x)|)dx < \infty. \tag{1}$$

(ii) Let $a < 0$. Then $L_\infty(\log L)_a(\Omega)$ is the set of all measurable functions $f : \Omega \to \mathbf{C}$ for which there exists a constant $\lambda > 0$ such that

$$\int_\Omega \exp\left\{(\lambda |f(x)|)^{-1/a}\right\} dx < \infty. \tag{2}$$

Remark 1 We follow [ET4]. When $p < \infty$ our notation resembles that of Bennett and Sharpley [BeS], pp. 252–3. The alternative notation $L_{\exp,-a}(\Omega)$ for $L_\infty(\log L)_a(\Omega)$ is closer to that employed in [BeS] for that case. The representation theorems in 2.6.2 justify our notation, which may seem somewhat complicated at first glance. We always assume that these spaces $L_\infty(\log L)_a(\Omega)$ are given the Luxemburg norm in the sense of Orlicz spaces, where the generating function $\exp(t^{-1/a})$ might be replaced by, say, $t^2 \exp(t^{-1/a})$. Background material for Orlicz spaces may be found in [BeS], Chapter 4.8, [KuJF], Chapter 3, [KrR] and [RaR]. A discussion of the Luxemburg norm in $L_\infty(\log L)_a(\Omega)$ may also be found in [Tri3], pp. 591–2. Even when $p < \infty$, the spaces $L_p(\log L)_a(\Omega)$ can, for certain values of p and a, be regarded as Orlicz spaces in the same way, but we find it more convenient to use the characterisation of these spaces in terms of the non-increasing rearrangement f^* of a function f on Ω. We recall that this is defined by

$$f^*(t) = \inf\{\tau > 0 : |\{x \in \Omega : |f(x)| > \tau\}| \le t\}. \tag{3}$$

Then if $0 < p < \infty, a \in \mathbf{R}$, or $p = \infty$ and $a < 0$, it is shown in [BeS], p. 252, that $f \in L_p(\log L)_a(\Omega)$ if, and only if,

$$\left(\int_0^{|\Omega|} \left[(1 + |\log t|)^a f^*(t)\right]^p dt\right)^{1/p} < \infty \tag{4}$$

(with the obvious modification if $p = \infty$). The expression in (4) is, in general, only a quasi-norm, but it can be shown, by the use of a convenient form of the Hardy maximal inequality, that for $1 < p < \infty$ and $a \in \mathbf{R}$, or $p = \infty$ and $a < 0$, the analogue of (4) with $f^*(t)$

replaced by $f^{**}(t) = t^{-1} \int_0^t f^*(s)ds$ defines a norm on $L_p(\log L)_a(\Omega)$ which is equivalent to the quasi-norm (4). Moreover, all these spaces are complete. Concerning the last remarks we refer to [BeS], p. 246 for a special case and [BeR], Corollary 8.2 for the general assertion. Henceforth we shall assume that $L_p(\log L)_a(\Omega)$ is provided with the quasi-norm (4), and we may regard it as a Banach space if $1 < p < \infty$ and $a \in \mathbf{R}$, or $p = \infty$ and $a < 0$. Furthermore, $L_1(\log L)_a(\Omega)$ with $a \geq 0$ may also be regarded as a Banach space; see in addition [BeR], Theorem 8.3.

Remark 2 The spaces $L_{\exp} = L_{\exp,1} = L_\infty(\log L)_{-1}$ and $L(\log L) = L_1(\log L)_1$ have a long history which goes back to the 1920s and early 1930s, when they were introduced by Zygmund, Titchmarsh, Hardy and Littlewood; see [BeS], pp. 288–9 and [BeR] for details. New interest in the spaces $L_{\exp,-a} = L_\infty(\log L)_a$ with $a < 0$ came from Trudinger's observation that they may serve as target spaces of limiting embeddings of, say, $H_p^{n/p}(\Omega)$ in $L_\infty(\log L)_a(\Omega)$ for some $a < 0$; see Trudinger [Tru] and Strichartz [Str]. For the further history of the spaces $L_p(\log L)_a$ and their generalisations $L_{p,q}(\log L)_a$ we refer to [BeS] and [BeR].

Proposition 1 *(i) Let $0 < \varepsilon < p < \infty$ and $-\infty < a_2 < a_1 < \infty$. Then*

$$L_{p+\varepsilon}(\Omega) \subset L_p(\log L)_{a_1}(\Omega) \subset L_p(\log L)_{a_2}(\Omega) \subset L_{p-\varepsilon}(\Omega) \qquad (5)$$

and

$$L_p(\log L)_\varepsilon(\Omega) \subset L_p(\Omega) \subset L_p(\log L)_{-\varepsilon}(\Omega). \qquad (6)$$

(ii) Let $-\infty < b_1 < b_2 < 0$. Then

$$L_\infty(\Omega) \subset L_\infty(\log L)_{b_2}(\Omega) \subset L_\infty(\log L)_{b_1}(\Omega). \qquad (7)$$

Proof Since $|\Omega| < \infty$ the proposition follows immediately from the above definition.

As has been observed, the spaces $L_p(\log L)_a(\Omega)$ with $1 < p < \infty, a \in \mathbf{R}$, or $p = \infty, a < 0$, or $p = 1, a \geq 0$, may be regarded as Banach spaces. They are Banach function spaces and the theory developed in [BeS], Chapters 1 and 2, may be applied. In particular one can ask for the associate and the dual spaces. In the case of the dual spaces we interpret them with respect to the dual pairing $\left(\mathscr{D}(\Omega), \mathscr{D}'(\Omega) \right)$, where $\mathscr{D}(\Omega) = C_0^\infty(\Omega)$ is the collection of all complex C^∞ functions in Ω with compact support in Ω, and $\mathscr{D}'(\Omega)$ stands for the complex distributions on Ω. In our situation this means $(f, g) = \int_\Omega f(x)g(x)dx$ with $fg \in L_1(\Omega)$ in agreement with associate norms and spaces. It can easily be seen that $\mathscr{D}(\Omega)$ is dense in the spaces of interest $L_p(\log L)_a(\Omega)$ if $1 < p < \infty, a \in \mathbf{R}$, or $p = 1, a \geq 0$. Let $L_\infty(\log L)_a^0(\Omega)$ with $a < 0$ be the completion of $L_\infty(\Omega)$ in $L_\infty(\log L)_a(\Omega)$. Then $\mathscr{D}(\Omega)$ is dense in $L_\infty(\log L)_a^0(\Omega)$. This follows from (4) (with $p = \infty$), but it is also a consequence of the representation given in Theorem 2.6.2/1. We remark that in a special case $L_\infty(\log L)_a^0$ coincides with the space Z_0^a in [BeR], 3, Theorem D(d). Hence for all these spaces, duality in the above interpretation makes sense.

Proposition 2 *(i) Let* $1 < p < \infty$ *and* $a \in \mathbf{R}$: *Then*

$$\left(L_p(\log L)_a(\Omega)\right)' = L_{p'}(\log L)_{-a}(\Omega), \text{ where} \frac{1}{p} + \frac{1}{p'} = 1. \qquad (8)$$

(ii) Let $a > 0$. *Then*

$$(L_1(\log L)_a(\Omega))' = L_\infty(\log L)_{-a}(\Omega). \qquad (9)$$

(iii) Let $a < 0$. *Then*

$$\left(L_\infty(\log L)_a^0(\Omega)\right)' = L_1(\log L)_{-a}(\Omega). \qquad (10)$$

Proof A proof of (9) with $a = 1$ may be found in [BeS], pp. 247–8. This proof can be extended immediately to all $a > 0$. In the same way one proves (8) for $1 < p < \infty$ and $a \in \mathbf{R}$. By the same proof it follows that $L_1(\log L)_a(\Omega)$ and $L_\infty(\log L)_{-a}(\Omega)$ with $a > 0$ are mutually associate. Finally, to prove (10) we apply [BeS], Theorem 5.5 on pp. 67–8 with $X = L_\infty(\log L)_a(\Omega)$. In the notation used there X' and X_b coincide with $L_1(\log L)_{-a}(\Omega)$ and $L_\infty(\log L)_a^0(\Omega)$, respectively; see also [BeS], Proposition 3.10 on p. 17. For the fundamental function we have $\varphi_X(t) \to 0$ if $t \downarrow 0$. This follows from (4) with $p = \infty$. Now (10) coincides with the assertion (iii) of the theorem mentioned above.

Remark 3 We refer to [BeR], Theorem 8.4, which covers parts (i) and (ii) of the proposition above. The representation theorem 2.6.2/1 for the spaces $L_p(\log L)_a(\Omega)$ with $a < 0$ and the above proposition are the basis for the representation theorem 2.6.2/2 for the spaces $L_p(\log L)_a(\Omega)$ with $a > 0$. This explains our interest in part (iii). It enables us to prove a representation theorem, for example for the interesting space $L(\log L)(\Omega) = L_1(\log L)_1(\Omega)$.

The spaces $L_{p,q}(\log L)_a\,(\Omega)$ The Lorentz-space version of $L_p(\log L)_a\,(\Omega)$ will be of interest to us later on. Let $0 < p < \infty, 0 < q \le \infty$ and $a \in \mathbf{R}$. Then $L_{p,q}(\log L)_a\,(\Omega)$ is defined to be the set of all measurable functions $f : \Omega \to \mathbf{C}$ such that

$$\left(\int_0^{|\Omega|} \left[t^{1/p} \left(1 + |\log t| \right)^a f^*(t) \right]^q \frac{dt}{t} \right)^{1/q} \tag{11}$$

(with the obvious modification if $q = \infty$) is finite. Just as for the $L_p(\log L)_a\,(\Omega)$ spaces, it can be shown that $L_{p,q}(\log L)_a\,(\Omega)$ is a quasi-Banach space with quasi-norm given by (11) or its modification when $q = \infty$, and that for $1 < p < \infty, 1 \le q \le \infty$ and $a \in \mathbf{R}$ the functional obtained from (11) by replacing f^* by f^{**} is a norm on $L_{p,q}(\log L)_a\,(\Omega)$ equivalent to the original quasi-norm. These spaces have been studied in great detail in [BeR]; see also [BeS], p. 253. Note that $L_{p,p}(\log L)_a\,(\Omega) = L_p(\log L)_a\,(\Omega)$ and that if $a = 0$, then $L_{p,q}(\log L)_0\,(\Omega)$ is the usual Lorentz space $L_{p,q}\,(\Omega)$ (see [BeS], pp. 216–7).

2.6.2 Basic theorems

In agreement with our notation in (2.4.3/3) we write

$$\frac{1}{p^\sigma} = \frac{1}{p} + \frac{\sigma}{n} > 0, \text{ where } 0 < p \le \infty \text{ and } \sigma \in \mathbf{R}. \tag{1}$$

To avoid clumsy notation we set

$$\sigma_j = 2^{-j}, \; \lambda_j = -2^{-j}, \; j \in \mathbf{N}, \tag{2}$$

and use p^{σ_j} and p^{λ_j} in the sense of (1). We recall that Ω is a fixed domain Ω in $\mathbf{R}^n$ with finite Lebesgue measure $|\Omega|$.

Theorem 1 *Let $0 < p \le \infty$ and $a < 0$. Then $L_p(\log L)_a(\Omega)$ is the set of all measurable functions $f : \Omega \to \mathbf{C}$ such that*

$$\left(\int_0^\varepsilon [v^{-a} \| f \mid L_{p^v}(\Omega) \|]^p \frac{d\sigma}{\sigma} \right)^{\frac{1}{p}} < \infty \tag{3}$$

(with the usual modification if $p = \infty$) for $\varepsilon > 0$, and (3) defines an equivalent quasi-norm on $L_p(\log L)_a(\Omega)$. Furthermore, (3) can be replaced

by the equivalent quasi-norm

$$\left(\sum_{j=J}^{\infty} 2^{jap} \|f \mid L_{p^{\sigma_j}}(\Omega)\|^p \right)^{\frac{1}{p}} < \infty \tag{4}$$

(with the usual modification if $p = \infty$) for $J \in \mathbf{N}$.

Proof Step 1 From the monotonicity of the $L_p(\Omega)$ spaces it follows immediately that the expressions in (3) and (4) for all $\varepsilon > 0$ and all $J \in \mathbf{N}$ are equivalent quasi-norms.

Step 2 Let $0 < p < \infty$. Then $D(\Omega)$ is dense both in $L_p(\log L)_a(\Omega)$ and in a space quasi-normed by the expression, say, in (3). Hence if $p < \infty$ it remains to prove that (3) is an equivalent quasi-norm on $L_p(\log L)_a(\Omega)$ for $f \in \mathcal{D}(\Omega)$. We need two steps. Here we modify (3) by

$$\left(\int_0^\varepsilon [\sigma^{-a} \|f \mid L_{p^\sigma,p}(\Omega)\|]^p \frac{d\sigma}{\sigma} \right)^{\frac{1}{p}} = \|f\|_{p,a}, \tag{5}$$

where $L_{p^\sigma,p}(\Omega)$ are the Lorentz spaces introduced at the end of 2.6.1. First we prove the desired equivalence for this modified quasi-norm. From (1),

$$\frac{p}{p^\sigma} - 1 = \frac{\sigma}{n}p, \tag{6}$$

and (2.6.1/11) we have

$$\|f \mid L_{p^\sigma,p}(\Omega)\| = \left(\int_0^{|\Omega|} t^{\frac{\sigma p}{n}} f^*(t)^p dt \right)^{\frac{1}{p}}. \tag{7}$$

Together with (5) this gives

$$\|f\|_{p,a}^p = \int_0^{|\Omega|} f^*(t)^p \int_0^\varepsilon \sigma^{-ap-1} t^{\frac{\sigma p}{n}} d\sigma dt. \tag{8}$$

For small t we write the inner integral as

$$\int_0^\varepsilon \sigma^{-ap-1} \exp\left(-\frac{\sigma p}{n} |\log t| \right) d\sigma = \left(\frac{p}{n} |\log t| \right)^{ap} \int_0^{\varepsilon p |\log t|/n} \tau^{-ap-1} e^{-\tau} d\tau \tag{9}$$

and observe that the final integral tends to $\Gamma(-ap)$ as $t \downarrow 0$. It follows that

$$\|f\|_{p,a}^p \sim \int_0^{|\Omega|} f^*(t)^p (1 + |\log t|)^{ap} dt \tag{10}$$

which by (2.6.1/4) gives the desired equivalence. If $p = \infty$, then there are no technical problems. Hence we may assume that for a given f we have either (2.6.1/4) or (3) with $p = \infty$ and $p^\sigma = n/\sigma$. There are obvious counterparts of (5) and (7). Then (8) is modified by

$$
\begin{aligned}
\|f\|_{\infty,a} &= \sup_{0<t<|\Omega|} f^*(t) \sup_{0<\sigma<\varepsilon} \sigma^{-a} t^{\sigma/n} \\
&\sim \sup_{0<t<|\Omega|} (1 + |\log t|)^a f^*(t),
\end{aligned}
\tag{11}
$$

which is the desired equivalence in this case.

Step 3 All that is left is to prove that (3) and (5) are equivalent quasi-norms when $0 < p \le \infty$ and $a < 0$. First we remark that a simple scaling argument shows that we can replace p^σ in (3) by $p^{c\sigma}$ with $c > 0$. Now it is clear that the expressions in (3) and (5) are equivalent if there are two positive constants c_1 and c_2 such that for all small $\sigma > 0$, and all $f \in \mathscr{D}(\Omega)$ when $p < \infty$ (with obvious modification if $p = \infty$),

$$
c_1 \left\| f \mid L_{p^{2\sigma}}(\Omega) \right\| \le \left\| f \mid L_{p^\sigma,p}(\Omega) \right\| \le c_2 \left\| f \mid L_{p^\sigma}(\Omega) \right\|.
\tag{12}
$$

Recall that $p > p^\sigma > p^{2\sigma}$; see (1). Then the right-hand inequality of (12) holds even with $c_2 = 1$; see [BeS], p. 217. As for the left-hand inequality, we choose $\kappa = \sigma/n$ and apply Hölder's inequality based on

$$
\frac{1}{p^{2\sigma}} = \frac{1}{p} + \frac{2\sigma}{n} = \frac{1}{p} + 2\kappa.
\tag{13}
$$

Then we obtain

$$
\begin{aligned}
\left\| f \mid L_{p^{2\sigma}}(\Omega) \right\| &= \left(\int_0^{|\Omega|} f^*(t)^{p^{2\sigma}} t^{\kappa p^{2\sigma}} t^{-\kappa p^{2\sigma}} dt \right)^{1/p^{2\sigma}} \\
&\le \left(\int_0^{|\Omega|} f^*(t)^p t^{\sigma p/n} dt \right)^{\frac{1}{p}} \left(\int_0^{|\Omega|} t^{-1/2} dt \right)^{2\kappa} \\
&\le c \left\| f \mid L_{p^\sigma,p}(\Omega) \right\|
\end{aligned}
\tag{14}
$$

by (7), for some c independent of σ (with obvious modification if $p = \infty$). The proof is complete.

Remark 1 As observed above, we can take $c_2 = 1$ in (12). Furthermore, 2σ in (12) can be replaced by $b\sigma$ for any $b > 1$. We followed [E14], where the corresponding assertion was proved by the same arguments under the (unnecessary) restriction $1 < p < \infty$. In that case it is reasonable to choose $\varepsilon > 0$ in (3) small and $J \in \mathbf{N}$ in (4) large such that $p^\sigma > 1$

and $p^{\sigma_j} > 1$, respectively. Then the expressions in (3) and (4) are norms and hence $L_p(\log L)_a(\Omega)$ are Banach spaces, as had also been mentioned in Remark 2.6.1/1. A different short direct proof for the case $p = \infty$ may be found in [Tri3], Proposition 2.2.4. See also [Gar], VI "Exercises and further results", 17. The natural incorporation of the case $p = \infty$ in the above theorem justifies our notation $L_\infty(\log L)_a(\Omega)$ compared with $L_{\exp,-a}(\Omega)$.

Remark 2 Let $1 < p < \infty$. Then the advantages of (3) and (4) compared with (2.6.1/1) are quite clear: assertions which hold for L_p spaces on Ω, such as properties of integral operators or pseudodifferential operators, can be carried over immediately to the spaces $L_p(\log L)_a(\Omega)$. As we see from Proposition 2.6.1/1, the spaces $L_p(\log L)_a(\Omega)$ provide a refined tuning of the L_p scale.

Remark 3 Just before Proposition 2.6.1/2 we introduced the space $L_\infty(\log L)_a^0(\Omega)$ where $a < 0$. Now it follows easily from the theorem above that $L_\infty(\log L)_a^0(\Omega)$ is the set of all

$$f \in L_\infty(\log L)_a(\Omega) \text{ with } \lim_{\sigma \downarrow 0} \sigma^{-a} \left\| f \mid L_{n/\sigma}(\Omega) \right\| = 0. \tag{15}$$

Of course, $\lim_{\sigma \downarrow 0}$ can be replaced by

$$\lim_{j \to \infty} 2^{ja} \left\| f \mid L_{n2^j}(\Omega) \right\| = 0. \tag{16}$$

Our next aim is to find appropriate characterisations for the spaces $L_p(\log L)_a(\Omega)$ with $a > 0$. The basic idea is to combine Theorem 1 and Proposition 2.6.1/2. This makes it clear that we must now restrict our considerations to the Banach space cases making their appearance as dual spaces in Proposition 2.6.1/2.

Theorem 2 *Let $1 \leq p < \infty$ and $a > 0$. Then $L_p(\log L)_a(\Omega)$ is the set of all measurable functions $g : \Omega \to \mathbf{C}$ which can be represented as*

$$g = \sum_{j=J}^{\infty} g_j, \ g_j \in L_{p^{\lambda_j}}(\Omega) , \tag{17}$$

for $J \in \mathbf{N}$ with $2^J > p/n$, such that

$$\left(\sum_{j=J}^{\infty} 2^{jap} \left\| g_j \mid L_{p^{\lambda_j}}(\Omega) \right\|^p \right)^{\frac{1}{p}} < \infty. \tag{18}$$

The infimum of the expression (18) taken over all admissible representations (17) is an equivalent norm on $L_p(\log L)_a(\Omega)$.

Proof Step 1 By (1) the condition $2^j > p/n$ ensures that $1 < p^{\lambda_j} < \infty$. Let $1 < p < \infty$. Then by (2.6.1/8) we have

$$(L_q(\log L)_{-a}(\Omega))' = L_p(\log L)_a(\Omega) \tag{19}$$

where

$$\frac{1}{p} + \frac{1}{q} = \frac{1}{p^{\lambda_j}} + \frac{1}{q^{\sigma_j}} = 1, \ j \geq J; \tag{20}$$

see (1) and (2). Let $l_q(L_{q^{\sigma_j}}(\Omega))$ be the Banach space of all sequences $F = (F_J, F_{J+1}, ...)$ with each $F_j \in L_{q^{\sigma_j}}(\Omega)$, normed in the natural way. With the usual interpretation the dual space is given by

$$[l_q(L_{q^{\sigma_j}}(\Omega))]' = l_p(L_{p^{\lambda_j}}(\Omega)). \tag{21}$$

By Theorem 1 and (4) the space $L_q(\log L)_{-a}(\Omega)$ can be identified with the subspace $l_q^{\text{sym}}(L_{q^{\sigma_j}}(\Omega))$ of $l_q(L_{q^{\sigma_j}}(\Omega))$ which consists of all elements $F = (F_J, F_{J+1}, ...)$ with $F_j = 2^{-ja}f$, where $f \in L_q(\log L)_{-a}(\Omega)$. By (21) and the Hahn–Banach theorem any $G \in [l_q^{\text{sym}}(L_{q^{\sigma_j}}(\Omega))]'$ can be represented as

$$G = (G_J, G_{J+1}, ...) \in l_p(L_{p^{\lambda_j}}(\Omega)), \tag{22}$$

where

$$\left\| G \mid [l_q^{\text{sym}}(L_{q^{\sigma_j}}(\Omega))]' \right\| = \left(\sum_{j=J}^{\infty} \left\| G_j \mid L_{p^{\lambda_j}}(\Omega) \right\|^p \right)^{\frac{1}{p}} \tag{23}$$

and the dual pairing is given by

$$(F, G) = \int_{\Omega} \sum_{j=J}^{\infty} F_j G_j dx = \int_{\Omega} \left(\sum_{j=J}^{\infty} 2^{-ja} G_j \right) f dx = \int_{\Omega} gf dx. \tag{24}$$

By (19) and (23) we finally have (17) and (18). This yields the desired result when $p > 1$.

Step 2 Let $p = 1$. Let $(A_J, A_{J+1}, ...)$ be a sequence of Banach spaces; then $l_{\infty}(A_j)$ has the usual meaning and $c_0(A_j)$ is the subspace of $l_{\infty}(A_j)$ consisting of all sequences $(a_J, a_{J+1}, ...)$ with $\|a_j \mid A_j\| \to 0$. With the usual interpretation we have

$$[c_0(A_j)]' = l_1(A_j'), \tag{25}$$

see [Triα], Lemma 1.11.1, p. 68 (generalised in an obvious way). Now (19) must be replaced by (2.6.1/10) with $-a$ in place of a. We have (20) with $p = 1, q = \infty$ and, hence, $q^{\sigma_j} = n2^j$. By Theorem 1 and Remark 3 the desired replacement of (21) is given by (25) with $A_j = L_{n2^j}(\Omega)$. The rest is the same as above.

Remark 4 Analogously to (3) in Theorem 1 one might ask for a continuous version of (17), (18). Neglecting measure-theoretical discussions, this can be done as follows: $g \in L_p(\log L)_a(\Omega)$ if, and only if, g can be represented as

$$g(x) = \int_0^\varepsilon g(\lambda, x)\frac{d\lambda}{\lambda}, \; g(\lambda, x) \in L_{p-\lambda}(\Omega), \tag{17'}$$

with

$$\left(\int_0^\varepsilon \lambda^{-ap} \left\| g(\lambda, \cdot) \mid L_{p-\lambda} \right\|^p \frac{d\lambda}{\lambda} \right)^{1/p} < \infty. \tag{18'}$$

If we have (17), (18) then we put (besides constants) $g(\lambda, x) = g_j$ if $\lambda \in (2^{-j}, 2^{-j+1}]$ and obtain (18'). Conversely, if we have (17'), (18') then we choose $g_j(x) = \int_{2^{-j}}^{2^{-j+1}} g(\lambda, x)\lambda^{-1}d\lambda$ and (18) follows easily from (18').

Remark 5 We again followed [ET4], now complemented by the case $p = 1$. It is just this limiting case for which the above representation theorem might be of peculiar interest. We give an example. Again let $L(\log L)(\Omega) = L_1(\log L)_1(\Omega)$. Then from (17), (18) with $a = p = 1$ and

$$\frac{1}{p^{\lambda_j}} = 1 - \frac{1}{2^j n} \text{ or } 2^j \sim \frac{1}{p^{\lambda_j} - 1}, \tag{26}$$

we see that g belongs to $L(\log L)(\Omega)$ if, and only if, it can be represented by (17) with

$$\sum_{j=1}^\infty \frac{1}{p^{\lambda_j} - 1} \left\| g_j \mid L_{p^{\lambda_j}}(\Omega) \right\| < \infty \tag{27}$$

(equivalent norms). Let Mf be the Hardy–Littlewood maximal function. Then there exists a positive constant c such that

$$\| Mf \mid L_p(\Omega) \| \le \frac{c}{p-1} \| f \mid L_p(\Omega) \|, \; p \downarrow 1; \tag{28}$$

see [BeS], p. 125, for an elementary proof of this well-known assertion; see also [Ste], pp. 5–8. Since $|\Omega| < \infty$ and M is sublinear, (27) and (28) prove that

$$\| Mf \mid L_1(\Omega) \| \le c \| f \mid L(\log L)(\Omega) \|. \tag{29}$$

This very classical assertion goes back to Hardy and Littlewood (1930); see [BeR], Theorem 3.2, [BeS], pp. 250, 288 and [Ste], p. 23, and the references given there. This result can easily be generalised. In case of $L_1(\log L)_a(\Omega)$ with $a > 0$ one has simply to replace $(p^{\lambda_j} - 1)^{-1}$ in (27) by $(p^{\lambda_j} - 1)^{-a}$. Then by (28) it follows immediately that

$$\|Mf \mid L_1(\log L)_a(\Omega)\| \leq c \|f \mid L_1(\log L)_{a+1}(\Omega)\|, \qquad (30)$$

where $a \geq 0$; see also [Ste], p. 23. Assertions of type (28) are by no means restricted to the maximal function. On the contrary, if a sublinear operator is of weak-type $(1,1)$ and of (strong or weak)-type (q,q) for some q with $1 < q \leq \infty$, then one has always an estimate of type (28) (see [Tor], Theorem IV, 4.1 on p. 87, Theorem IV, 4.3 on p. 91), and hence of type (30). This also applies, for example, to the Hilbert transform. New light is thus shed on the classical results of Hardy, Littlewood, Titchmarsh, Riesz, Zygmund and Stein, together with the more recent extensions in [BeS] and, in particular, [BeR] (where one also finds careful references).

2.6.3 Logarithmic Sobolev spaces

As we remarked at the beginning of 2.6.1, to deal with a few sophisticated limiting cases in the spectral theory of some degenerate elliptic operators a refined tuning of the scale of Sobolev spaces $H_p^s(\Omega)$, $s \in \mathbf{R}$, $0 < p < \infty$, is desirable. Recall that we introduced the spaces

$$H_p^s(\mathbf{R}^n) = F_{p,2}^s(\mathbf{R}^n), s \in \mathbf{R}, 0 < p < \infty, \qquad (1)$$

in (2.2.2/7) and (2.3.3/11). Even in this generality we shall call them Sobolev spaces. This justifies our calling the spaces which we introduce below logarithmic Sobolev spaces. Let Ω be a domain in $\mathbf{R}^n$; then $H_p^s(\Omega)$ is the restriction of $H_p^s(\mathbf{R}^n)$ on Ω, in agreement with Definition 2.5.1/1. In 2.5.1 we also discussed some problems connected with this definition.

In this subsection Ω stands for a bounded domain in $\mathbf{R}^n$ with C^∞ boundary. Then there exists a bounded linear extension operator from $H_p^s(\Omega)$ to $H_p^s(\mathbf{R}^n)$; see [Triγ], 5.1.3, p. 239. Besides $H_p^s(\Omega)$ we introduce the closed subspaces of $H_p^s(\mathbf{R}^n)$

$$\widetilde{H}_p^s(\Omega) = \{f \subset H_p^s(\mathbf{R}^n) : \text{supp } f \subset \overline{\Omega}\}, s \in \mathbf{R}, 0 < p < \infty \qquad (2)$$

Let $0 < p < \infty$ and let p^σ, σ_j and λ_j be given by (2.6.2/1) and (2.6.2/2).

Definition *Let $0 < p < \infty$ and $s \in \mathbf{R}$.*

(i) Let $a < 0$. Then $H_p^s (\log H)_a (\Omega)$ consists of all $f \in \mathscr{D}' (\Omega)$ such that

$$\left(\sum_{j=J}^{\infty} 2^{jap} \left\| f \mid H_{p}^{s_{\sigma_j}} (\Omega) \right\|^p \right)^{\frac{1}{p}} < \infty \tag{3}$$

for $J \in \mathbf{N}$.

(ii) Let $a > 0$. Then $H_p^s (\log H)_a (\Omega)$ consists of all $g \in \mathscr{D}' (\Omega)$ which can be represented as

$$g = \sum_{j=J}^{\infty} g_j, \;\; g_j \in H_{p^{\lambda_j}}^s (\Omega) \; for \; J \in \mathbf{N}, \;\; 2^J > \frac{p}{n}, \tag{4}$$

with

$$\left(\sum_{j=J}^{\infty} 2^{jap} \left\| g_j \mid H_{p^{\lambda_j}}^s (\Omega) \right\|^p \right)^{\frac{1}{p}} < \infty, \tag{5}$$

and quasi-normed by the infimum of all expressions in (5) over all admissible representations (4).

(iii) Let $0 \neq a \in \mathbf{R}$. Then $\widetilde{H}_p^s \left(\log \widetilde{H} \right)_a (\Omega)$ is defined as in (i) and (ii) with $\widetilde{H}$ instead of H.

Remark 1 If $a = 0$ then we put $H_p^s (\log H)_0 (\Omega) = H_p^s (\Omega)$ and $\widetilde{H}_p^s \left(\log \widetilde{H} \right)_0 (\Omega) = \widetilde{H}_p^s (\Omega)$. To justify the above definition we must be sure that it is independent of $J \in \mathbf{N}$ (up to equivalent quasi-norms). Let $0 < p_1 \leq p_2 < \infty$, $s \in \mathbf{R}$ and let B be a ball in $\mathbf{R}^n$; then there exists a positive constant c such that

$$\left\| f \mid H_{p_1}^s (\mathbf{R}^n) \right\| \leq c \left\| f \mid H_{p_2}^s (\mathbf{R}^n) \right\| \text{ for all } f \in H_{p_2}^s (\mathbf{R}^n) \tag{6}$$

with supp$f \subset B$. This follows from Hölder's inequality and the characterisation of these spaces via local means; see [Triγ], 2.4.6, p. 122. Then we have by restriction

$$H_{p_2}^s (\Omega) \subset H_{p_1}^s (\Omega), \, 0 < p_1 \leq p_2 < \infty. \tag{7}$$

Note that (7) also follows from the atomic representations of these spaces described in 2.5.2 and 2.5.3, since any (s, p_2)-atom according to (2.5.2/8) is also an (s, p_1)-atom. Armed with (7) and the respective quasi-norm inequalities it follows easily that the spaces $H_p^s (\log H)_a (\Omega)$ and their modifications in (iii) are independent of $J \in \mathbf{N}$. By the same argument one obtains the counterpart of Proposition 2.6.1/1: Let $s \in \mathbf{R}$, $0 < \varepsilon < p < \infty$ and $-\infty < a_2 < a_1 < \infty$. Then

$$H_{p+\varepsilon}^s (\Omega) \subset H_p^s (\log H)_{a_1} (\Omega) \subset H_p^s (\log H)_{a_2} (\Omega) \subset H_{p-\varepsilon}^s (\Omega) \qquad (8)$$

and

$$H_p^s (\log H)_\varepsilon (\Omega) \subset H_p^s (\Omega) \subset H_p^s (\log H)_{-\varepsilon} (\Omega). \qquad (9)$$

Finally, it follows by standard arguments that all those spaces are quasi-Banach spaces.

Remark 2 We again followed [ET4]. If $1 < p < \infty$ and $a \in \mathbf{R}$ then we have, by Theorems 2.6.2/1 and 2.6.2/2,

$$H_p^0 (\log H)_a (\Omega) = \widetilde{H}_p^0 \left(\log \widetilde{H}\right)_a (\Omega) = L_p (\log L)_a (\Omega). \qquad (10)$$

It is clear that we used these theorems as the motivation for the introduction of the spaces above, for denoting them by $H_p^s (\log H)_a (\Omega)$ and for calling them "logarithmic Sobolev spaces".

Remark 3 We mentioned that the constant c in (6) comes from the classical Hölder inequality for L_p spaces. We have a similar situation in (7) and hence for all spaces involved in (3) and (5). Then the quasi-norm in (3) can be replaced by the (equivalent) quasi-norm

$$\left(\int_0^\varepsilon [\sigma^{-a} \|f \mid H_{p^\sigma}^s (\Omega)\|]^p \frac{d\sigma}{\sigma}\right)^{\frac{1}{p}} < \infty \text{ for } \varepsilon > 0, \qquad (11)$$

which corresponds to (2.6.2/3). In a similar way one has a continuous counterpart of (4), (5) corresponding to (2.6.2/17') and (2.6.2/18').

The case $1 < p < \infty$

If $1 < p < \infty$, then we have (10) and it can be expected that other known relations between $H_p^s (\Omega)$ and $L_p (\Omega)$ have appropriate counterparts for $H_p^s (\log H)_a (\Omega)$. First we collect what is known. Recall that Ω is a bounded domain in $\mathbf{R}^n$ with C^∞ boundary. Besides $\widetilde{H}_p^s (\Omega)$, defined in (2), we introduce $\overset{0}{H}_p^s (\Omega)$ as the completion of $\mathscr{D}(\Omega) = C_0^\infty(\Omega)$ in $H_p^s (\Omega)$, where $s \in \mathbf{R}$, $1 < p < \infty$. (Of course this makes sense for all $p \in (0, \infty)$, but this is not of interest for us here.) To clarify the situation we compare $H_p^s (\Omega)$, $\widetilde{H}_p^s (\Omega)$ and $\overset{0}{H}_p^s (\Omega)$. We have

$$\overset{0}{H}_p^s (\Omega) = H_p^s (\Omega) \text{ if } 1 < p < \infty \text{ and } -\infty < s \leq \tfrac{1}{p}, \qquad (12)$$

$$\widetilde{H}_p^s(\Omega) = \overset{0}{H_p^s}(\Omega) \text{ if } 1 < p < \infty, \frac{1}{p} - 1 < s < \infty, s - \frac{1}{p} \notin \mathbf{N}_0. \tag{13}$$

Furthermore, $D(\Omega)$ is dense in $\widetilde{H}_p^s(\Omega)$ if $1 < p < \infty$ and $s \in \mathbf{R}$. By (12) and (13) we have, in particular,

$$H_p^s(\Omega) = \widetilde{H}_p^s(\Omega) \text{ if } 1 < p < \infty \text{ and } \frac{1}{p} - 1 < s < \frac{1}{p}. \tag{14}$$

For details we refer to [Triα], 4.3.2, p. 318, where further discussion is to be found, in particular for the exceptional case in which $s - 1/p \in \mathbf{N}_0$. We now describe lifting properties for these spaces. Let

$$A_m f = (-\Delta + \mathrm{id})^m f, \, m \in \mathbf{N}, \tag{15}$$

where Δ is the Laplacian. Then $A_{m,D}$ and $A_{m,N}$, defined by

$$\begin{aligned} A_{m,D} f &= A_m f, \, \mathrm{dom}\left(A_{m,D}\right) \\ &= \left\{ f \in H_p^{2m}(\Omega) : \frac{\partial^j f}{dv^j} \mid_{\partial\Omega} = 0 \text{ if } j = 0, ..., m - 1 \right\} \end{aligned} \tag{16}$$

and

$$\begin{aligned} A_{m,N} f &= A_m f, \, \mathrm{dom}\left(A_{m,N}\right) \\ &= \left\{ f \in H_p^{2m}(\Omega) : \frac{\partial^{j+m} f}{\partial v^{j+m}} \mid_{\partial\Omega} = 0 \text{ if } j = 0, ..., m - 1 \right\} \end{aligned} \tag{17}$$

stand for the Dirichlet and Neumann realisations, respectively, of the operator A_m. Since $1 < p < \infty$, we can follow [Triα], 4.9.2, p. 335, and have for the fractional powers $A_{m,D}^\tau$ and $A_{m,N}^\tau$,

$$\mathrm{dom}\left(A_{m,D}^\tau\right) = \widetilde{H}_p^{2m\tau}(\Omega), \, \mathrm{dom}\left(A_{m,N}^\tau\right) = H_p^{2m\tau}(\Omega) \text{ for } 0 \le \tau \le \frac{1}{2}. \tag{18}$$

If now we interpret $A_{m,D}^\tau$ and $A_{m,N}^\tau$, with $0 \le \tau \le \frac{1}{2}$, as mappings between $\mathrm{dom}\left(A_{m,D}^\eta\right)$ and $\mathrm{dom}\left(A_{m,N}^\eta\right)$, always with (18) in mind, without indicating the initial and target spaces and using the duality arguments in [Triα], 4.9.2, p. 336 we have (up to isomorphisms)

$$A_{m,N}^\tau H_p^{s+2m\tau}(\Omega) = H_p^s(\Omega), 0 \le s \le s + 2m\tau \le m, \tag{19}$$

and

$$A_{m,D}^\tau \overline{H}_p^{s+2m\tau}(\Omega) = \overline{H}_p^s(\Omega), \, -m \le s \le s + 2m\tau \le m, 0 \le \tau \le \frac{1}{2}, \tag{20}$$

where $\overline{H}_p^\kappa(\Omega) = \widetilde{H}_p^\kappa(\Omega)$ if $\kappa \ge 0$ and $\overline{H}_p^\kappa(\Omega) = H_p^\kappa(\Omega)$ if $\kappa \le 0$. In that way $A_{m,N}^\tau$ and $A_{m,D}^\tau$, now even with $|\tau| \le \frac{1}{2}$, provide isomorphic maps in the way indicated. For further explanations and proofs we refer to [Triα], 4.9.2, pp. 335–6.

Theorem *Let Ω be a bounded domain in $\mathbf{R}^n$ with C^∞ boundary and let $1 < p < \infty$, $a \in \mathbf{R}$ and $m \in \mathbf{N}$.*
(i) Let $0 \le \tau \le \frac{1}{2}$. Then

$$A_{m,N}^{-\tau} L_p (\log L)_a (\Omega) = H_p^{2m\tau} (\log H)_a (\Omega), \tag{21}$$

$$A_{m,D}^{-\tau} L_p (\log L)_a (\Omega) = \widetilde{H}_p^{2m\tau} \left(\log \widetilde{H}\right)_a (\Omega) \tag{22}$$

and

$$A_{m,D}^{\tau} L_p (\log L)_a (\Omega) = H_p^{-2m\tau} (\log H)_a (\Omega). \tag{23}$$

(ii) For all $s \in \mathbf{R}$, $\mathscr{D}(\Omega)$ is dense in $\widetilde{H}_p^s \left(\log \widetilde{H}\right)_a (\Omega)$.
(iii) If $s \ge 0$, then in the sense of the dual pairing $\left(\mathscr{D}(\Omega), \mathscr{D}'(\Omega)\right)$,

$$\left[\widetilde{H}_p^s \left(\log \widetilde{H}\right)_a (\Omega)\right]' = H_{p'}^{-s} (\log H)_{-a} (\Omega), \quad \frac{1}{p} + \frac{1}{p'} = 1. \tag{24}$$

(iv) If $s \in \mathbf{N}_0$, then

$$H_p^s (\log H)_a (\Omega) = \{f \in \mathscr{D}'(\Omega) : D^\alpha f \in L_p (\log L)_a (\Omega) \text{ if } |\alpha| \le s\} \tag{25}$$

with the equivalent norm

$$\sum_{|\alpha| \le s} \|D^\alpha f \mid L_p (\log L)_a (\Omega)\|. \tag{26}$$

Proof Step 1 Let $1 < p_1 < p_2 < \infty$ and $\frac{1}{p} = \frac{1-\theta}{p_1} + \frac{\theta}{p_2}$ with $0 < \theta < 1$. We have the complex interpolation formula

$$H_p^s (\Omega) = \left[H_{p_1}^s (\Omega), H_{p_2}^s (\Omega)\right]_\theta \tag{27}$$

and a corresponding formula with $\widetilde{H}$ instead of H; see [Triα], Theorems 4.3.1/1 and 4.3.2/2 (complemented by (2.4.2/11) in the formulation of that theorem), pp. 317–19. By (19) and (20) and the interpolation property it follows that all the equivalence constants connected with the isomorphisms (19) and (20) with $p_1 < p < p_2$ can be estimated independently of p. Now part (i) of the theorem follows immediately from the above definition, (19), (20) and Theorems 2.6.2/1 and 2.6.2/2.

Step 2 We prove part (ii). Recall that the density of $\mathscr{D}\left(\mathbf{R}_+^n\right)$ in $\widetilde{H}_p^s\left(\mathbf{R}_+^n\right)$ is proved by standard arguments: translation in the x_n-direction, multiplication by cut-off functions, mollification, which applies uniformly to all p^{σ_j} and p^{λ_j} near p (and this is sufficient), see [Triα], 2.10.3, p. 231. The corresponding assertion for $\widetilde{H}_p^s(\Omega)$ follows afterwards by the technique of local coordinates and hence we have a similar assertion for the spaces $\widetilde{H}_p^s(\Omega), \widetilde{H}_{p^{\sigma_j}}^s(\Omega)$ and $\widetilde{H}_{p^{\lambda_j}}^s(\Omega)$; see [Triα], 4.3.2, p. 318. Now (ii) is a consequence of these observations.

Step 3 By (ii) the left-hand side of (24) makes sense within $\mathscr{D}'(\Omega)$. By (22) and (23), and the self-adjointness of $A_{m,D}$ and its fractional powers, (24) can be reduced to the case $s = 0$. But this coincides with (2.6.1/8).

Step 4 Let $s \in \mathbf{N}$ and $1 < p_1 < p_2 < \infty$. Then

$$\left\| f \mid H_p^s(\Omega) \right\| \sim \sum_{|\alpha| \leq s} \left\| D^\alpha f \mid L_p(\Omega) \right\|, \tag{28}$$

where the equivalence constants can be estimated uniformly with respect to p provided that $p_1 \leq p \leq p_2$. This is well known and also can be ensured by the interpolation arguments used in the first step. When $a < 0$, (iv) then follows from (3) and Theorem 2.6.2/1. By the same argument one obtains in the case of $a > 0$,

$$\sum_{|\alpha| \leq s} \left\| D^\alpha f \mid L_p(\log L)_a(\Omega) \right\| \leq c \left\| f \mid H_p^s(\log H)_a(\Omega) \right\| \tag{29}$$

for some $c > 0$. To prove the opposite assertion we recall the integral representation formula

$$f = \sum_{|\alpha| \leq s} K_\alpha(D^\alpha f), \tag{30}$$

where the K_α are weakly singular integral operators, which is the basis for Calderón's extension method for Sobolev spaces from domains with the cone-property to $\mathbf{R}^n$; see [Triα], pp. 312–13. If one inserts optimal decompositions of $D^\alpha f \in L_p(\log L)_a(\Omega)$ according to Theorem 2.6.2/2, then (30) yields a decomposition of f in $H_p^s(\log H)_a(\Omega)$. By the arguments in connection with Calderón's extension (boundedness of singular integrals in L_p spaces, $1 < p < \infty$), we obtain the opposite estimate to (29), which completes the proof in that case also.

Remark 4 The equivalent norm (26) supports our intention expressed at the beginning of this subsection to call these spaces "logarithmic Sobolev spaces".

Remark 5 In [JawM], B. Jawerth and M. Milman developed an extrapolation theory as a complement to the well-known interpolation theory; see also the recent book [Mil]. They use abstract versions of the constructions in the above definition and in the two theorems in 2.6.2. We refer especially to [JawM], 5.1 and 5.3, where one finds assertions related to the two theorems in 2.6.2 and to (26) in the limiting cases $p = 1$ and $p = \infty$.

2.7 Limiting embeddings

2.7.1 Extremal functions

The spaces B_{pq}^s and F_{pq}^s with $s = \frac{n}{p}$ on $\mathbf{R}^n$ and on domains play a peculiar role. In the $(\frac{1}{p}, s)$-diagram in Fig. 2.1 on p. 52 they correspond to the ray $s = \frac{n}{p}$ with $(0,0)$ as an end point. If one looks for embeddings along this ray in spaces related to $(0,0)$, then there are several possible candidates: besides $B_{\infty,q}^0$ the spaces L_∞, C, bmo and (in the case of bounded domains) $L_\infty (\log L)_a (\Omega) = L_{\exp,-a} (\Omega)$ with $a < 0$. A first discussion was given in 2.3.3(iii). In the case of bounded domains one may ask whether these embeddings, if they exist, are compact. This is not the case if the target space is L_∞, C or bmo (we return to this point in 2.7.3). But the situation is different when the $L_\infty (\log L)_a (\Omega)$ are target spaces. The study of this phenomenon, and its connection with entropy numbers and spectral theory, is one of the main subjects of this book. Section 2.7 provides the necessary background, and it turns out that the "extremal functions" discussed now play a decisive role. To simplify our notation we use temporarily in this subsection the abbreviations

$$\mathcal{B}_p = B_{pp}^{n/p} (\mathbf{R}^n) \text{ and } \mathcal{H}_p = H_p^{n/p} (\mathbf{R}^n), \ 0 < p < \infty; \tag{1}$$

see 2.2.1, 2.2.2 and (2.3.3/11). Recall that $L_p = L_p (\mathbf{R}^n)$ are the Lebesgue spaces (here with $p = j \in \mathbf{N}$) and $L_\infty (\log L)_a = L_{\exp,-a}$ with $a < 0$ are the spaces introduced in Definition 2.6.1(ii) with respect to, say, Ω taken to be the unit ball (but this is immaterial in this subsection). Finally, let $\psi \in C^\infty$ be a function on $\mathbf{R}^n$ with

$$\text{supp } \psi \subset \{y \in \mathbf{R}^n : |y| < \delta\}, \ \psi (x) = 1 \text{ if } |x| \leq \delta/2 \tag{2}$$

for some small $\delta > 0$.

Theorem *Let $1 < p < \infty$ and $\sigma > 1/p$. Then*

$$f_{p\sigma}(x) = |\log|x||^{1-1/p} \left(\log\left(1 - \log|x|\right)\right)^{-\sigma} \psi(x) \in \mathscr{B}_p \cap \mathscr{H}_p, \qquad (3)$$

$$f_{p\sigma} \in L_\infty (\log L)_{\frac{1}{p}-1} \qquad (4)$$

and

$$\|f_{p\sigma} \mid L_j\| \sim j^{1-\frac{1}{p}} \text{ if } j \in \mathbf{N}. \qquad (5)$$

Proof Step 1 Of course, "$\sim$" in (5) means that the corresponding equivalence constants are independent of $j \in \mathbf{N}$. Recall that

$$\|f \mid L_\infty (\log L)_a\| \sim \sup_{j\in\mathbf{N}} j^a \|f \mid L_j\|, \ a < 0; \qquad (6)$$

see (2.6.2/4) with ∞ and $n2^j$ in place of p and p^{σ_j}, respectively, and obvious modifications. Now (4) follows from (5) with $a = \frac{1}{p} - 1$. Hence we have to prove (3) and (5). For that purpose we generalise (3) by

$$f^t_{p\sigma}(x) = |\log|x||^{1-\frac{1}{p}+it} \left(\log|\log\varepsilon|x||\right)^{-\sigma+it} \psi(x), \qquad (7)$$

where $t \in \mathbf{R}$ and $\varepsilon > 0$ is small. Of course the replacement of $1 - \log|x|$ in (3) by $\log\varepsilon|x|$ is immaterial. Since $1 < p < \infty$ and $\sigma > \frac{1}{p}$ are fixed we omit p and σ in (7) in what follows and write $f^t(x)$ instead of $f^t_{p\sigma}(x)$.

Step 2 Let $n = 1$ and $1 < p < \infty$. We prove in this step that

$$f^t \in \mathscr{B}_p \left(= B^{1/p}_{pp}\right) \text{ and } \|f^t \mid \mathscr{B}_p\| \le c \left(1 + |t|\right), \qquad (8)$$

where c is independent of $t \in \mathbf{R}$. Since $\mathscr{B}_p$ is a classical Besov space, we have

$$\|f^t \mid \mathscr{B}_p\|^p \sim \|f^t \mid L_p\|^p + \int_{-1}^1 \int_{-1}^1 \frac{|f^t(x) - f^t(y)|^p}{|x-y|^2} dxdy; \qquad (9)$$

see (2.2.2/12). We may assume that f^t is an even function. Then

$$\frac{|f^t(x)-f^t(y)|^p}{|x-y|^2} \le \frac{|f^t(x)-f^t(-y)|^p}{|x-(-y)|^2}, \ x > 0 \text{ and } y < 0. \qquad (10)$$

It follows that the integration in (9) over $(-1,1)\times(-1,1)$ can be reduced to an integration over $0 < x < y < 1$. Thus to prove (8) it is sufficient to show that

$$I = \left(\int_0^1 \int_0^1 h^{-2} |f^t(x+h) - f^t(x)|^p dxdh\right)^{1/p} \le c\left(1 + |t|\right). \qquad (11)$$

Since

$$f^t(x+h) - f^t(x) = h \int_0^1 (f^t)'(x+h\lambda)\,d\lambda, \tag{12}$$

we have

$$I \leq \int_0^1 d\lambda \left(\int_0^1 \int_0^1 h^{-2+p} \left| (f^t)'(x+h\lambda) \right|^p dxdh \right)^{1/p}. \tag{13}$$

The remainder terms $+\ldots$ in

$$(f^t)'(x) = -\left(1 - \frac{1}{p} + it\right) |\log x|^{-\frac{1}{p}+it} x^{-1} \left(\log |\log \varepsilon x|\right)^{-\sigma+it} \psi(x) + \ldots \tag{14}$$

can be estimated from above near the origin by the absolute value of the first term on the right-hand side. We insert (14) in (13), replace $h\lambda$ by h and obtain

$$I \leq c\left(1 + |t|\right) \int_0^1 \lambda^{\frac{1}{p}-1} d\lambda$$

$$\times \left(\int_0^\delta \int_0^\delta h^{-2+p} |\log(x+h)|^{-1} (x+h)^{-p} \left(\log |\log(\varepsilon x + \varepsilon h)|\right)^{-\sigma p} dxdh \right)^{1/p}, \tag{15}$$

where δ has the same meaning as in (2), and is assumed to be small. With $u = x + h$ we obtain

$$I^p \leq c\left(1 + |t|\right)^p \int_0^\delta h^{-2+p} \int_h^{\delta+h} \frac{-1}{u^p \log u} \left(\log |\log \varepsilon u|\right)^{-\sigma p} dudh. \tag{16}$$

We have

$$\left(\frac{\left(\log |\log \varepsilon u|\right)^{-\sigma p}}{u^{p-1} \log u} \right)' = \frac{1-p}{u^p \log u} \left(\log |\log \varepsilon u|\right)^{-\sigma p} (1 + o(1)) \tag{17}$$

as $u \downarrow 0$. Hence using (17) in the inner integral in (16) we obtain

$$I^p \leq c\left(1 + |t|\right)^p \left| \int_0^\delta h^{-2+p} \frac{\left(\log |\log \varepsilon h|\right)^{-\sigma p}}{h^{p-1} \log h} dh \right| \leq c'\left(1 + |t|\right)^p, \tag{18}$$

on using the fact that $\sigma p > 1$. Hence we have (11) and (8) and, in particular,

$$f_{p\sigma} \in \mathscr{B}_p, \quad 1 < p < \infty, \quad n = 1. \tag{19}$$

Step 3 Let $1 < n < p < \infty$. Here we prove the n-dimensional counterpart of (8),

$$f^t \in \mathscr{B}_p \left(= B_{pp}^{n/p}\right) \quad \text{and} \quad \left\| f^t \mid \mathscr{B}_p \right\| \leq c\left(1 + |t|\right), \tag{20}$$

where c is independent of $t \in \mathbf{R}$. By the Fubini property of $\mathscr{B}_p = B_{pp}^{n/p}$ (see [Triβ], 2.5.13, p. 114) we have

$$\|f^t \mid \mathscr{B}_p\|^p \sim \|f^t \mid L_p\|^p + \int_{\mathbf{R}^{n-1}} \int_{\mathbf{R} \times \mathbf{R}} \frac{\left|f^t\left(x', x_n\right) - f^t\left(x', y_n\right)\right|^p}{\left|x_n - y_n\right|^{n+1}} \, dx_n dy_n dx' + \dots$$
(21)

where $x' = (x_1, \dots, x_{n-1})$ and $+\dots$ stands for $n-1$ similar summands with $x_1, \dots, x_{n-1}$ in place of x_n. For fixed $x' \in \mathbf{R}^{n-1}$ the inner integral in (21) is of the same type as the integral in (9) with $n+1$ instead of 2 and can be treated in the same way as in (10)–(16), which results in the following counterpart of (16),

$$I^p \le c \left(1 + |t|\right)^p \int_{|x'| \le \delta} dx' \int_0^\delta h^{-n-1+p} \int_{|x'|+h}^{\delta + |x'| + h} \frac{-1}{u^p \log u} \left(\log |\log \varepsilon u|\right)^{-\sigma p} dudh.$$
(22)

We use (17) and obtain as the counterpart of (18),

$$I^p \le c \left(1 + |t|\right)^p \int_{|x'| \le \delta} dx' \int_0^\delta h^{-n-1+p} \frac{\left[\log \left|\log \varepsilon \left(|x'| + h\right)\right|\right]^{-\sigma p}}{\left(|x'| + h\right)^{p-1} \left|\log \left(|x'| + h\right)\right|} dh$$

$$\le c \left(1 + |t|\right)^p \int_0^\delta h^{-n-1+p} dh \left(\int_{|x'| \le h} \dots + \int_{h \le |x'| \le \delta} \dots\right)$$

$$= c \left(1 + |t|\right)^p \left(I_1^p + I_2^p\right).$$
(23)

In the first integral where $|x'| \le h$ we replace $|x'| + h$ by h and obtain

$$I_1^p \le c \int_0^\delta h^{-2+p} \frac{\left[\log |\log \varepsilon h|\right]^{-\sigma p}}{h^{p-1} |\log h|} dh < \infty,$$
(24)

since $\sigma p > 1$. In the second integral we replace $|x'| + h$ by $|x'|$ and find

$$I_2^p \le c \int_0^\delta h^{-n-1+p} dh \int_{h \le |x'| \le \delta} \frac{\left[\log |\log \varepsilon |x'||\right]^{-\sigma p}}{|x'|^{p-1} |\log |x'||} dx'.$$
(25)

We substitute $|x'| = hr$ in the inner integral and use an appropriate modification of (17). Then we have

$$I_2^p \le c_1 \int_0^\delta h^{-2+p} h^{-p+1} dh \int_1^{\delta h^{-1}} \frac{\left[\log |\log \varepsilon h r|\right]^{-\sigma p} r^{n-2}}{r^{p-1} |\log h r|} dr$$

$$\le c_2 \int_0^\delta h^{-1} \frac{\left[\log |\log \varepsilon h|\right]^{-\sigma p}}{|\log h|} dh + c_2 \int_0^\delta h^{-1} h^{p-n} dh < \infty,$$
(26)

where we used $\sigma p > 1$ in the first integral and $p > n$ in the second one. We insert (24) and (26) in (23) and obtain (20); in particular, taking account of (19), we have

$$f_{p\sigma} \in \mathscr{B}_p, \ 1 \le n < p < \infty, \ \sigma > \frac{1}{p}. \tag{27}$$

Step 4 In this step we prove (20) under the assumption that $1 < p < n$. Recall that $f^t \in \mathscr{B}_p$ if, and only if,

$$f^t \in L_p \text{ and } \frac{\partial f^t}{\partial x_m} \in B_{pp}^{\frac{n}{p}-1} \text{ where } m = 1, ..., n. \tag{28}$$

The first assertion is obvious. To prove the second we use the atomic decomposition theorem 2.2.3. Since $\frac{n}{p} - 1 > 0$ no moment conditions are needed ($L = -1$ in the formulation given there). Let $v \in \mathbf{N}$ and let $y = (y_1, ..., y_n)$ in what follows. We divide the corridors

$$\{y \in \mathbf{R}^n : |y_m| < 2^{-v+1}, \ m = 1, ..., n\} \setminus \{y \in \mathbf{R}^n : |y_m| < 2^{-v}, \ m = 1, ..., n\} \tag{29}$$

in a natural way into $4^n - 2^n$ congruent cubes with side length 2^{-v}, which are appropriate translates of the cube $\{y \in \mathbf{R}^n : 0 < y_m < 2^{-v}\}$ (we do not care about the faces of the cubes involved, which do not play any role). We denote these cubes by Q_{vl} with $v \in \mathbf{N}$ and $l = 1, ..., L$, where $L = 4^n - 2^n$. Let

$$\sum_{l=1}^{L} \sum_{v \in \mathbf{N}} \psi_{vl}(x) = 1 \text{ if } x = (x_1, ..., x_n) \text{ with } |x_m| < 1, \ x \ne 0, \tag{30}$$

be a corresponding resolution of unity, where the ψ_{vl} are C^∞ functions and

$$\text{supp } \psi_{vl} \subset rQ_{vl} \text{ with } 1 < r < \frac{3}{2}; \tag{31}$$

see 2.2.3 for the explanation of rQ_{vl}. According to (2.2.3/20) we ask for the atomic decomposition

$$\frac{\partial f^t}{\partial x_m} = \sum_{l=1}^{L} \sum_{v=2}^{\infty} \psi_{vl} \frac{\partial f^t}{\partial x_m} = \sum_{l=1}^{L} \sum_{v=2}^{\infty} \lambda_{vl} a_{vl}. \tag{32}$$

The remainder terms $+...$ in

$$\frac{\partial f^t}{\partial x_m}(x) = -\left(1 - \frac{1}{p} + it\right) |\log|x||^{it-\frac{1}{p}} \frac{x_m}{|x|^2} (\log|\log \varepsilon |x||)^{-\sigma+it} \psi(x) + ... \tag{33}$$

are less singular near the origin than the term indicated. We have

$$\left|\frac{\partial f^t}{\partial x_m}(x)\right| \le c \left(1 + |t|\right) 2^v v^{-1/p} (\log v)^{-\sigma} \text{ if } |x| \sim 2^{-v}. \tag{34}$$

By $(2.2.3/6_K)$ with $s = \frac{n}{p} - 1$ and (34) we may choose $\lambda_{vl} = c\left(1 + |t|\right)$ $v^{-1/p} (\log v)^{-\sigma}$ in (32). Since $\sigma p > 1$, we have

$$\sum_{l=1}^{L} \sum_{v=2}^{\infty} |\lambda_{vl}|^p = c\left(1 + |t|\right)^p \sum_{v=2}^{\infty} \frac{1}{v} (\log v)^{-\sigma p} \le c'\left(1 + |t|\right)^p. \qquad (35)$$

Hence by Theorem 2.2.3 it follows that (32) yields an atomic representation in $B_{pp}^{\frac{n}{p}-1}$. By (28) and (35) we obtain (20) now for $1 < p < n$; in particular, taking account of (27),

$$f_{p\sigma} \in \mathcal{B}_p, \ 1 < p < \infty, p \neq n, \sigma > 1/p. \qquad (36)$$

Step 5 The reason for extending $f_{p\sigma}$ to $f_{p\sigma}^t$ is that we wish to apply complex interpolation. For that purpose we also have to consider the limiting case $p = \infty$. More precisely, now let

$$f^t(x) = |\log |x||^{1+it} \left(\log |\log \varepsilon |x||\right)^{it} \psi(x), \ t \in \mathbf{R}; \qquad (37)$$

then we prove in this step that

$$f^t \in \text{bmo and } \|f^t \mid \text{bmo}\| \le c\left(1 + |t|\right), \qquad (38)$$

where c is independent of t. For bmo we refer to (2.2.2/16). It is sufficient to prove that for each cube Q with $|Q| \le 1$ there exists a constant c_Q with

$$|Q|^{-1} \int_Q |f^t(x) - c_Q| \, dx \le c\left(1 + |t|\right), \qquad (39)$$

where c is independent of Q. This follows from (2.2.2/16) where one can replace f in the last summand by $f - c_Q$ and then use the triangle inequality. Of course, only cubes near the origin are of interest. Without restriction of generality we may assume that

$$Q = \{y \in \mathbf{R}^n : |y_m| < 2^{-\mu}, m = 1, ..., n\} \qquad (40)$$

for some $\mu \in \mathbf{N}$. We subdivide this cube as in Step 4, first into corridors and the latter then into cubes Q_{vl} now with $v > \mu$. In (39) we choose $c_Q = f^t\left(x^Q\right)$ with, say, $|x^Q| = 2^{-\mu}$. Then we obtain

$$|Q|^{-1} \int_Q |f^t(x) - f^t\left(x^Q\right)| \, dx = c2^{\mu n} \sum_{l=1}^{L} \sum_{v>\mu} \int_{Q_{vl}} |f^t(x) - f^t\left(x^Q\right)| \, dx. \qquad (41)$$

If $x \in Q_{vl}$ then we have $|x| \sim 2^{-v}$ and

$$|f^t(x) - f^t\left(x^Q\right)| \le c\left(1 + |t|\right)(v - \mu) \qquad (42)$$

for some $c > 0$. Inserting (42) in (41) we obtain

$$|Q|^{-1} \int_Q |f^t(x) - f^t(x^Q)| \, dx \le c (1 + |t|) \sum_{\nu > \mu} 2^{-(\nu-\mu)n} (\nu - \mu). \qquad (43)$$

This proves (39) and, in turn, (38).

Step 6 In this step we prove that

$$f_{p\sigma} \in \mathscr{H}_p, \ 1 < p < \infty, \sigma > 1/p. \qquad (44)$$

If $1 < p < 2$ then (44) follows from (36) and $\mathscr{B}_p \subset \mathscr{H}_p$; see (2.3.3/1) with $q = 2, u = p$ and $s = n/p$. To extend this assertion to $2 \le p < \infty$ we use the classical complex interpolation method $[\cdot, \cdot]_\theta$, see e.g. [Triα], 1.9, p. 55. In Remark 1 below we prove that

$$[\text{bmo}, \mathscr{H}_p]_\theta = \mathscr{H}_r \text{ for } 1 < p < \infty, 0 < \theta < 1, \frac{1}{r} = \frac{\theta}{p}. \qquad (45)$$

In agreement with the complex interpolation method we generalise (7) by

$$f(z, x) = |\log|x||^{1 - \frac{z}{p}} \left(\log|\log \varepsilon |x||\right)^{-\sigma z} \psi(x), \qquad (46)$$

where

$$z \in \mathbf{C}, 0 \le \text{Re}\, z \le 1, 1 < p < 2 \text{ and } \sigma > 1/p. \qquad (47)$$

Then (37) yields

$$f(it, \cdot) \in \text{bmo}, \ \|f(it, \cdot) \mid \text{bmo}\| \le c (1 + |t|) \text{ where } t \in \mathbf{R}. \qquad (48)$$

Furthermore, by (20), which by Step 4 also holds for $1 < p < 2$, and the above-mentioned embedding $\mathscr{B}_p \subset \mathscr{H}_p$, we have

$$f(1 + it, \cdot) \in \mathscr{H}_p, \|f(1 + it, \cdot) \mid \mathscr{H}_p\| \le c (1 + |t|), 1 < p < 2, t \in \mathbf{R}. \ (49)$$

Now by (45)–(49) and the complex interpolation method as described in [Triα], 1.9, p. 55 we obtain

$$f(\theta, x) = |\log|x||^{1 - \frac{1}{r}} \left(\log|\log \varepsilon |x||\right)^{-\sigma\theta} \psi(x) \in \mathscr{H}_r. \qquad (50)$$

We have $\sigma\theta > 1/r$ and, of course, any number larger than $1/r$ can be represented in that way; any r with $1 < r < \infty$ can be obtained in the way indicated. The proof of (44) is complete.

Step 7 To remove the restriction in (36) we remark that $\mathscr{H}_p \subset \mathscr{B}_p$ if $2 \le p < \infty$; see (2.3.3/1) with $q = 2, v = p$ and $s = n/p$. Hence, by (44), we have $f_{p\sigma} \in \mathscr{B}_p$ in the remaining case $p = n \geq 2$. The proof of (3) is now complete.

Step 8 It remains to prove (5). Let

$$I(j,\kappa) = \int_0^1 |\log r|^\kappa r^{n-1} \left(\log |\log \varepsilon r|\right)^{-\sigma j} dr, \tag{51}$$

where $\varepsilon > 0$ is small, $\sigma > 0$ is fixed and $\kappa > 0$. Then (5) is covered by

$$I(j,\kappa j)^{1/j} \sim j^\kappa, \quad j \in \mathbf{N}, \tag{52}$$

where "$\sim$" indicates that the equivalence constants are independent of $j \in \mathbf{N}$. Let $\kappa j > 1$. Since

$$\frac{1}{n}\left(|\log r|^{\kappa j} r^n\right)' = |\log r|^{\kappa j} r^{n-1} - \frac{\kappa j}{n}|\log r|^{\kappa j-1} r^{n-1} \tag{53}$$

we have

$$I(j,\kappa j) = \frac{1}{n}\int_0^1 \left(|\log r|^{\kappa j} r^n\right)' \left(\log |\log \varepsilon r|\right)^{-\sigma j} dr + \frac{\kappa j}{n}I(j,\kappa j - 1). \tag{54}$$

The integral in (54) can be calculated and estimated from above by

$$\frac{\sigma j}{n}\int_0^1 |\log r|^{\kappa j} r^n \left(\log |\log \varepsilon r|\right)^{-\sigma j-1} r^{-1} |\log \varepsilon r|^{-1} dr$$

$$\leq \eta j \int_0^1 |\log r|^{\kappa j-1} r^{n-1} \left(\log |\log \varepsilon r|\right)^{-\sigma j} dr = \eta j I(j,\kappa j - 1) \tag{55}$$

where we may assume that η is small (in dependence on ε). From (54) and (55) we obtain

$$I(j,\kappa j) \sim \kappa j I(j,\kappa j - 1). \tag{56}$$

Let $\kappa j = [\kappa j] + \lambda$ with $0 \leq \lambda < 1$. We obtain by iteration

$$c_1^{\kappa j}\Gamma(\kappa j) I(j,\lambda) \leq I(j,\kappa j) \leq c_2^{\kappa j}\Gamma(\kappa j) I(j,\lambda) \tag{57}$$

where c_1 and c_2 are two approriate positive numbers and Γ stands for the gamma function. By (51) the integral $I(j,\lambda)$ can be estimated from above and from below by c_3^j and c_4^j, respectively, where c_3 and c_4 are appropriate positive numbers. Then (57) and

$$\Gamma(\kappa j)^{\frac{1}{j}} \sim j^\kappa \tag{58}$$

prove (52). The proof is complete.

Remark 1 We now prove the interpolation formula in (45). For that purpose we use duality, both for the complex interpolation method and for bmo; see [Triα], 1.11.3, p. 72 and [Triβ], 2.5.8 and 2.11.2, pp. 93, 178, respectively. We have

$$
\begin{aligned}
\left[\text{bmo}, \mathscr{H}_p\right]_\theta &= \left[\text{bmo}, H_p^{n/p}\right]_\theta = \left[h_1', \left(H_{p'}^{-n/p}\right)'\right]_\theta \\
&= \left[h_1, H_{p'}^{-n/p}\right]_\theta' = \left[F_{1,2}^0, F_{p',2}^{-n/p}\right]_\theta' = \left(F_{r',2}^{-n/r}\right)' \\
&= F_{r,2}^{n/r} = \mathscr{H}_r,
\end{aligned}
\tag{59}
$$

where, of course, $\frac{1}{p} + \frac{1}{p'} = \frac{1}{r} + \frac{1}{r'} = 1$. We used [Triα], 2.4.2, p. 185, extended to $F_{1,2}^0$.

Remark 2 We followed essentially [Tri3], Theorem 3.1.2 and its proof. By the same techniques as in (59) one can also prove that

$$
[\text{bmo}, h_1]_\theta = L_p, \quad \theta = 1/p, \; 0 < \theta < 1,
\tag{60}
$$

see [Tri3], Proposition 3.1.4 and its proof.

2.7.2 Embedding constants

We continue to study the limiting embeddings as described at the beginning of 2.7.1. Again let $\mathscr{B}_p$ and $\mathscr{H}_p$ be given by (2.7.1/1). Since these spaces and also all the other spaces in this subsection are defined on $\mathbf{R}^n$ we omit "$\mathbf{R}^n$" in what follows. Let $1 \le p \le q < \infty$ and let $\text{id}_{p,q}$ be one of the following embedding operators:

$$
\text{id}_{p,q} : \mathscr{B}_p \to L_q \text{ or } \text{id}_{p,q} : \mathscr{H}_p \to L_q.
\tag{1}
$$

We are interested in the dependence of $\|\text{id}_{p,q}\|$ on q, where p is assumed to be fixed and $q \to \infty$. These continuous embeddings exist; see the beginning of 2.7.1, Fig. 2.1 on p. 52 and 2.3.3. If $p = 1$ then by (2.3.3/9) and (2.3.3/10),

$$
\text{id} : \mathscr{B}_1 \to L_\infty \text{ and } \mathscr{H}_1 \to L_\infty
\tag{2}
$$

and the question of the dependence on q of the embedding constants in (1) is not of interest. If $p > 1$, then again by (2.3.3/9) and (2.3.3/10) the situation is different and it is to be expected that $\|\text{id}_{p,q}\| \to \infty$ if $q \to \infty$.

Theorem *Let* $1 < p < \infty$. *There exist two positive constants* c_1 *and* c_2 *(depending on* p*) such that*

$$c_1 q^{1-\frac{1}{p}} \leq \|\text{id}_{p,q}\| \leq c_2 q^{1-\frac{1}{p}} \tag{3}$$

for every q *with* $p \leq q < \infty$.

Proof Step 1 Let $\{\varphi_k\}_{k=0}^{\infty}$ be the dyadic resolution of unity introduced in (2.2.1/5) and let $\frac{1}{p} + \frac{1}{p'} = 1$. By (2.2.1/6) and Nikol'skii's inequality (see [Triβ], 1.3.2, Remark 1, p. 18), we have

$$
\begin{aligned}
\|f \mid L_q\| &\leq \sum_{k=0}^{\infty} \left\| \left(\varphi_k \hat{f} \right)^{\vee} \mid L_q \right\| \\
&\leq c \sum_{k=0}^{\infty} 2^{kn\left(\frac{1}{p}-\frac{1}{q}\right)} \left\| \left(\varphi_k \hat{f} \right)^{\vee} \mid L_p \right\| \\
&\leq c \left(\sum_{k=0}^{\infty} 2^{-knp'/q} \right)^{1/p'} \left(\sum_{k=0}^{\infty} 2^{kn} \left\| \left(\varphi_k \hat{f} \right)^{\vee} \mid L_p \right\|^p \right)^{1/p} \\
&\leq c' q^{1/p'} \|f \mid \mathscr{B}_p\|.
\end{aligned}
\tag{4}
$$

Here we used Hölder's inequality and the fact that $\mathscr{B}_p = B_{pp}^{n/p}$. This proves the right-hand side of (3) in the case of $\mathscr{B}_p$.

Step 2 By Step 7 of the proof of Theorem 2.7.1 we also have the right-hand inequality of (3) in the case of $\mathscr{H}_p$ if $2 \leq p < \infty$. To prove the corresponding estimate if $1 < p < 2$ we interpolate

$$\|f \mid L_q\| \leq c \|f \mid \mathscr{H}_1\| \tag{5}$$

(see (2)) and

$$\|f \mid L_q\| \leq cq^{1/2} \|f \mid \mathscr{H}_2\|, \quad 2 \leq q < \infty. \tag{6}$$

We use the complex interpolation

$$[\mathscr{H}_1, \mathscr{H}_2]_\theta = \left[H_1^n, H_2^{n/2} \right]_\theta = H_p^{n/p} = \mathscr{H}_p, \quad 0 < \theta < 1, \tag{7}$$

$\frac{1}{p} = 1 - \theta + \frac{\theta}{2}$; see [Tri$\alpha$], (2.4.2/11), p. 185 extended to $p_0 = 1$. By (5)–(7) and the interpolation property we have

$$\|f \mid L_q\| \leq cq^{\theta/2} \|f \mid \mathscr{H}_p\|, \quad 2 \leq q < \infty. \tag{8}$$

But this is just what we want, since $\frac{\theta}{2} = 1 - \frac{1}{p}$, and the extension of (8) to $p \leq q < 2$ is obvious.

Step 3 The left-hand inequality is an easy consequence of Theorem 2.7.1.

Remark 1 We followed [Tri3], Propositions 4.3.1 and 4.3.2. Now it is clear why the functions $f_{p\sigma}$ in (2.7.1/3) deserve to be called "extremal".

For our later applications we need a refinement of the right-hand inequality of (3).

Corollary *Let* $1 < p < \infty$. *There exists a positive constant* c *(depending on* p) *such that*

$$\|f \mid L_q\| \leq c q^{1-\frac{1}{p}} 2^{-\frac{Nn}{q}} \|f \mid \mathscr{B}_p\| \tag{9}$$

for all q *with* $p \leq q < \infty$, *all* $N \in \mathbf{N}$ *and all* $f \in \mathscr{B}_p$ *with*

$$\operatorname{supp} \hat{f} \cap \{y : |y| \leq 2^{N+2}\} = \emptyset, \tag{10}$$

and such that

$$\|f \mid L_q\| \leq c q^{1-\frac{1}{p}} 2^{-\frac{Nn}{q}} \|f \mid \mathscr{H}_p\| \tag{11}$$

for all q *with* $p \leq q < \infty$ *and all* $f \in \mathscr{H}_p$ *with (10)*.

Proof We prove (9) by homogeneity. Let $\{\varphi_k\}_{k=0}^{\infty}$ be as above. By standard arguments we have

$$\left(\varphi_k f\left(2^{-N}\cdot\right)^{\wedge}\right)^{\vee}(\xi) = \left(\varphi_{k+N} \hat{f}\right)^{\vee}\left(2^{-N}\xi\right) \tag{12}$$

which by (2.2.1/6), with $s = n/p$ and $q = p$, gives

$$\left\|f\left(2^{-N}\cdot\right) \mid \mathscr{B}_p\right\| \sim \|f \mid \mathscr{B}_p\|. \tag{13}$$

Then (4) applied to $f\left(2^{-N}\cdot\right)$ yields (9). To prove (11) we use the following counterpart of (13):

$$\left\|f\left(2^{-N}\cdot\right) \mid \mathscr{H}_p\right\| \sim \left\|\left(|\xi|^{\frac{n}{p}} f\left(2^{-N}\cdot\right)^{\wedge}\right)^{\vee} \mid L_p\right\|$$
$$\sim \left\|\left(|\xi|^{\frac{n}{p}} \hat{f}\right)^{\vee} \mid L_p\right\| \sim \|f \mid \mathscr{H}_p\|. \tag{14}$$

Now (14) and (4) with $\mathscr{H}_p$ in place of $\mathscr{B}_p$ applied to $f\left(2^{-N}\cdot\right)$ give (11).

Remark 2 Our interest in (3), (9) and (11) comes from the later applications, where always p is fixed and $q \to \infty$. The differential dimension (as introduced in 2.3.3(ii)) of the initial space $\mathcal{H}_p$ or $\mathcal{B}_p$ is 0 whereas the differential dimension of the target space L_q is $-n/q$. Usually one asks for best constants for embeddings between spaces with the same differential dimensions, see 2.3.3(ii). The starting point here is

$$\|f \mid L_q\| \leq C \|\nabla f \mid L_p\|, \ 1 < p < n, \ 1 - \frac{n}{p} = -\frac{n}{q}, \tag{15}$$

where the best possible constant C is known; see [Aub], pp. 39–40, [GiT], p. 151 or [Tal]. From the explicit value of this constant C, it follows that

$$\|f \mid L_q\| \leq cq^{1-\frac{1}{p}} \|f \mid W_p^1\|, \ 1 < p < n, \ 1 - \frac{n}{p} = -\frac{n}{q}, \tag{16}$$

where c depends only on n. In [EET] we extended this result in the following way. Let $n \geq 2$, $p^* > 1$ and

$$p^* < p < n, \ 1 < s < \frac{n}{p}, \ s - \frac{n}{p} = -\frac{n}{q}. \tag{17}$$

Then there is a constant c independent of p, q, s (but which may depend upon n and p^*) such that for all $f \in H_p^s$,

$$\|f \mid L_q\| \leq cq^{1-\frac{1}{p}} \|f \mid H_p^s\| \tag{18}$$

and for all $f \in B_{pp}^s$,

$$\|f \mid L_q\| \leq cq^{1-\frac{1}{p}} \|f \mid B_{pp}^s\|. \tag{19}$$

On the other hand, by the restrictions (17) we have

$$\mathcal{B}_p \subset B_{pp}^s \text{ and } \mathcal{H}_p \subset H_p^s. \tag{20}$$

Now by (3) and (18) it follows under the above restrictions (17) that

$$\left\|\text{id} : B_{pp}^s \to L_q\right\| \sim q^{1-\frac{1}{p}}, \tag{21}$$

and a corresponding assertion holds with H_p^s in place of B_{pp}^s. This is an affirmative answer to the conjectures in [EET], Remark 3 on p. 127.

2.7.3 Embeddings

In this subsection Ω stands for a bounded domain in $\mathbf{R}^n$ with C^∞ boundary. In agreement with (2.7.1/1) we put

$$\mathcal{B}_p(\Omega) = B_{pp}^{n/p}(\Omega) \text{ and } \mathcal{H}_p(\Omega) = H_p^{n/p}(\Omega), \ 0 < p < \infty; \tag{1}$$

see 2.2.1, 2.2.2, (2.3.3/11) and 2.5.1. Armed with the results of the two preceding subsections we return to the problem described at the beginning of 2.7.1: limiting embeddings of $\mathscr{B}_p(\Omega)$ and $\mathscr{H}_p(\Omega)$ in the potential target spaces: $L_\infty(\Omega)$ (or $C(\Omega)$: there is no difference since $C^\infty(\overline{\Omega})$ is dense in the initial spaces); bmo (Ω), which is the restriction of bmo $(\mathbf{R}^n)$ defined in 2.2.2 to Ω; or $L_\infty(\log L)_a(\Omega)$ with $a < 0$, introduced in 2.6.1. Since Ω is bounded the question whether (an existing) embedding is compact is reasonable. As for the formulation below we note that in our context an embedding between two function spaces is automatically continuous if it exists.

Theorem *Let Ω be a bounded domain in $\mathbf{R}^n$ with C^∞ boundary.*
(i) Let $1 < p < \infty$ and $a < 0$. The embedding

$$\mathrm{id} : \mathscr{H}_p(\Omega) \to L_\infty(\log L)_a(\Omega) \tag{2}$$

exists if, and only if, $a \le \dfrac{1}{p} - 1$; $\qquad$ (3)

is compact if, and only if, $a < \dfrac{1}{p} - 1$. $\qquad$ (4)

Furthermore, the assertions in (3) and (4) also hold for the embedding (2) with $\mathscr{B}_p(\Omega)$ in place of $\mathscr{H}_p(\Omega)$.
(ii) Let $0 < p \le 1$. The embeddings

$$\mathrm{id} : \mathscr{H}_p(\Omega) \to L_\infty(\Omega) \quad and \quad \mathscr{B}_p(\Omega) \to L_\infty(\Omega) \tag{5}$$

exist, but they are not compact.
(iii) Let $0 < p < \infty$. The embeddings

$$\mathrm{id} : \mathscr{H}_p(\Omega) \to \mathrm{bmo}(\Omega) \quad and \quad \mathscr{B}_p(\Omega) \to \mathrm{bmo}(\Omega) \tag{6}$$

exist, but they are not compact.

Proof Step 1 Recall that $L_\infty(\log L)_a(\Omega)$ can be normed by (2.6.2/4) or, more explicitly, by (2.7.1/6). Then Theorem 2.7.2 (see also the explicit version (2.7.2/4)), shows that (2) exists (and is continuous) if $a \le \frac{1}{p} - 1$. The same holds with $\mathscr{B}_p(\Omega)$ in place of $\mathscr{H}_p(\Omega)$. On the other hand the extremal functions in Theorem 2.7.1 show that there is no embedding (2) (also with $\mathscr{B}_p(\Omega)$ in place of $\mathscr{H}_p(\Omega)$) if $a > \frac{1}{p} - 1$. Hence the proof of (2), (3), also with $\mathscr{B}_p(\Omega)$, is complete.

Step 2 We prove assertion (4). The embedding

$$\mathscr{H}_p(\Omega) \to L_q(\Omega), \; p \le q < \infty, \tag{7}$$

is compact (this will be discussed in great detail in Chapter 3); see also [Triβ], 4.3.2, Remark 1, p. 233. Then it follows again by Theorem 2.7.2 (see also (2.7.1/6)) that the embedding (2) is compact if $a < \frac{1}{p} - 1$. The same holds with $\mathscr{B}_p(\Omega)$ in place of $\mathscr{H}_p(\Omega)$. Let $a = \frac{1}{p} - 1$. Let

$$Q_v = \left\{ x \in \mathbf{R}^n : 2^{-v} < x_l < 2^{-v+1} \text{ for } l = 1, ..., n \right\}, \; v \in \mathbf{N}_0. \tag{8}$$

We assume, without loss of generality, that $Q_v \subset \Omega$. Let x^v be the centre of Q_v and let

$$f_{p\sigma}^v(x) = f_{p\sigma}(2^v(x - x^v)), \; v \in \mathbf{N}_0, \tag{9}$$

where $f_{p\sigma}(x)$ is given by (2.7.1/3). (Of course, these functions should not be confused with the functions $f_{p\sigma}^t$ introduced in (2.7.1/7).) By (2.7.1/5) and (2.7.1/6) we have

$$\left\| f_{p\sigma}^v \mid L_j(\Omega) \right\| \sim 2^{-vn/j} j^{1-\frac{1}{p}} \tag{10}$$

and

$$\left\| f_{p\sigma}^v \mid L_\infty (\log L)_{\frac{1}{p}-1}(\Omega) \right\| \ge c > 0 \tag{11}$$

for some c which is independent of v. By Proposition 2.3.1/1 the sequence $\left\{ f_{p\sigma}^v \right\}_{v \in \mathbf{N}_0}$ is bounded in both $\mathscr{H}_p(\Omega)$ and $\mathscr{B}_p(\Omega)$. Since the supports of $f_{p\sigma}^v$ and $f_{p\sigma}^\mu$ with $\mu \ne v$ are disjoint we have (11) also for $f_{p\sigma}^v - f_{p\sigma}^\mu$. In particular, the sequence $\left\{ f_{p\sigma}^v \right\}_{v \in \mathbf{N}_0}$ is not pre-compact in $L_\infty (\log L)_{\frac{1}{p}-1}(\Omega)$. Hence the embedding (2) and its $\mathscr{B}$-counterpart is not compact if $a = \frac{1}{p} - 1$. The proof of (i) is complete.

Step 3 We prove (ii). The continuity of (5) is covered by (2.7.2/2) and the monotonicity of the spaces $\mathscr{B}_p$ and $\mathscr{H}_p$; see also (2.3.3/7). By an argument similar to, but simpler than, that of Step 2 it follows that these embeddings cannot be compact.

Step 4 We prove (iii). The continuity of the embeddings (6) is covered by (2.3.3/12). It remains to prove that these embeddings are not compact. Let f be a C^∞ function with

$$\operatorname{supp} f \subset Q_0, \; \|f \mid L_1(\Omega)\| = 1 \text{ and } \int_{Q_0} f(x)\,dx = 0. \tag{12}$$

Then we have supp $f(2^v \cdot) \subset Q_v, v \in \mathbf{N}_0$, where Q_v is given by (8). We construct the following sequence of maps:

$$l_p \xrightarrow{A} \mathscr{B}_p(\Omega) \xrightarrow{\text{id}} \text{bmo}(\Omega) \xrightarrow{B} l_\infty, \tag{13}$$

where id is given by (6), and where $\mathscr{B}_p(\Omega)$ can be replaced immediately by $\mathscr{H}_p(\Omega)$. The operators A and B are defined as follows:

$$A : \{a_v\}_{v \in \mathbf{N}_0} \to \sum_{v=0}^{\infty} a_v f(2^v x) \tag{14}$$

and

$$B : g \to \left\{ 2^{vn} \int_{Q_v} \text{sgn}\, f(2^v x)\, (g(x) - g_{Q_v})\, dx \right\}_{v=0}^{\infty} \tag{15}$$

where

$$g_{Q_v} = 2^{-vn} \int_{Q_v} g(y)\, dy \tag{16}$$

is the mean value of g in Q_v (we corrected in (15) a bad misprint in [Tri3], p. 615, formula (25)). By the atomic representation theorem 2.2.3 it follows that in both cases, $\mathscr{B}_p(\Omega)$ and $\mathscr{H}_p(\Omega)$, A is a continuous map. The continuity of B follows from (2.2.2/16). Finally by (12) we have

$$B \circ \text{id} \circ A : \{a_v\}_{v=0}^{\infty} \to \{a_v\}_{v=0}^{\infty}, \tag{17}$$

the identity map from l_p to l_∞. But this map is not compact, and hence id is also not compact. The proof is complete.

Remark 1 We partly followed [Tri3].

Remark 2 In connection with the above theorem the question arises as to how the target spaces are related to each other. Of course, $L_\infty(\Omega)$ is the smallest space. Furthermore from (4) and the non-compactness of (6) it follows that there is no embedding of $L_\infty(\log L)_a(\Omega)$ in bmo (Ω). On the other hand by (6) and (3) there is also no embedding of bmo (Ω) in $L_\infty(\log L)_a(\Omega)$ if $a > -1$. As for the remaining cases $a \le -1$ we refer to [KuJF], Theorem 4.8.3, p. 231, where one finds a proof that bmo (Ω) is continuously embedded in $L_\infty(\log L)_a(\Omega)$ if $a \le -1$. In other words,

$$\text{bmo}(\Omega) \subset L_\infty(\log L)_a(\Omega) \text{ if, and only if, } a \le -1. \tag{18}$$

3

Entropy and Approximation Numbers of Embeddings

3.1 Introduction

The theme of this chapter is the analysis, from the standpoint of entropy and approximation numbers, of compact embeddings between function spaces on bounded domains in $\mathbf{R}^n$. Initially we focus on the case in which the function spaces lie in the scales B_{pq}^s and F_{pq}^s introduced in Chapter 2 and the underlying domain Ω has C^∞ boundary. In such a situation we give upper and lower estimates for the entropy and approximation numbers which are optimal in all cases for the entropy numbers and in many cases for the approximation numbers. These results include all those previously known; unlike earlier work by other authors, which was based upon the piecewise-polynomial approach pioneered by Birman and Solomyak and upon more refined spline approximations, our approach uses Fourier-analytical techniques and the characterisations of the function spaces given in Chapter 2. It also involves a reduction of the problems to the study of mappings between finite-dimensional sequence spaces.

The embeddings mentioned above are of the non-limiting kind, in a sense which we make precise later. After dealing with them, we go on to handle limiting embeddings, typified by the celebrated embedding of a Sobolev space with critical exponent in an Orlicz space of exponential type. However, this is but an end point case of a whole range of limiting embeddings which are covered, involving embeddings between logarithmic Sobolev spaces, and for all of these cases we provide upper and lower estimates for the entropy and approximation numbers.

To conclude the chapter we examine the position when the bounded domain Ω may have a non-smooth boundary. Here we rely on material in 2.5 on the corresponding function spaces.

3.2 The embedding of ℓ_p^m in ℓ_q^m

3.2.1 The spaces ℓ_p^m

Let $m \in \mathbf{N}$ and $0 < p \le \infty$. By ℓ_p^m we shall mean the linear space of all complex m-tuples $y = (y_j)$, endowed with the quasi-norm

$$\|y \,|\, \ell_p^m\| = \left(\sum_{j=1}^m |y_j|^p \right)^{1/p}, \qquad \text{if } 0 < p < \infty,$$

with the usual modification if $p = \infty$. Let

$$U_p^m = \{ y \in \ell_p^m : \|y \,|\, \ell_p^m\| \le 1 \}$$

be the closed unit ball in ℓ_p^m. Since $\mathbf{C}^m$ may be naturally identified with $\mathbf{R}^{2m}$, we shall understand by the volume of U_p^m the Lebesgue $2m$-measure of $\left\{ (x_1, ..., x_{2m}) \in \mathbf{R}^{2m} : \sum_{j=1}^m (x_{2j-1}^2 + x_{2j}^2)^{p/2} \le 1 \right\}$. A formula for this volume is provided by the following result.

Proposition (i) *If* $0 < p \le \infty$, *then the volume of the unit ball in* ℓ_p^m *is*

$$\mathrm{vol}\, U_p^m = \pi^m \{ \Gamma(1 + 2/p) \}^m / \Gamma(1 + 2m/p). \tag{1}$$

(ii) *There is a function* $\theta : (0, \infty) \to \mathbf{R}$, *with* $0 < \theta(t) < 1/12$ *for all* $t > 0$, *such that for all* $p \in (0, \infty)$,

$$\mathrm{vol}\, U_p^m = 2^{m-1} \pi^{\frac{1}{2}(3m-1)} p^{-(m-1)/2} m^{-\frac{2m}{p}-\frac{1}{2}} \exp\{ m\theta(2/p)p/2 - \theta(2m/p)p/(2m) \}. \tag{2}$$

Proof If $0 < p < \infty$, then

$$\mathrm{vol}\, U_p^m = \int dx,$$

where the integration is over $\left\{ x = (x_j) \in \mathbf{R}^{2m} : \sum_{j=1}^m \left(x_{2j-1}^2 + x_{2j}^2 \right)^{p/2} \le 1 \right\}$. The change of variable given by $x_{2j-1} = r_j \cos \varphi_j, x_{2j} = r_j \sin \varphi_j, j = 1, ..., m$ shows that

$$\mathrm{vol}\, U_p^m = (2\pi)^m \int \prod_{j=1}^m r_j \, dr_1 ... dr_m,$$

where the integration is over $r_1^p + ... + r_m^p \leq 1$. The further substitution $r_j^p = t_j$, $j = 1, ..., m$ leads to

$$\text{vol} U_p^m = \left(\frac{2\pi}{p}\right)^m \int \prod_{j=1}^m t_j^{\frac{2}{p}-1} dt,$$

where the integration is over $\{t : \sum_1^m t_k \leq 1, t_k > 0\}$. Hence by a result of Dirichlet (see [WiW], 12.5),

$$\text{vol} U_p^m = (2\pi/p)^m \{\Gamma(2/p)\}^m \{\Gamma(2m/p)\}^{-1} \int_0^1 \tau^{2m/p-1} d\tau,$$

and (1) follows. When $p = \infty$, (1) is obvious.

To prove (2) we use Stirling's formula in the version

$$\Gamma(t) = e^{-t} t^{t-\frac{1}{2}} (2\pi)^{1/2} e^{\theta(t)/t}, \quad 0 < t < \infty,$$

where $0 < \theta(t) < 1/12$ for all $t > 0$ (see [WiW], 12.33). This gives (2) immediately. Note that the exponential factor in (2) is bounded above and below by $e^{pm/24}$ and $e^{-pm/24}$ respectively.

3.2.2 Entropy numbers

The entropy numbers have been introduced in an abstract setting in Definition 1.3.1/1. Now Proposition 3.2.1 enables us to estimate from above the entropy numbers of embeddings between finite-dimensional sequence spaces.

Proposition *Let* $0 < p_1 \leq p_2 \leq \infty$ *and for each* $k \in \mathbf{N}$ *let* e_k *be the* k*th entropy number of the embedding* id $: \ell_{p_1}^m \to \ell_{p_2}^m$. *Then*

$$e_k \leq c \times \begin{cases} 1 & \text{if} \quad 1 \leq k \leq \log_2(2m), \\ \left(k^{-1}\log_2(1+2m/k)\right)^{1/p_1-1/p_2} & \text{if} \quad \log_2(2m) \leq k \leq 2m, \\ 2^{-k/2m}(2m)^{1/p_2-1/p_1} & \text{if} \quad k \geq 2m, \end{cases}$$

(1)

where c *is a positive constant which is independent of* m *(and* k*) but may depend upon* p_1 *and* p_2.

Proof Step 1 First we deal with the situation in which k is large and $0 < p_1 \leq p_2 \leq 1$. Put $r = 2^{-k/2m}(2m)^{1/p_2-1/p_1}$ with $k \geq 2m$, and let K_r be the maximal number of points $y^\ell \in U_{p_1}^m$ with $\|y^\ell - y^n | \ell_{p_2}^m\| > r$ if $\ell \neq n$.

When $z \in U_{p_2}^m$, Hölder's inequality and this choice of r show that

$$
\begin{aligned}
\left\| y^\ell + rz \,|\ell_{p_1}^m \|\right\|^{p_1} &\leq 1 + r^{p_1} \left\| z \,|\ell_{p_1}^m \|\right\|^{p_1} \\
&\leq 1 + r^{p_1} \left\| z \,|\ell_{p_2}^m \|\right\|^{p_1} m^{p_1(1/p_1 - 1/p_2)} \\
&\leq 2.
\end{aligned}
\tag{2}
$$

Let $\{ y^\ell : \ell = 1, ..., K_r \}$ be a set which is maximal in the above sense. Then

$$
U_{p_1}^m \subset \bigcup_\ell \{ y^\ell + r U_{p_2}^m \} \subset 2^{1/p_1} U_{p_1}^m.
\tag{3}
$$

Moreover, the balls $y^\ell + 2^{-1/p_2} r U_{p_2}^m$, $\ell = 1, ..., K_r$, are pairwise disjoint, for if z belongs to two of these balls, indexed by ℓ and n, say, then

$$
\left\| y^\ell - y^n \,|\ell_{p_2}^m \|\right\|^{p_2} \leq \left\| y^\ell - z \,|\ell_{p_2}^m \|\right\|^{p_2} + \left\| z - y^n \,|\ell_{p_2}^m \|\right\|^{p_2} \leq r^{p_2},
$$

and we have a contradiction. Together with (3) this gives

$$
K_r 2^{-2m/p_2} r^{2m} \mathrm{vol} U_{p_2}^m \leq 2^{2m/p_1} \mathrm{vol} U_{p_1}^m.
\tag{4}
$$

From (3.2.1/2) we see that

$$
\mathrm{vol} U_{p_1}^m \leq c^{2m} (2m)^{-2m(1/p_1 - 1/p_2)} \mathrm{vol} U_{p_2}^m
\tag{5}
$$

for some positive constant c which depends only on p_1 and p_2. The choice of r, plus (4) and (5), now gives

$$
K_r \leq 2^{k+cm} \quad \text{if} \quad k \geq 2m,
\tag{6}
$$

and hence

$$
e_{k+cm} \leq 2^{-k/2m} (2m)^{1/p_2 - 1/p_1} \quad \text{if} \quad k \geq 2m,
\tag{7}
$$

where c is independent of k and m: here, and henceforth, we adopt the useful convention that $e_\lambda = e_{[\lambda]+1}$ if $\lambda \geq 1$. This proves (1) provided that $k \geq c_1 m$ for some $c_1 > 1$ which is independent of m and k, always supposing that $0 < p_1 \leq p_2 \leq 1$.

Step 2 Routine modifications of the above argument, involving the triangle inequality, show that (1) holds for all p_1, p_2 with $0 < p_1 \leq p_2 \leq \infty$, if $k \geq c_1 m$. In particular, if $0 < p_1 = p_2 \leq \infty$, then the proposition follows immediately if we use the obvious fact that $e_k \leq 1$ for all $k \in \mathbf{N}$.

Step 3 Here we suppose that $0 < p = p_1 < \infty$, $p_2 = \infty$ and $1 \leq k \leq c_1 m$, where c_1 has the same meaning as above. We choose

$$c_2 > \left\{ \frac{1}{c_1} \log_2 \left(1 + \frac{1}{c_1} \right) \right\}^{-1/p}$$

and note that

$$\sigma := c_2 \left\{ k^{-1} \log_2 \left(\frac{m}{k} + 1 \right) \right\}^{1/p} = c_2 m^{-1/p} \left\{ \frac{m}{k} \log_2 \left(\frac{m}{k} + 1 \right) \right\}^{1/p} > m^{-1/p}. \tag{8}$$

Let m_σ be the maximal number of components y_n which any point $y = (y_1, ..., y_m) \in U_p^m$ may have with $|y_n| > \sigma$. By (8) we have $m_\sigma < m$ and

$$m_\sigma \sigma^p \leq 1, \quad \text{that is,} \quad m_\sigma \leq \sigma^{-p}. \tag{9}$$

We may, and shall, assume that $\sigma^{-p} \in \mathbf{N}$ and that $m_\sigma = \sigma^{-p}$. Let

$$e_k^{(\sigma)} = e_k \left(\mathrm{id} : \ell_p^{m_\sigma} \to \ell_\infty^{m_\sigma} \right).$$

By Step 1, with $k = c_1 \sigma^{-p}$, we have

$$e_{c_1 \sigma^{-p}}^{(\sigma)} \leq c_3 m_\sigma^{-1/p} = c_3 \sigma, \qquad c_1 \geq 1, \ c_3 \geq 1. \tag{10}$$

Hence $2^{c_1 \sigma^{-p}}$ balls in $\ell_\infty^{m_\sigma}$ with radius $c_3 \sigma$ cover $U_p^{m_\sigma}$. As there are $\binom{m}{m_\sigma}$ ways of selecting m_σ coordinates out of m, it follows by construction that $2^{c_1 \sigma^{-p}} \binom{m}{\sigma^{-p}}$ balls in ℓ_∞^m with radius $c_3 \sigma$ cover U_p^m. Since

$$\begin{aligned}
\log_2 \binom{m}{m_\sigma} &\leq m_\sigma \log_2 m - \sum_{j=1}^{m_\sigma} \log_2 j \\
&\leq m_\sigma \log_2 m - m_\sigma \log_2 m_\sigma + c m_\sigma \\
&\leq c' m_\sigma \log_2 \left(\frac{m}{m_\sigma} + 1 \right)
\end{aligned}$$

for some positive numbers c and c', it follows that U_p^m is covered by

$$2^{c_4 \sigma^{-p} \log_2(m\sigma^p + 1)} \tag{11}$$

ℓ_∞^m-balls of radius $c_3 \sigma$. We see from (8) that

$$\log_2 (m\sigma^p + 1) \leq c \log_2 \left(\frac{m}{k} + 1 \right),$$

and hence the expression in (11) can be estimated from above by $2^{c_5 k}$ for some positive constant c_5. Together with (8) this shows that

$$e_{c_5 k} \leq c_6 \left\{ k^{-1} \log_2 \left(\frac{m}{k} + 1 \right) \right\}^{1/p} \quad \text{if} \quad 1 \leq k \leq c_1 m \tag{12}$$

for some numbers $c_5 \geq 1$ and $c_6 > 0$ which are independent of m and k. The proposition for the case in which $0 < p_1 < \infty$ and $p_2 = \infty$ now follows from (12) and the fact that $e_k \leq 1$ for all $k \in \mathbf{N}$.

Step 4 Finally we deal with the case in which $0 < p_1 < p_2 < \infty$ and $1 \leq k \leq c_1 m$, where c_1 has the same meaning as in Step 1. To deal with this we use Theorem 1.3.2(i) with

$$A = B_0 = \ell_{p_1}, \; B_1 = \ell_\infty, \; B_\theta = \ell_{p_2}, \text{ where } 1/p_2 = (1 - \theta)/p_1. \quad (13)$$

Then we have

$$e_k \left(\mathrm{id} : \ell_{p_1}^m \to \ell_{p_2}^m \right) \leq c e_k^\theta \left(\mathrm{id} : \ell_{p_1}^m \to \ell_\infty^m \right). \quad (14)$$

We now use (12) with $p = p_1$ and $\theta/p_1 = 1/p_1 - 1/p_2$ on the right-hand side of (14). The same argument as at the end of Step 3 completes the proof for the remaining cases.

Remark 1 If $1 \leq p_1 < p_2 \leq \infty$, the estimate (1) is even an equivalence: see Schütt [Schü] and König [Kön], 3.c.8, pp. 190–1.

Remark 2 The constant c in (1) may in fact be chosen independently of p_2. This follows from Theorem 1.3.2(i).

3.2.3 Approximation numbers

We shall also need accurate information about the approximation numbers of the natural embedding

$$\mathrm{id} : \ell_p^m \to \ell_q^m \quad (1)$$

when $m \in \mathbf{N}$, $0 < p \leq \infty$ and $0 < q \leq \infty$. The sequence spaces involved here are complex ones; we shall denote by α_k the kth approximation number of id, in the sense of Definition 1.3.1/2. The corresponding *real* sequence spaces will be denoted by $\ell_p^{m,R}$, etc. and α_k^R will stand for the approximation numbers of the embedding between real spaces corresponding to (1). We mention these real spaces and embeddings because the calculation of the α_k^R when $p, q \in [1, \infty]$ has attracted much attention, the question being finally given a complete solution by Gluskin [Glu]. The survey article by R. Linde [Lin] gives an account of the history of this topic and provides detailed references. We are, of course, principally interested in the complex case, but first state the real results in the form given by Mynbaev and Otelbaev [MyOt] and then discuss how they relate to the complex situation.

Theorem 1 *Let $k \le m$ and put*

$$\Phi(m,k,p,q) =$$

$$\begin{cases} \left(\min \left\{ 1, m^{1/q} k^{-1/2} \right\} \right)^{(1/p-1/q)/(1/2-1/q)} & \text{if } 2 \le p < q \le \infty, \\ \max \left\{ m^{1/q-1/p}, \min \left\{ 1, m^{1/q} k^{-1/2} \right\} \sqrt{(1-k/m)} \right\} & \text{if } 1 \le p < 2 \le q \le \infty, \\ \max \left\{ m^{1/q-1/p}, (\sqrt{(1-k/m)})^{(1/p-1/q)/(1/p-1/2)} \right\} & \text{if } 1 \le q < q \le 2; \end{cases}$$

(2)

set

$$\Psi(m,k,p,q) = \begin{cases} \Phi(m,k,p,q) & \text{if } 1 \le p < q \le p', \\ \Phi(m,k,q',p') & \text{if } \max\{p,p'\} < q \le \infty. \end{cases}$$

(3)

(i) Suppose that $1 \le p < q \le \infty$ and that $(p,q) \ne (1,\infty)$. Then there are positive constants c_1, c_2 depending only on p and q, such that

$$c_1 \le \alpha_k^R / \Psi(m,k,p,q) \le c_2.$$

(4)

(ii) If $1 \le p \le q \le 2$ or $2 \le p \le q \le \infty$, then

$$\alpha_k^R \ge \sqrt{(1-k/m)}.$$

(5)

Remark 1 Inspection of the proof of (4) shows that c_1 and c_2 depend boundedly on $q \in (1,\infty]$, and that they can be replaced by positive constants which depend only on p, for q in the interval $(1,\infty]$.

The relationship between the real and complex cases is given by the following

Proposition *Let $k \in \mathbf{N}$, $k \le m$ and suppose that $p, q \in [1,\infty]$. Then*

$$\alpha_{2k-1}^R \le \alpha_k \le 2\alpha_k^R$$

(6)

(with $\alpha_k^R = 0$ if $\ell > m$).

Proof Let T be a complex $m \times m$ matrix with $\operatorname{rank} T < k$; then $\operatorname{rank} \operatorname{Re}(T) < 2k-1$, where $\operatorname{Re}(T) = \frac{1}{2}(T+\overline{T})$. We identify such matrices with the corresponding linear operators. Then the left-hand side of (6) is a consequence of

$$\left\| x - \operatorname{Re}(T)x \, | \ell_q^m \right\| \le \left\| x - Tx \, | \ell_q^m \right\|, \quad x \in \ell_p^{m,R}.$$

(7)

Now let T be a real $m \times m$ matrix. Then the right-hand side of (6) follows from the inequality

$$\left\|x - Tx \,|\ell_q^m\right\| \leq \left\|\mathrm{Re}x - T(\mathrm{Re}x)\,|\ell_q^m\right\| + \left\|\mathrm{Im}x - T(\mathrm{Im}x)\,|\ell_q^m\right\| \qquad (8)$$

for $x \in \ell_p^m$. Here $\mathrm{Re}x$ and $\mathrm{Im}x$ are the real and imaginary parts of x, in the obvious sense.

Remark 2 If $1 \leq q \leq p \leq \infty$ it is known (see [Pi2], p. 109) that

$$\alpha_k = (m - k + 1)^{1/q - 1/p}, \quad k = 1, ..., m. \qquad (9)$$

We next give some immediate consequences of the results described above. In these, the statement that $\alpha_k \sim K$ is to be interpreted as meaning that α_k/K is bounded above and below by positive constants which depend only upon p. Moreover, given any $p \in (0, \infty]$, we define p' by $1/p' + 1/p = 1$ if $1 \leq p \leq \infty$, and by $p' = \infty$ if $0 < p < 1$.

The following corollary is a direct consequence of Theorem 1 and the above proposition.

Corollary *Let $m \in \mathbf{N}$.*

(i) If $1 \leq p < q \leq 2$ and $k \leq m/4$, then

$$\alpha_k \sim 1. \qquad (10)$$

(ii) If $1 \leq p < 2 \leq q < p'$ and $k \leq m/4$, then

$$\alpha_k \sim \min\left(1, m^{1/q} k^{-1/2}\right). \qquad (11)$$

(iii) If $1 \leq p \leq 2 \leq p' \leq q \leq \infty$, with $(p, q) \neq (1, \infty)$, and $k \leq m/4$, then

$$\alpha_k \sim \min\left(1, m^{1/p'} k^{-1/2}\right). \qquad (12)$$

(iv) If $2 \leq p \leq q \leq \infty$ and $k \leq m/4$, then

$$\alpha_k \sim 1. \qquad (13)$$

These results will now be extended to allow for the possibility that p or q may belong to $(0, 1)$.

Theorem 2 *(i) Let $0 < p \leq q \leq 2$ and put $k \leq m/4$. Then*

$$\alpha_k \sim 1. \qquad (14)$$

(ii) Let $0 < q \leq p \leq \infty$ and let $m = 2k$. Then

$$\alpha_k \geq 2^{-1/q} m^{1/q - 1/p}. \tag{15}$$

(iii) Let $0 < p < 2 \leq q < p'$ and $k \leq m/4$. Then

$$\alpha_k \sim \min\left(1, m^{1/q} k^{-1/2}\right). \tag{16}$$

Proof Step 1 Here we prove (16): all that is necessary is to extend (11) to values of p in $(0,1)$. Let β_k be the kth approximation number of id : $\ell_1^m \to \ell_q^m$, and let $x^r \in \ell_1^m$ have jth component δ_{rj}, $j, r = 1, ..., m$. The unit ball in ℓ_1^m is

$$\left\{x = \sum_{r=1}^{m} \lambda_r x^r : \sum_{r=1}^{m} |\lambda_r| \leq 1\right\}.$$

Let $T : \ell_1^m \to \ell_q^m$ be linear and with $\dim \operatorname{im} T < k$. Then

$$x - Tx = \sum_{r=1}^{m} \lambda_r (x^r - Tx^r) = \sum_{r=1}^{m} \lambda_r w^r,$$

say, and thus

$$\beta_k \leq \sup \left\| x - Tx \left| \ell_q^m \right\| \right. \leq \sup_{r=1,...,m} \left\| w^r \left| \ell_q^m \right\| \right., \tag{17}$$

where the first supremum is taken over the unit ball in ℓ_1^m. Now take the infimum over all admissible T: (17) and (11) give

$$\min\left(1, m^{1/q} k^{-1/2}\right) \sim \beta_k \leq \alpha_k. \tag{18}$$

The opposite inequality follows from the fact that $p < 1$ and from the monotonicity of the ℓ_p spaces. This completes the proof of (16).

Step 2 We prove (14), and in view of (10), we shall assume that $0 < p < 1$. Let $q > 1$; then we have (17) and (18) with 1 on the left-hand side; see (10). Since $\alpha_k \leq \beta_k$, it follows that $\alpha_k \sim 1$. Next suppose that $0 < p \leq q \leq 1$ and replace β_k in (17) by the kth approximation number of the identity map in ℓ_q^m, which is 1. Then the analogue of (17) is

$$1 \leq \left\| \sum_{r=1}^{m} \lambda_r w^r \left| \ell_q^m \right\| \right\|^q \leq \sum_{r=1}^{m} |\lambda_r|^q \left\| w^r \left| \ell_q^m \right\| \right\|^q$$
$$\leq \sup_{r=1,...,m} \left\| w^r \left| \ell_q^m \right\| \right\|^q. \tag{19}$$

Hence $\alpha_k \geq 1$; and by the argument above we finally obtain $\alpha_k \sim 1$.

Step 3 Let T be represented as an $m \times m$ matrix with rank less than k. Then $\dim \ker T > k$. By V.D. Milman's lemma in the form proved in [Pi2], p. 108, it follows that there is an element $x = (x_1, ..., x_m) \in \ker T$ with $|x_j| \leq 1$ for all j and, say, $|x_1| = \cdots = |x_{m/2}| = 1$. For this element x we have

$$\left\| x \left| \ell_q^m \right\| \right. \geq (m/2)^{1/q} \quad \text{and} \quad \left\| x \left| \ell_p^m \right\| \right. \leq m^{1/p}.$$

Then

$$\alpha_{k+1} \geq \left\| x \left| \ell_q^m \right\| \right. / \left\| x \left| \ell_p^m \right\| \right. \geq 2^{-1/q} m^{1/q - 1/p},$$

which gives (15) and completes the proof of the theorem.

3.3 Embeddings between function spaces

3.3.1 Notation

Throughout this section Ω will be a bounded domain in $\mathbf{R}^n$ with C^∞ boundary and the spaces $B^s_{pq}(\Omega)$, $F^s_{pq}(\Omega)$ will be as in Definition 2.5.1/1. Given any $a \in \mathbf{R}$ we shall write $a_+ = \max(a, 0)$. Let

$$-\infty < s_2 < s_1 < \infty, \quad p_1, p_2, q_1, q_2 \in (0, \infty] \tag{1}$$

and suppose that

$$\delta_+ := s_1 - s_2 - n \left(\frac{1}{p_1} - \frac{1}{p_2} \right)_+ > 0. \tag{2}$$

We shall let $A^s_{pq}(\Omega)$ stand for either $B^s_{pq}(\Omega)$ or $F^s_{pq}(\Omega)$, with the understanding that for the F spaces we must have $p < \infty$. With this convention, let

$$\mathrm{id} : A^{s_1}_{p_1 q_1}(\Omega) \to A^{s_2}_{p_2 q_2}(\Omega) \tag{3}$$

be the natural embedding. The existence of this embedding is guaranteed by 2.5.1.

3.3.2 Entropy numbers: upper estimates

Let e_k be the kth entropy number of any of the embeddings (3.3.1/3).

Theorem *Under the above hypotheses, there is a positive constant c such that for all $k \in \mathbf{N}$,*

$$e_k \leq c k^{-(s_1 - s_2)/n}. \tag{1}$$

Proof We know that if $s \in \mathbf{R}$, $0 < p < \infty$ and $0 < q \leq \infty$, then in the sense of continuous embedding,

$$B_{pu}^s(\Omega) \subset F_{pq}^s(\Omega) \subset B_{pv}^s(\Omega),$$

where $u = \min(p,q)$ and $v = \max(p,q)$; see (2.3.3/1) and 2.5.1. It is therefore sufficient to prove the theorem for the B spaces, and initially we assume that $p_1 \leq p_2$. Then $\delta_+ = \delta = s_1 - \frac{n}{p_1} - \left(s_2 - \frac{n}{p_2}\right)$.

Step 1 Let

$$Q_r = \{x = (x_j) \in \mathbf{R}^n : |x_j| \leq r \quad \text{for} \quad j = 1, ..., n\}, \quad r > 0. \tag{2}$$

Without loss of generality we shall assume that

$$\Omega \subset Q_1. \tag{3}$$

Now let $\psi \in \mathscr{S}(\mathbf{R}^n)$ be such that

$$\operatorname{supp}\psi \subset Q_2 \quad \text{and} \quad \psi(x) = 1 \quad \text{for all} \quad x \in Q_1. \tag{4}$$

By [Triγ], 5.1.3, p. 238 we may assume that

$$B_{p_1q_1}^{s_1}(\Omega) \xrightarrow{E} \{f \in B_{p_1q_1}^{s_1}(\mathbf{R}^n) : \operatorname{supp}f \subset Q_1\} \subset B_{p_2q_2}^{s_2}(\mathbf{R}^n) \xrightarrow{R} B_{p_2q_2}^{s_2}(\Omega) \tag{5}$$

where E is an extension operator and R is the restriction operator. It is thus enough to prove the theorem for the middle embedding. In particular, we have $f = \psi f$ and hence

$$f = \psi \sum_{j=1}^{N} \left(\phi_j \widehat{f}\right)^{\vee} + \psi \sum_{j=N+1}^{\infty} \left(\phi_j \widehat{f}\right)^{\vee} := f_N + f^N \tag{6}$$

for any $N \in \mathbf{N}$. Here the ϕ_j are as in 2.2.1. Let $\|f \,|\, B_{p_1q_1}^{s_1}(\mathbf{R}^n)\| \leq 1$. If $q_2 < \infty$, then

$$\left\|f^N \,\big|\, B_{p_2q_2}^{s_2}(\mathbf{R}^n)\right\| \leq c_1 \left\|\sum_{j=N+1}^{\infty} \left(\phi_j \widehat{f}\right)^{\vee} \,\bigg|\, B_{p_2q_2}^{s_2}(\mathbf{R}^n)\right\|$$

$$\leq c_2 \left(\sum_{j=N-1}^{\infty} 2^{js_2q_2} \left\|\left(\phi_j \widehat{f}\right)^{\vee} \,\Big|\, L_{p_2}(\mathbf{R}^n)\right\|^{q_2}\right)^{1/q_2}$$

$$\leq c_3 \left(\sum_{j=N-1}^{\infty} 2^{j(s_1-\delta)q_2} \left\|\left(\phi_j \widehat{f}\right)^{\vee} \,\Big|\, L_{p_1}(\mathbf{R}^n)\right\|^{q_2}\right)^{1/q_2}$$

$$\leq c_4 2^{-N\delta} \left\|f \,|\, B_{p_1q_1}^{s_1}(\mathbf{R}^n)\right\|$$

$$\leq 2^{-N\delta} c_4. \tag{7}$$

The first inequality comes from pointwise multiplier properties (see [Triβ], 2.8.2, p. 140), the second from the definition of the norm $B^{s_2}_{p_2 q_2}(\mathbf{R}^n)$ and Fourier multiplier properties (see [Triβ], 2.3.2, p. 46), and the crucial third inequality follows from the definition of $\delta_+ = \delta$ in (3.3.1/2) and Nikol'skij's inequality (see [Triβ], (1.3.2/5), p. 18). Obvious modifications of this argument establish (7) when $q_2 = \infty$.

Step 2 To handle the term f_N in (6) we expand $\phi_j \hat{f}$ in the cube $Q_{2^j \pi}$ in a trigonometric series (in the sense of periodic distributions)

$$\left(\phi_j \hat{f}\right)(\xi) = \sum_{m \in \mathbf{Z}^n} a_m \exp\left(-2^{-j} im \cdot \xi\right), \quad \xi \in Q_{2^j \pi}, \qquad (8)$$

where

$$a_m = c_1 2^{-jn} \int_{Q_{2^j \pi}} \exp\left(2^{-j} im \cdot \xi\right) \left(\phi_j \hat{f}\right)(\xi) d\xi$$

$$= c_2 2^{-jn} \left(\phi_j \hat{f}\right)^{\vee}\left(2^{-j} m\right). \qquad (9)$$

As usual, $\mathbf{Z}^n$ is the lattice of all points in $\mathbf{R}^n$ with integer-valued coordinates and $m \cdot \xi$ is the scalar product of $m \in \mathbf{Z}^n$ and $\xi \in \mathbf{R}^n$. Note that $\operatorname{supp}\phi_j \subset Q_{2^j \pi}$. Let ψ be as in (4) and put $\psi_\lambda(\xi) = \psi(2^\lambda \xi)$, where $\lambda = \lambda(n)$ is a positive number so chosen that

$$(\psi - \psi_\lambda)\left(2^{-j-1} \xi\right) \phi_j(\xi) = \phi_j(\xi) \quad \text{if} \quad \xi \in \mathbf{R}^n \quad \text{and} \quad j \geq 1. \qquad (10)$$

Also $\phi_0(\xi) = \phi_0(\xi)\psi(2^{-1}\xi)$. Thus multiplication of (8) by $\psi(2^{-1}\xi)$ (if $j = 0$) or by $(\psi - \psi_\lambda)\left(2^{-j-1}\xi\right)$ (if $j \neq 0$) extends (8) from $Q_{2^j \pi}$ to $\mathbf{R}^n$. Application of the inverse Fourier transform gives

$$\left(\phi_j \hat{f}\right)^{\vee}(x) = c_2 2^{-jn} \sum_{m \in \mathbf{Z}^n} \left(\phi_j \hat{f}\right)^{\vee}\left(2^{-j} m\right)$$

$$\times \left(\exp\left(-2^{-j} im \cdot \xi\right)(\psi - \psi_\lambda)\left(2^{-j-1}\xi\right)\right)^{\vee}(x)$$

$$= c_3 \sum_{m \in \mathbf{Z}^n} \left(\phi_j \hat{f}\right)^{\vee}\left(2^{-j} m\right)(\psi - \psi_\lambda)^{\vee}\left(2^{j+1} x - 2m\right) \qquad (11)$$

if $j \geq 1$, and a corresponding formula when $j = 0$. By the theory developed in [Triβ], 1.4.1 and 1.3.3, pp. 22, 19, respectively, we have, if $p_1 < \infty$,

$$\left(\sum_{m \in \mathbf{Z}^n} \left|\left(\phi_j \hat{f}\right)^{\vee}\left(2^{-j} m\right)\right|^{p_1}\right)^{1/p_1} \leq c \left\|\left(\phi_j \hat{f}\right)^{\vee}\left(2^{-j} \cdot\right)\right| L_{p_1}(\mathbf{R}^n)\right\|$$

$$= c 2^{jn/p_1} \left\|\left(\phi_j \hat{f}\right)^{\vee}\right| L_{p_1}(\mathbf{R}^n)\right\|; \qquad (12)$$

a natural modification holds when $p_1 = \infty$. Here the constant c is independent of j.

Step 3 Here we introduce a further decomposition. Let ψ be as in (4), and suppose additionally that for every multi-index $\gamma \in \mathbf{N}_0^n$,

$$|D^\gamma \widehat{\varphi}(x)| \leq c_{\gamma,a} 2^{-\sqrt{|x|}} |x|^{-a} \quad \text{if} \quad |x| \geq 1, \tag{13}$$

where a is a positive number at our disposal and the $c_{\gamma,a}$ are appropriate positive constants: see [SchmT], 1.2.1, 1.2.2, pp. 14–16. With the estimation of f_N in mind, let

$$f_N^j(x) = c\psi(x) \sum_{m \in \mathbf{Z}^n, |m| \leq N_j} \left(\phi_j\widehat{f}\right)^\vee (2^{-j}m)(\psi - \psi_\lambda)^\vee (2^{j+1}x - 2m), \tag{14}$$

for $j = 0, 1, ..., N$ (there being the natural modification for $j = 0$). Here

$$N_j = \max\left(N^2\delta^2, 2^{j+2}\sqrt{n}\right), \tag{15}$$

$c = c_3$ and λ have the same meaning as in (11). By (6) and (11),

$$f_N = \sum_{j=0}^N f_N^j + f_{N,2} := f_{N,1} + f_{N,2}, \tag{16}$$

where

$$f_{N,2}(x) = c\psi(x) \sum_{j=1}^N \sum_{|m|>N_j} \left(\phi_j\widehat{f}\right)^\vee (2^{-j}m)(\psi - \psi_\lambda)^\vee (2^{j+1}x - 2m)$$

$$+ c\psi(x) \sum_{|m|>N_0} \left(\phi_0\widehat{f}\right)^\vee (m) \, \overset{\vee}{\psi} (2x - 2m). \tag{17}$$

To estimate $f_{N,2}$ we use the pointwise multiplier property from [Triβ], 2.8.2, p. 140, together with (12) and the fact that $\left\|f \mid B_{p_1q_1}^{s_1}(\mathbf{R}^n)\right\| \leq 1$, to obtain

$$\left\|f_{N,2} \mid B_{p_2q_2}^{s_2}(\mathbf{R}^n)\right\| \leq c \left\|\psi \mid B_{p_2q_2}^{s_2}(\mathbf{R}^n)\right\|$$

$$\times \sum_{j=0}^N \sum_{|m|>N_j} 2^{j\rho} \sup\left\{\left|D_x^\gamma \overset{\vee}{\psi} (2^{j+1}x - 2m)\right|\right.$$

$$\left. + \left|D_x^\gamma \overset{\vee}{\psi_\lambda} (2^{j+1}x - 2m)\right|\right\}, \tag{18}$$

where the supremum is taken over all x with $|x| \leq 2\sqrt{n}$ and all $\gamma \in \mathbf{N}_0^n$ with $|\gamma| \leq M$; c and ρ are positive numbers and M is an appropriate natural number. In view of (15) and (13), which holds for ψ_λ as well as for

ψ, the supremum in (18) can be estimated from above by $c2^{-N\delta}|m|^{-a}2^{-jb}$, where the positive numbers a and b are at our disposal. Thus from (18) we obtain

$$\left\| f_{N,2} \,\middle|\, B^{s_2}_{p_2 q_2}(\mathbf{R}^n) \right\| \le 2^{-N\delta}c, \tag{19}$$

where c is independent of N.

Step 4: the N-core A closer examination of $f_{N,1}$ is now desirable. We discuss the geometrical structure of the N-core, which is defined to be

$$\bigcup_{j=0}^{N} \left\{ f_N^j : f \in B^{s_1}_{p_1 q_1}(\mathbf{R}^n), \; \left\| f \,\middle|\, B^{s_1}_{p_1 q_1}(\mathbf{R}^n) \right\| \le 1, \; \mathrm{supp} f \subset Q_1 \right\}, \tag{20}$$

considered as a subset of $B^{s_2}_{p_2 q_2}(\mathbf{R}^n)$. Here f_N^j is given by (14), with the usual modification if $j = 0$. Let $K, L \in \mathbf{N}$ be such that

$$2^{K+2}\sqrt{n} \sim \delta^2 N^2 \quad \text{and} \quad L(s_1 - s_2) \sim \delta N, \tag{21}$$

where '$\sim$' means that K and L are chosen in an optimal way. Note that

$$2^{L+2}\sqrt{n} \ge c2^{\delta N/(s_1-s_2)} > 2\delta^2 N^2 > 2^{K+2}\sqrt{n} \text{ if } N \ge N_0, \text{ say}; \tag{22}$$

in what follows we shall assume that $N \ge N_0$. Thus $N \ge L > K$ and we split $f_{N,1}$ as follows:

$$f_{N,1} = \sum_{j=0}^{K} f_N^j + \sum_{j=K+1}^{L} f_N^j + \sum_{j=L+1}^{N} f_N^j := f_{N,3} + f_{N,4} + f_{N,5}. \tag{23}$$

We cover each of the j-components of the N-core (20) separately. Let $j > K$, so that the corresponding terms f_N^j belong to $f_{N,4}$ and $f_{N,5}$. Then in (14) we have $|m| \le 2^{j+2}\sqrt{n}$. We factorise the map $F_j : f \mapsto f_N^j$ as follows:

$$
\begin{array}{ccc}
\{f \in B^{s_1}_{p_1 q_1}(\mathbf{R}^n) : \mathrm{supp} f \subset Q_1\} & \overset{S_j}{\longrightarrow} & \ell^{M_j}_{p_1} \\
F_j \downarrow & & \downarrow \text{ id} \\
B^{s_2}_{p_2 q_2}(\mathbf{R}^n) & \overset{T_j}{\longleftarrow} & \ell^{M_j}_{p_2}
\end{array}
\tag{24}
$$

with $M_j = \mathrm{rank} F_j \le c2^{nj}$. The maps S_j and T_j are given by

$$S_j f = \left\{ \left(\phi_j \hat{f}\right)^{\vee} (2^{-j}m) \right\}_{|m| \le 2^{j+2}\sqrt{n}} \tag{25}$$

and

$$T_j \{a_m\}_{|m| \leq 2^{j+2}\sqrt{n}}(x) = c\psi(x) \sum_{|m| \leq 2^{j+2}\sqrt{n}} a_m (\psi - \psi_\lambda)^\vee (2^{j+1}x - 2m)$$

$$:= c\psi(x)g_j (2^{j+1}x). \tag{26}$$

By (14) we have

$$F_j = T_j \circ \mathrm{id} \circ S_j, \tag{27}$$

and from (12) we obtain

$$\|S_j\| \leq 2^{-j(s_1 - n/p_1)}c, \tag{28}$$

where c is independent of j. Moreover,

$$\begin{aligned}
\|T_j \{a_m\} \,|\, B_{p_2 q_2}^{s_2}(\mathbf{R}^n)\| &\leq c_1 \|g_j (2^{j+1} \cdot) \,|\, B_{p_2 q_2}^{s_2}(\mathbf{R}^n)\| \\
&\leq c_2 2^{js_2} \|g_j (2^{j+1} \cdot) \,|\, L_{p_2}(\mathbf{R}^n)\| \\
&\leq c_3 2^{j(s_2 - n/p_2)} \|g_j \,|\, L_{p_2}(\mathbf{R}^n)\|,
\end{aligned} \tag{29}$$

where the first inequality follows from the pointwise multiplier property in $B_{p_2 q_2}^{s_2}(\mathbf{R}^n)$, and the second is a consequence of the fact that the Fourier transform of $g_j (2^{j+1} \cdot)$ has support in the annulus $\{\xi \in \mathbf{R}^n : 2^{j-\mu} \leq |\xi| \leq 2^{j+\nu}\}$ for some $\mu > 0$ and $\nu > 0$. Since $\mathrm{supp}\, \widehat{g}_j \subset Q_2$, we have for some $\kappa > 0$,

$$\|g_j \,|\, L_{p_2}(\mathbf{R}^n)\| \leq c \left(\sum_{\ell \in \mathbf{Z}^n} |g_j(\kappa\ell)|^{p_2} \right)^{1/p_2} \tag{30}$$

(appropriately modified if $p_2 = \infty$): see [Triß], 1.3.3, 1.4.1, pp. 19, 22. In view of the decay properties of $(\psi - \psi_\lambda)^\vee$ it follows from (26) that

$$\begin{aligned}
|g_j(\kappa\ell)| &\leq c_1 \sum_m |a_m| \left(|\kappa\ell - 2m| + 1\right)^{-a} \\
&\leq c_2 \left(\sum_m |a_m|^{p_2} \left(1 + |\kappa\ell - 2m|\right)^{-b} \right)^{1/p_2},
\end{aligned} \tag{31}$$

where $a > 0$ and $b > 0$ are at our disposal. Hence

$$\left(\sum_{\ell \in \mathbf{Z}^n} |g_j(\kappa\ell)|^{p_2} \right)^{1/p_2} \leq c \left(\sum_m |a_m|^{p_2} \right)^{1/p_2} \tag{32}$$

where c is independent of j. Together with (29) and (30) this gives

$$\|T_j\| \leq 2^{j(s_2 - n/p_2)}c. \tag{33}$$

Now let $e_k^{(j)}$ be the kth entropy number of F_j, and let $\varepsilon_k^{(j)}$ be the kth entropy number of id : $\ell_{p_1}^{M_j} \to \ell_{p_2}^{M_j}$. From (27), (28) and (33) it follows that

$$e_k^{(j)} \le c2^{-j\delta} \varepsilon_k^{(j)}, \quad j \ge K; \tag{34}$$

we remind the reader that $\delta = \delta_+$ is given by (3.3.1/2). In view of the estimates (7) and (19), we look for the smallest $k \in \mathbf{N}$ such that (see (21))

$$2^{-j\delta} \varepsilon_k^{(j)} \le c2^{-N\delta} \sim c2^{-L(s_1-s_2)}. \tag{35}$$

Of course, the numbers $\varepsilon_k^{(j)}$ are simply the numbers e_k from (3.2.2/1) with $m = M_j \sim c2^{nj}$. What will emerge is that the term with $j = L$ is dominant and that all the other terms can be regarded as perturbations. Some care is necessary because we have to cover not only the various components of the N-core from (20), but the sum of all these components: it is at this point that the dyadic structures both of the dimensions and of the radii which are involved come into their own.

Step 5: the terms $f_{N,4}$ First we handle the f_N^j for which $K < j \le L$: see (14), (21) and (23). We may assume that $M_j = c_1 2^{nj}$ (see (24)) and choose

$$k_j = c_2 2^{nj}(1 + L - j) \ge 2M_j, \quad c_2 > c_1, \tag{36}$$

where c_2 will be determined later. From (34) with $k = k_j$, and the last line in (3.2.2/1), we find that

$$
\begin{aligned}
e_{k_j}^{(j)} &\le c2^{-j\delta} 2^{-k_j/2M_j} M_j^{1/p_2-1/p_1} \\
&= cc_1^{1/p_2-1/p_1} 2^{-j(s_1-s_2)} 2^{-c_3(1+L-j)};
\end{aligned} \tag{37}
$$

the final line follows from the definition (3.3.1/2) of $\delta = \delta_+$. Choose $c_3 = c_2/c_1 > \max(1, s_1 - s_2)$. Then

$$e_{k_j}^{(j)} \le c_4 2^{-L(s_1-s_2)-c_5(L-j)} \tag{38}$$

for some positive numbers c_4, c_5 which are independent of j and L. This means that 2^{k_j} balls in $B_{p_2 q_2}^{s_2}(\mathbf{R}^n)$ of radius $c_4 2^{-L(s_1-s_2)-c_5(L-j)}$ cover the j-component of the N-core (20), where $K < j \le L$. Let $f \in B_{p_1 q_1}^{s_1}(\mathbf{R}^n)$ be such that $\|f \,|\, B_{p_1 q_1}^{s_1}(\mathbf{R}^n)\| \le 1$ and supp $f \subset Q_1$, and let $\rho = \min(p_2, q_2, 1)$. An optimal choice of the centres $G_i = G_i(f)$ of the corresponding balls in $B_{p_2 q_2}^{s_2}(\mathbf{R}^n)$ gives, together with (21), the estimate

$$\left\| f_{N,4} - \sum_{j=K+1}^{L} G_j \,\big|\, B_{p_2 q_2}^{s_2}(\mathbf{R}^n) \right\|^{\rho} \leq \sum_{j=K+1}^{L} \left\| f_N^j - G_j \,\big|\, B_{p_2 q_2}^{s_2}(\mathbf{R}^n) \right\|^{\rho}$$

$$\leq c_4^{\rho} 2^{-L\rho(s_1-s_2)} \sum_{j=K+1}^{L} 2^{-c_5\rho(L-j)}$$

$$\leq c_6 2^{-L\rho(s_1-s_2)}$$

$$\sim c_6 2^{-N\rho\delta}. \tag{39}$$

The number of balls needed to ensure that (39) holds is at most $\displaystyle\prod_{j=K+1}^{L} 2^{k_j}$;
and

$$\sum_{j=K+1}^{L} k_j \leq c_2 2^{nL} \sum_{j=K+1}^{L} 2^{-n(L-j+1)}(L-j+1) \leq c_7 2^{nL}. \tag{40}$$

Step 6: the terms $f_{N,5}$ Here we deal with the term f_N^j with $L < j \leq N$.
When $p_1 = p_2$ there are no terms of this type; thus we shall assume that
$p_1 < p_2$. As before we may suppose that $M_j = c_1 2^{nj}$. Choose

$$k_j = c_1 2^{nL-(j-L)\kappa} \text{ with } 0 < \kappa < \delta \left(1/p_1 - 1/p_2\right)^{-1}, \tag{41}$$

and apply (34) with $k = k_j$. By the middle line in (3.2.2/1), which also
covers the first line in the sense of an estimate from above, we obtain

$$e_{k_j}^{(j)} \leq c_2 2^{-j\delta} 2^{-nL(1/p_1-1/p_2)} 2^{(j-L)\kappa(1/p_1-1/p_2)} \left(\log_2 2^{n(j-L)+\kappa(j-L)}\right)^{1/p_1-1/p_2}$$

$$\leq c_3 2^{-L(s_1-s_2)} 2^{-\mu(j-L)} (j-L)^{1/p_1-1/p_2}, \tag{42}$$

where

$$\mu = \delta - \kappa \left(1/p_1 - 1/p_2\right) > 0. \tag{43}$$

The position is now as in (39) and (40): $\displaystyle\prod_{j=L+1}^{N} 2^{k_j}$ balls $\left(\text{with } \displaystyle\sum_{j=L+1}^{N} k_j \leq 2^{nL} c_1\right)$
in $B_{p_2 q_2}^{s_2}(\mathbf{R}^n)$ of radius $2^{-L(s_1-s_2)} c_2$ cover

$$\left\{ f_{N,5} : f \in B_{p_1 q_1}^{s_1}(\mathbf{R}^n),\ \left\| f \,\big|\, B_{p_1 q_1}^{s_1}(\mathbf{R}^n) \right\| \leq 1,\ \mathrm{supp} f \subset Q_1 \right\}. \tag{44}$$

Step 7: the terms $f_{N,3}$ Suppose that $f \in B^{s_1}_{p_1 q_1}(\mathbf{R}^n)$, with $\left\| f \mid B^{s_1}_{p_1 q_1}(\mathbf{R}^n) \right\| \le 1$ and supp$f \subset Q_1$, and let $v = \min(p_2, q_2, 1)$. Then

$$\left\| f_{N,3} \mid B^{s_2}_{p_2 q_2}(\mathbf{R}^n) \right\|^v = \left\| \sum_{j=0}^{K} f_N^j \mid B^{s_2}_{p_2 q_2}(\mathbf{R}^n) \right\|^v \le \sum_{j=0}^{K} \left\| f_N^j \mid B^{s_2}_{p_2 q_2}(\mathbf{R}^n) \right\|^v$$

$$\le c + c \sum_{j=0}^{K} \left\| \psi \sum_{m \in \mathbf{Z}^n} \left(\phi_j \widehat{f} \right)^{\vee} (2^{-j} m) (\psi - \psi_\lambda)^{\vee} (2^{j+1} \cdot -2m) \mid B^{s_2}_{p_2 q_2}(\mathbf{R}^n) \right\|^v,$$

(45)

with suitable modification of the term with $j = 0$: see (14) and the estimate of $f_{N,2}$ in (18) and (19). From (11) and the same arguments as in (7) we obtain

$$\left\| f_{N,3} \mid B^{s_2}_{p_2 q_2}(\mathbf{R}^n) \right\| \le c \tag{46}$$

for some c which is independent of N. Since $K \sim \log_2 N \sim \log_2 L$, we cover

$$\left\{ f_{N,3} : f \in B^{s_1}_{p_1 q_1}(\mathbf{R}^n), \ \left\| f \mid B^{s_1}_{p_1 q_1}(\mathbf{R}^n) \right\| \le 1, \ \text{supp} f \subset Q_1 \right\} \tag{47}$$

in $B^{s_2}_{p_2 q_2}(\mathbf{R}^n)$ with balls of radius $c2^{-\delta N} \sim c2^{-(s_1 - s_2)L}$; by (46) we need at most

$$c_1 \left(2^{\delta N} \right)^{c_2 N^{2n+1}} = c_1 2^{c_3 N^{2n+2}} \tag{48}$$

such balls. We now use the crude estimate $c_3 N^{2n+2} \le c_4 2^{nL}$.

Step 8: the entropy numbers We aim to cover

$$\left\{ f_{N,1} : f \in B^{s_1}_{p_1 q_1}(\mathbf{R}^n), \ \left\| f \mid B^{s_1}_{p_1 q_1}(\mathbf{R}^n) \right\| \le 1, \ \text{supp} f \subset Q_1 \right\} \tag{49}$$

in $B^{s_2}_{p_2 q_2}(\mathbf{R}^n)$ with balls of radius $c_1 2^{-(s_1 - s_2)L}$. By Steps 5–7 the corresponding sets with $f_{N,3}$, $f_{N,4}$ and $f_{N,5}$ instead of $f_{N,1}$ can be covered by 2^m balls of this radius, with $m \le c_2 2^{nL}$. Thus by (23) the set in (49) can also be covered by 2^m balls of radius $c_3 2^{-(s_1 - s_2)L}$, where $m \le c_4 2^{nL}$. Since $f = f^N + f_{N,1} + f_{N,2}$ (see (6) and (16)), and since we have the estimates (7) and (19), with $2^{-N\delta} \sim 2^{-(s_1 - s_2)L}$, it follows that the set corresponding to (49), but with f instead of $f_{N,1}$, can be covered in the desired way. This shows that

$$e_{c_5 2^{nL}} \le c_6 2^{-(s_1 - s_2)L}$$

with $(s_1 - s_2)L \sim N\delta$ and $N \in \mathbf{N}$. Hence

$$e_k \le c_7 k^{-(s_1 - s_2)/n}$$

for all $k \in \mathbf{N}$.

Step 9 To complete the proof of the theorem we merely have to remove the initial assumption that $p_1 \leq p_2$. To extend the estimate (1) to the case in which $p_1 > p_2$, it is plainly enough to show that

$$B^{s_2}_{p_1 q_2}(\Omega) \subset B^{s_2}_{p_2 q_2}(\Omega), \quad 0 < p_2 \leq p_1 \leq \infty, \tag{50}$$

in the sense of continuous embeddings. To do this we use the fact, established in 2.3.2 (see especially (2.3.2/13)) that

$$\|K_0(1, g) \,|L_{p_1}(\mathbf{R}^n)\| + \left(\int_0^1 t^{-s_2 q_2} \|K(t, g) \,|L_{p_1}(\mathbf{R}^n)\|^{q_2} \frac{dt}{t} \right)^{1/q_2}, \tag{51}$$

with $g \in B^{s_2}_{p_1 q_2}(\mathbf{R}^n)$, is a quasi-norm on $B^{s_2}_{p_1 q_2}(\mathbf{R}^n)$ equivalent to the usual one. We recall that K_0 is a function in $C_0^\infty(\mathbf{R}^n)$, that $K = \triangle^N K_0$ for some suitably large $N \in \mathbf{N}$, and that $K(t, g)$ and $K_0(1, g)$ are local means: see (2.3.2/12).

Now let $f \in B^{s_2}_{p_1 q_2}(\Omega)$ and let $g = \psi(x) E f$, where E is a bounded linear extension operator from $B^{s_2}_{p_1 q_2}(\Omega)$ to $B^{s_2}_{p_1 q_2}(\mathbf{R}^n)$ and ψ is given by (4): see Remark 2.5.1/1 and [Triγ], 5.1.3, p. 239. Then all the local means in (51) vanish outside some compact subset of $\mathbf{R}^n$ and hence, by Hölder's inequality (and (51), also with p_2 instead of p_1), we have

$$\left\| g \,\big| B^{s_2}_{p_2 q_2}(\mathbf{R}^n) \right\| \leq c \left\| g \,\big| B^{s_2}_{p_1 q_2}(\mathbf{R}^n) \right\|$$

for some positive constant c, independent of g. The desired embedding (50) now follows by the usual procedures of extension and restriction.

3.3.3 Entropy numbers: lower estimates

We remind the reader that the assumptions made in 3.3.1 will still hold here, and that we shall continue to use the notation of that subsection.

Theorem 1 *Under the hypotheses of 3.3.1, there is a positive constant c such that for all $k \in \mathbf{N}$,*

$$e_k \geq c k^{-(s_1 - s_2)/n}. \tag{1}$$

Proof As in 3.3.2 it is enough to deal with the B-spaces. The main idea is to study more closely the N-core (3.3.2/20). To do this we modify the function in (3.3.2/20) in a suitable way. Let

$$G(x) = \sum_{|m| \leq b 2^j} a_m \overset{\vee}{\chi} \left(2^j x - \kappa m \right) \exp \left(i 2^j \lambda \mathbf{1} \cdot x \right) \tag{2}$$

and

$$f(x) = h(x)G(x). \tag{3}$$

Here $j \in \mathbf{N}$ and $x \in \mathbf{R}^n$; b, κ and λ are positive numbers, with $\lambda\kappa(2\pi)^{-1} \in \mathbf{N}$, to be chosen later; $\mathbf{1} = (1, ..., 1)$ and $\mathbf{1} \cdot x = \sum_{k=1}^{n} x_k$. The summation in (2) extends over all those lattice points $m \in \mathbf{Z}^n$ with $|m| \leq b2^j$, and the a_m are non-negative numbers. Lastly, the function h in (3) is a cut-off function which we shall specify presently, and χ is a function in $C^\infty(\mathbf{R}^n)$, supported by the unit ball and with $\hat{\chi}(x) > 0$ for $x \in \mathbf{R}^n$. Such a function exists. To verify this, let χ_0 be a real-valued, even function in $C_0^\infty(\mathbf{R}^n)$. Then $\hat{\chi}_0(x)$ is real and $(\chi_0 * \chi_0)^\wedge(x) \sim (\hat{\chi}(x))^2 \geq 0$. If the support of χ_0 is chosen appropriately, then

$$\chi(x) := (\chi_0 * \chi_0)(x)e^{-|x|^2/2}$$

has the desired properties, since

$$\hat{\chi}(x) = c \int_{\mathbf{R}^n} e^{-|x-y|^2/2} (\hat{\chi}_0(y))^2 \, dy$$

for some $c > 0$.

Now that these preliminaries are over, we can embark on the proof.

Step 1 Let $s \in \mathbf{R}$ and suppose that $p, q \in (0, \infty]$. We shall prove that

$$\left\| G \mid B_{pq}^s(\mathbf{R}^n) \right\| \sim 2^{j(s-n/p)} \left\| \{a_m\} \mid \ell_p^{N_j} \right\|, \quad N_j \sim 2^{jn}, \tag{4}$$

where $\ell_p^{N_j}$ has the same meaning as in 3.2.1. Assertions of the form of (4) are just those considered in connection with localisation: see 2.3.2. Here we can give a short, direct proof. We remind the reader that the equivalence in (4) is to be understood in the sense that the positive constants in the related inequalities are independent of $j \in \mathbf{N}$ and of the a_m. From the definition (2) of G we see that

$$\hat{G}(x) = \sum_{|m| \leq b2^j} a_m 2^{-jn} \left(\overset{\vee}{\chi}(\xi - \kappa m) \exp(i\lambda \mathbf{1} \cdot \xi) \right)^\wedge (2^{-j}x)$$

$$= 2^{-jn} \sum_{|m| \leq b2^j}{}' a_m \left(\overset{\vee}{\chi}(\xi - \kappa m) \right)^\wedge (2^{-j}x - \lambda\mathbf{1})$$

$$= 2^{-jn}\chi(2^{-j}x - \lambda\mathbf{1}) \sum_{|m| \leq b2^j} a_m \exp(-i\kappa 2^{-j}x \cdot m) \tag{5}$$

and $\operatorname{supp}\widehat{G} \subset \{x \in \mathbf{R}^n : |x - 2^j \lambda \mathbf{1}| \le 2^j\}$. From the definition (2.2.1/6) of the norm on $B^s_{pq}(\mathbf{R}^n)$ we saw that, for large enough λ, we have the first of the following two equivalences:

$$\left\| G \,\big|\, B^s_{pq}(\mathbf{R}^n) \right\| \sim 2^{js} \left\| G \,|\, L_p(\mathbf{R}^n) \right\| \sim 2^{j(s-n/p)} \left\| \{ G(2^{-j}\kappa M) \}_{M \in \mathbf{Z}^n} \,\big|\, \ell_p \right\|. \tag{6}$$

The second equivalence in (6) is an inequality of Plancherel–Pólya–Nikol'skii type, for the validity of which we must choose $\kappa > 0$ to be small enough: see [Triβ], 1.4.1, 1.4.2 and 1.3.3, pp. 22, 23, 19 respectively combined with a homogeneity argument with respect to 2^j. Now (2) and (6) give

$$\left\| G \,\big|\, B^s_{pq}(\mathbf{R}^n) \right\| \sim 2^{j(s-n/p)} \left\| \left\{ \sum_{|m| \le b2^j} a_m \overset{\vee}{\chi}(\kappa M - \kappa m) \right\}_M \,\bigg|\, \ell_p \right\|. \tag{7}$$

Since $\overset{\vee}{\chi}$ decays rapidly,

$$\left\| G \,\big|\, B^s_{pq}(\mathbf{R}^n) \right\| \le c2^{j(s-n/p)} \left\| \{a_m\}_{|m| \le b2^j} \,\big|\, \ell_p \right\|. \tag{8}$$

The reverse estimate follows from our assumptions that $a_m \ge 0$ and $\overset{\vee}{\chi} > 0$; the proof of (4) is complete.

Step 2 Here we show that, under the same assumptions as in Step 1,

$$\left\| f \,\big|\, B^s_{pq}(\mathbf{R}^n) \right\| \sim 2^{j(s-n/p)} \left\| \{a_m\} \,\big|\, \ell_p^{N_j} \right\|, \quad N_j \sim 2^{jn}, \tag{9}$$

where f is given by (3), and also show that a corresponding assertion holds with $\mathbf{R}^n$ replaced by Ω. To do this, we choose the cut-off function h so that it is in $C^\infty(\mathbf{R}^n)$ and satisfies

$$\operatorname{supp} h \subset \Omega \text{ and } h(x) = 1 \text{ if } |x| \le 1. \tag{10}$$

We shall suppose that these requirements (10) are not contradictory: if they are, then some routine modifications need to be implemented. Cover $\{y \in \mathbf{R}^n : |y| > 1\}$ with congruent balls K_ℓ centred at, say, $x^\ell = \varepsilon \ell$, with $|x^\ell| \ge 1$, $\ell \in \mathbf{Z}^n$, ε small and positive; each K_ℓ is to have radius less than $\frac{1}{2}$. Now choose functions $\rho_\ell \in C^\infty(\mathbf{R}^n)$ which form a corresponding resolution of unity; more precisely,

$$1 = \sum_\ell \rho_\ell(x) \text{ if } |x| \ge 1, \ \operatorname{supp}\rho_\ell \subset K_\ell, \tag{11}$$

and $\rho_\ell(x) = \rho(x - x^\ell)$ for some suitable standard function ρ. We write

$$f(x) = G(x) + (h(x) - 1)G(x) \tag{12}$$

and observe that

$$\left\| (1 - h(x))G(x) \,\big|\, B_{pq}^{s}(\mathbf{R}^{n}) \right\| \leq \left\| (1 - h(x))G(x) \,\big|\, B_{pp}^{\sigma}(\mathbf{R}^{n}) \right\|, \quad \sigma > s. \quad (13)$$

In view of (10), (11) and the localisation principle for $B_{pp}^{\sigma}(\mathbf{R}^{n})$ (see [Triγ], 2.4.7, p. 124), we have

$$\left\| (1 - h(x))G(x) \,\big|\, B_{pp}^{\sigma}(\mathbf{R}^{n}) \right\|^{p} \sim \sum_{\ell} \left\| \rho_{\ell}(1 - h)G \,\big|\, B_{pp}^{\sigma}(\mathbf{R}^{n}) \right\|^{p}, \quad (14)$$

where $\sum_{\ell}$ now, and for the time being, is the same as in (11). Think of $(1 - h)G$ as a pointwise multiplier in $B_{pp}^{\sigma}(\mathbf{R}^{n})$ (see [Triγ], 4.2.2, p. 203); by (13), (14) and the properties of ρ_{ℓ} we obtain

$$\left\| (1 - h)G \,\big|\, B_{pq}^{s}(\mathbf{R}^{n}) \right\|^{p} \leq c \sum_{\ell} \sup_{|\gamma| \leq L, x \in K_{\ell}} |D^{\gamma}(1 - h)(x)G(x)|^{p} \quad (15)$$

for some $L \in \mathbf{N}$ which depends on s, p and q. If $|\gamma| \leq L$ and $x \in K_{\ell}$, then (2) and the decay properties of $\overset{\vee}{\chi}$ give

$$|D^{\gamma}(1 - h)(x)G(x)| \leq 2^{jL} \sum_{|m| \leq b2^{j}} |a_{m}| \, \big|2^{j}x - \kappa m\big|^{-N}, \quad (16)$$

where $N \in \mathbf{N}$ is at our disposal. We now choose b small and positive and replace $|a_{m}|$ by $\left(\sum_{m} |a_{m}|^{p} \right)^{1/p}$: the right-hand side of (16) can thus be estimated from above by

$$c2^{-j\rho} \, |x^{\ell}|^{-\alpha} \left(\sum_{|m| \leq b2^{j}} |a_{m}|^{p} \right)^{1/p}, \quad (17)$$

where $\rho > 0$ and $\alpha > 0$ are at our disposal. Use of these estimates in (15) gives

$$\left\| (1 - h)G \,\big|\, B_{pq}^{s}(\mathbf{R}^{n}) \right\| \leq c2^{-j\rho} \left(\sum_{|m| \leq b2^{j}} |a_{m}|^{p} \right)^{1/p}, \quad (18)$$

where again $\rho > 0$ is at our disposal. The desired equivalence (9) follows immediately from (4), (17) and (18). Moreover, (10) leads to

$$\left\| f \,\big|\, B_{pq}^{s}(\Omega) \right\| \sim 2^{j(s-n/p)} \left\| \{a_{m}\} \,\big|\, \ell_{p}^{N_{j}} \right\|, \quad N_{j} \sim 2^{jn}, \quad (19)$$

as required.

Step 3 We can now establish the lower bound (1). Let f be given by (2) and (3), and suppose that $\left\| f \,|\, B^{s_1}_{p_1 q_1}(\Omega) \right\| \leq 1$; we recall that $a_m \geq 0$. In view of (19), the family V of all such functions generates a set $V^{N_j}_{p_1}$ of elements $\{a_m\} \in \mathbf{R}^{N_j}$, and

$$\mathrm{vol} V^{N_j}_{p_1} \sim 2^{-j(s_1 - n/p_1)N_j} \, \mathrm{vol} U^{N_j/2}_{p_1} \, 2^{-N_j}, \tag{20}$$

where $U^{N_j/2}_{p_1}$ stands for the unit ball in $\ell^{N_j/2}_{p_1}$. Let K be a ball of radius $\delta 2^{-j(s_1 - s_2)}$ in $B^{s_2}_{p_2 q_2}(\Omega)$. Then via (2) and (3), $K \cap V$ generates a subset $(K \cap V)^{N_j}$ of $V^{N_j}_{p_1}$ with

$$
\begin{aligned}
\mathrm{vol}\,(K \cap V)^{N_j} &\leq \delta^{N_j} 2^{-j(s_1 - s_2)N_j} 2^{-j(s_2 - n/p_2)N_j} \mathrm{vol} U^{N_j/2}_{p_2} \\
&= \left(\delta 2^{-j(s_1 - n/p_1)} 2^{-jn(1/p_1 - 1/p_2)} \right)^{N_j} \mathrm{vol} U^{N_j/2}_{p_2}. \tag{21}
\end{aligned}
$$

This means that the number M of $B^{s_2}_{p_2 q_2}(\Omega)$-balls of radius $\delta 2^{-j(s_1 - s_2)}$ which are needed to cover V is at least $\mathrm{vol} V^{N_j}_{p_1}$ divided by the right-hand side of (21); thus

$$M \geq c 2^{-N_j} \delta^{-N_j} 2^{jn(1/p_1 - 1/p_2)N_j} \mathrm{vol} U^{N_j/2}_{p_1} \Big/ \mathrm{vol} U^{N_j/2}_{p_2}. \tag{22}$$

By (3.2.1/2) and the last line in 3.2.1,

$$\left(\mathrm{vol} U^{N_j/2}_{p_1} \right)^{1/N_j} \sim \left(\mathrm{vol} U^{N_j/2}_{p_2} \right)^{1/N_j} N_j^{-1/p_1 + 1/p_2}, \tag{23}$$

where $\sim$ means that the associated constants are independent of N_j. Since $N_j \sim 2^{jn}$, (22) and (23) show that

$$M^{1/N_j} \geq c/\delta \quad \text{for some } c > 0. \tag{24}$$

Now choose δ small enough and obtain

$$M \geq 2^{jn}. \tag{25}$$

Thus at least $2^{2^{jn}}$ balls of radius $\delta 2^{-j(s_1 - s_2)}$ in $B^{s_2}_{p_2 q_2}(\Omega)$ are needed to cover the unit ball in $B^{s_1}_{p_1 q_1}(\Omega)$. It follows that

$$e_{2^{jn}} \geq \delta 2^{-j(s_1 - s_2)}. \tag{26}$$

This proves (1) for all k of the form 2^{jn}, and hence for all $k \in \mathbf{N}$.

We summarise, for later convenience, Theorem 1 and Theorem 3.3.2 as follows:

Theorem 2 *Under the hypotheses of 3.3.1, there are positive constants c_1 and c_2 such that for all $k \in \mathbf{N}$,*

$$c_1 k^{-(s_1 - s_2)/n} \leq e_k \leq c_2 k^{-(s_1 - s_2)/n}. \tag{27}$$

3.3.4 Approximation numbers

We remind the reader that the assumptions and conventions of 3.3.1 still hold here. By a_k we shall mean the kth approximation number of any of the embeddings (3.3.1/3). Also recall that $\delta_+ = \delta = s_1 - \frac{n}{p_1} - \left(s_2 - \frac{n}{p_2} \right)$ if $p_2 \geq p_1$.

Theorem *Let the general hypotheses of 3.3.1 be satisfied.*
(i) Suppose in addition that

$$0 < p_1 \leq p_2 \leq 2, \text{ or } 2 \leq p_1 \leq p_2 \leq \infty, \text{ or } 0 < p_2 \leq p_1 \leq \infty. \tag{1}$$

Then there are positive numbers c_1 and c_2 such that for all $k \in \mathbf{N}$,

$$c_1 k^{-\delta_+/n} \leq a_k \leq c_2 k^{-\delta_+/n}. \tag{2}$$

(ii) Suppose that in addition to the general hypotheses,

$$0 < p_1 \leq 2 \leq p_2 < \infty \text{ and } \lambda = \frac{s_1 - s_2}{n} - \max \left(\frac{1}{2} - \frac{1}{p_2}, \frac{1}{p_1} - \frac{1}{2} \right) > \frac{1}{2}. \tag{3}$$

Then there are positive constants c_1 and c_2 such that for all $k \in \mathbf{N}$,

$$c_1 k^{-\lambda} \leq a_k \leq c_2 k^{-\lambda}. \tag{4}$$

(iii) Suppose that in addition to the general hypotheses,

$$0 < p_1 \leq 2 \leq p_2 \leq \infty. \tag{5}$$

Then there are positive constants c_1 and c_2 such that for all $k \in \mathbf{N}$,

$$c_1 k^{-\lambda} \leq a_k \leq c_2 k^{-\delta_+/n}, \tag{6}$$

where λ is as in (3).

Proof As in the case of the entropy numbers, it is enough to deal with the *B* spaces. Our arguments rely to a large extent on those already given for the entropy numbers in 3.3.2 and 3.3.3, thus enabling us to present upper and lower estimates in a single theorem, without an excessively long proof. We begin with a general upper estimate, and use the terminology introduced in 3.3.2 and 3.3.3

Step 1 We shall show that under the general hypotheses of 3.3.1,

$$a_k \leq c k^{-\delta_+/n}, \quad k \in \mathbf{N}, \tag{7}$$

for some constant $c > 0$, independent of k. In fact, if $p_1 \leq p_2$, then by
(3.3.2/6), (3.3.2/16) and the estimates (3.3.2/7) and (3.3.2/19), we have

$$\left\| f - f_{N,1} \left| B_{p_2 q_2}^{s_2}(\mathbf{R}^n) \right\| \right. \leq 2^{-N\delta} c \tag{8}$$

for some c which is independent of N. Recall that $\delta_+ = \delta$ in this case. In
view of (3.3.2/14) and (3.3.2/16) the linear map $f \longmapsto f_{N,1}$ has rank less
than $2^{nN}c$. Thus (8) shows that for all $N \in \mathbf{N}$,

$$a_{2^{nN}c_1} \leq 2^{-N\delta} c_2,$$

and hence (7) follows, when $p_1 \leq p_2$. To remove the assumption that
$p_1 \leq p_2$ we proceed as in Step 9 of the proof of Theorem 3.3.2. We
therefore have the upper estimates claimed in (2) and (6).

Step 2 Here we prove the sharper upper estimate given in (4), under the
additional assumptions (3). Let $N \in \mathbf{N}$ be given and put

$$M = N\delta/n\lambda, \tag{9}$$

where δ and λ are as just before the theorem and as in (3) respectively. We
may assume that $M \in \mathbf{N}$; for if not, routine modifications overcome any
difficulties caused. By (3.3.2/14) and (3.3.2/23) it follows that $f \longmapsto f_{N,3}$,
$f \longmapsto f_{N,4}$ and $f \longmapsto f_{N,5}$ are linear maps of finite rank P_3, P_4 and P_5
respectively. Moreover, (3.3.2/21) and (3.3.2/23) show that

$$P_3 \leq c \left(\delta^2 N^2 \right)^n K, \text{ where } K \sim \log N,$$

and hence

$$P_3 \leq c 2^{nM} \tag{10}$$

where c is independent of N (see (9)). Similar arguments apply to P_4, P_5
and the ranks of the maps $f \longmapsto f_N^j$; see (3.3.2/14). We now see, from
(3.3.2/7), (3.3.2/19), (10), knowledge of the ranks of $f \longmapsto f_N^j$ and the
explicit formulae for f_N^j, that

$$a_{c2^{nM}} \leq c' 2^{-\delta N} + c'' \inf \sup \left\| \sum_{j=M+2}^{N} \psi(x) \sum_{|m| \leq 2^{j+2}\sqrt{n}} \left(\phi_j \hat{f} \right)^{\vee} \left(2^{-j} m \right) \right.$$
$$\times (\psi - \psi_\lambda)^{\vee} \left(2^{j+1} x - 2m \right) - Lf \left| B_{p_2 q_2}^{s_2}(\mathbf{R}^n) \right\|, \tag{11}$$

where the supremum is taken over all f with $\left\| f \left| B_{p_1 q_1}^{s_1}(\mathbf{R}^n) \right\| \right. \leq 1$, supp$f \subset$
Q_1, and the infimum is taken over all linear operators L from $B_{p_1 q_1}^{s_1}(\mathbf{R}^n)$
to $B_{p_2 q_2}^{s_2}(\mathbf{R}^n)$ with rank less than $c2^{nM}$.

The next idea is to reduce (11) to an estimate which involves the approximation numbers of maps acting between finite-dimensional sequence spaces. To do this, we split L in the form $L = L_{M+2} + \cdots + L_N$, with $\operatorname{rank} L_j = r_j$ and $r_{M+2} + \cdots + r_N = \operatorname{rank} L \sim 2^{nM}$; and then approximate the corresponding terms in (11) separately. We use the commutative diagram (3.3.2/24), now applied to approximation numbers rather than entropy numbers. As a direct counterpart of (3.3.2/34) we have

$$a_k^{(j)} \leq c 2^{-j\delta} \alpha_k^{(j)}, \quad j > M, \tag{12}$$

where $a_k^{(j)}$ is the kth approximation number of $f \longmapsto f_N^j$ and $\alpha_k^{(j)}$ is the kth approximation number of $\operatorname{id} : \ell_{p_1}^{M_j} \to \ell_{p_2}^{M_j}$ with $M_j \sim 2^{nj}$. Let $\rho = \min(1, p_2, q_2)$; then (11), the indicated splitting and (12) show that

$$a_{c2^{nM}}^\rho \leq c' 2^{-\delta N \rho} + \sum_{j=M+2}^N 2^{-j\delta\rho} \alpha_{r_j}^{(j)\rho}. \tag{13}$$

To complete the proof of the upper estimate we finally use Theorem 3.2.3/2(iii) and Corollary 3.2.3(iii). We remind the reader that we have the additional assumption (3), and with this in mind we first suppose that

$$0 < p_1 < 2 \leq p_2 < p_1' \text{ and } s_1 - s_2 > n/p_1. \tag{14}$$

Then

$$\delta = s_1 - s_2 - \frac{n}{p_1} + \frac{n}{p_2} > \frac{n}{p_2} \tag{15}$$

and

$$n\lambda = s_1 - s_2 + \frac{1}{2}n - n/p_1. \tag{16}$$

Choose two positive numbers κ and μ such that

$$\delta > \frac{n}{p_2} + \frac{1}{2}n\kappa, \quad n\mu\lambda/\delta = 1 + \kappa, \tag{17}$$

and take the numbers r_j in (13) to be

$$r_j = 2^{nN\mu} 2^{-j\kappa n}; \tag{18}$$

we suppose that each r_j is an integer. Then

$$\sum_{j=M+2}^N r_j \sim 2^{nN\mu} 2^{-M\kappa n} \sim 2^{nM} \tag{19}$$

since $N\mu - M\kappa = M\mu n\lambda/\delta - M\kappa = M$, in view of (9) and (17). Thus by Theorem 3.2.3/2(iii) with $m \sim 2^{nj}$, $j = M + 2, ..., N$, we have

$$\alpha_{r_j}^{(j)} \le c2^{nj/p_2}r_j^{-1/2} \sim 2^{nj/p_2 - nN\mu/2 + j\kappa n/2}. \tag{20}$$

To estimate the last exponent of 2 we need some identities. From (15) and (16),

$$\delta - \frac{n}{p_2} + \frac{1}{2}n = n\lambda \tag{21}$$

and hence

$$\delta - \frac{1}{2}n\delta(1 + \kappa)/n\lambda = \delta \left(\delta - \frac{n}{p_2} - \frac{1}{2}n\kappa \right) \Big/ n\lambda.$$

This gives, by (17) and (9),

$$N \left(\delta - \frac{1}{2}n\mu \right) = M \left(\delta - \frac{n}{p_2} - \frac{1}{2}n\kappa \right), \tag{22}$$

from which we see that

$$nM/p_2 - \frac{1}{2}nN\mu + \frac{1}{2}nM\kappa = \delta(M - N). \tag{23}$$

Hence the last exponent of 2 in (20) can be rewritten as

$$nj/p_2 - \frac{1}{2}nN\mu + \frac{1}{2}j\kappa n = j\delta - N\delta - (j - M)\sigma, \tag{24}$$

where

$$\sigma = \delta - n/p_2 - \frac{1}{2}n\kappa > 0$$

(see (17)). Substitution of (20), with the exponent (24), in (13) shows that

$$a_{c2^{nM}} \le c'2^{-\delta N} = c'2^{-Mn\lambda}, \tag{25}$$

in view of (9). From this the upper estimate in (4) follows immediately, under the assumption (14). All that remains is to establish the estimate in the case in which

$$1 < p_1 \le 2 \le p_1' \le p_2 < \infty, \quad s_1 - s_2 > n\left(1 - 1/p_2\right). \tag{26}$$

The counterparts of (15) and (16) are now

$$\delta = s_1 - s_2 - n/p_1 + n/p_2 > n/p_1' \tag{27}$$

and

$$n\lambda = s_1 - s_2 + \frac{1}{2}n - n/p_2'. \tag{28}$$

Comparison of (15) and (16) with (27) and (28) shows that we have to replace p_2 by p_1' and p_1 by p_2'. We can then follow the calculations above, now based on (3.2.3/12), to obtain (25), and with it the upper estimate in (4).

Step 3 When extra clarity will be gained, we shall denote by $a_k(T)$ the kth approximation number of a compact linear map T. Here we shall prove that there is a positive number c such that for all $j, k \in \mathbf{N}$,

$$a_k \geq c 2^{-j(s_1 - s_2) + jn(1/p_1 - 1/p_2)} a_k(\mathrm{id}^\ell), \qquad (29)$$

where $N_j = 2^{jn}$, id^ℓ is the identity map from $\ell_{p_1}^{N_j}$ to $\ell_{p_2}^{N_j}$; a_k as usual stands for the kth approximation number of the natural embedding id of $B_{p_1 q_1}^{s_1}(\Omega)$ in $B_{p_2 q_2}^{s_2}(\Omega)$. We construct a commutative diagram with the help of maps A and B such that

$$\mathrm{id}^\ell = B \circ \mathrm{id} \circ A : \qquad (30)$$

$$
\begin{array}{ccc}
B_{p_1 q_1}^{s_1}(\Omega) & \xrightarrow{\ \mathrm{id}\ } & B_{p_2 q_2}^{s_2}(\Omega) \\[6pt]
A \uparrow & & \downarrow B \\[6pt]
\ell_{p_1}^{N_j} & \xrightarrow{\ \mathrm{id}^\ell\ } & \ell_{p_2}^{N_j}
\end{array}
$$

This means that

$$a_k \geq \|A\|^{-1} \|B\|^{-1} a_k\left(\mathrm{id}^\ell\right), \qquad (31)$$

and from this we shall obtain (29) if we are able to estimate $\|A\|$ and $\|B\|$ suitably. To construct A, we assume without loss of generality that Ω contains the unit cube in $\mathbf{R}^n$, and divide the unit cube in the usual way into 2^{jn} congruent cubes with side length 2^{-j} and centres x^r. Let $\Phi \in C^\infty(\mathbf{R}^n)$ have support near the origin, put $\phi(x) = \Phi(2^j x)$ and $\phi_r(x) = \phi(x - x^r)$. Then A is given by

$$A\{\lambda_r\} = \sum_{r=1}^{N_j} \lambda_r \phi_r(x). \qquad (32)$$

We may, and shall, choose Φ in such a way that Theorem 2.3.2 can be applied. We have

$$\left\| A\{\lambda_r\} \,\big|\, B_{p_1 q_1}^{s_1}(\Omega) \right\| \leq c \left(\sum_{r=1}^{N_j} |\lambda_r|^{p_1} \right)^{1/p_1} 2^{j(s_1 - n/p_1)} \qquad (33)$$

for some c which is independent of j. The map B is defined by $Bf = \{\mu_r\}$, where

$$\mu_r = 2^{jn} \left(\Delta^M_{2^{-j}h}f, \phi_r\right)_{L_2}, \quad r = 1, ..., N_j = 2^{nj}; \tag{34}$$

here ϕ is as above, $(\cdot, \cdot)_{L_2}$ is the usual L_2 inner product,

$$\Delta^M_{2^{-j}h}f(x) = \sum_{\ell=0}^{M} f\left(x + \ell 2^{-j}h\right) \binom{M}{\ell} (-1)^{M-\ell} \tag{35}$$

are the usual differences (to be interpreted appropriately if f is a distribution), $M \in \mathbf{N}$ will be chosen later and $h \in \mathbf{R}^n$ is such that $|h| \sim 1$. We may assume that Ω is large compared with the unit cube and that there are no difficulties in (34) if $f \in B^{s_2}_{p_2 q_2}(\Omega)$.

Our first objective is to prove (30). We may suppose that $2^{jn} \|\phi_r |L_2\|^2 = c$ for some positive c which is independent of j and r. Substitute $f = A\{\lambda_r\}$ (defined by (32)) in (34) and choose h in (34) suitably: it follows that $\mu_r = c\lambda_r$. We thus have (30), possibly with cid^ℓ instead of id^ℓ, but this is of no consequence. Now we estimate $\|B\|$, using local means. Let Φ be as above and let $\Psi(x) = \Delta^M_{-h}\Phi(x)$, where Δ^M_{-h} is the difference given by (35) and $2M > \max\left(s_2, n\left(1/p_2 - 1\right)_+\right)$. Let $\Psi(t, f)$ be the means constructed from $\Psi(x)$ as $K(t, f)$ is constructed from $K(x)$ in (2.3.2/12), and let

$$\Psi^*\left(2^{-j}, f\right)(x) = \sup \frac{\left|\Psi\left(2^{-j}, f\right)(x+z)\right|}{1 + |2^j z|^a}, \quad a > n/p_2, \tag{36}$$

be a maximal function in the sense of [Triγ], (2.4.1/55), p. 109; here the supremum is taken over all $z \in \mathbf{R}^n$. By [Triγ], Corollary 2.5.1/2, p. 134, we have

$$\left(\sum_{j=1}^{\infty} 2^{js_2 q_2} \left\|\Psi^*\left(2^{-j}, f\right)(\cdot) |L_{p_2}(\mathbf{R}^n)\right\|^{q_2}\right)^{1/q_2} \leq c \|f |B^{s_2}_{p_2 q_2}(\mathbf{R}^n)\| \tag{37}$$

for some $c > 0$. Here we have, without loss of generality, assumed that f has been extended to $\mathbf{R}^n$. As before, let the x^r be the centres of the sub-cubes with side length 2^{-j}. Then (36) and (37) give

$$\left(\sum_{j=1}^{\infty} 2^{j(s_2 - n/p_2)q_2} \left(\sum_{j=1}^{N_j} \left|\Psi\left(2^{-j}, f\right)(x^r)\right|^{p_2}\right)^{q_2/p_2}\right)^{1/q_2} \leq c \|f |B^{s_2}_{p_2 q_2}(\Omega)\|.$$

$$\tag{38}$$

Now observe that since $\phi(x) = \Phi\left(2^j x\right)$ and $\phi_r(x) = \phi\left(x - x^r\right)$, the definition of Ψ shows that

$$
\begin{aligned}
\Psi\left(2^{-j}, f\right)(x) &= \int_{\mathbf{R}^n} \Delta_{-h}^M \Phi(y) f\left(x + 2^{-j} y\right) dy \\
&= \int_{\mathbf{R}^n} \Phi(y) \Delta_{2^{-j}h}^M f\left(x + 2^{-j} y\right) dy \\
&= 2^{jn} \int_{\mathbf{R}^n} \Phi(2^j y) \Delta_{2^{-j}h}^M f\left(x + y\right) dy \\
&= 2^{jn} \int_{\mathbf{R}^n} \phi(y - x) \Delta_{2^{-j}h}^M f(y) dy.
\end{aligned}
\tag{39}
$$

With $x = x^r$ we find that (see (34))

$$
\mu_r = \Psi\left(2^{-j}, f\right)\left(x^r\right).
\tag{40}
$$

In view of (38) and (40) we obtain

$$
\left\| Bf \left| \ell_{p_2}^{N_j} \right. \right\| = \left\| \{\mu_r\} \left| \ell_{p_2}^{N_j} \right. \right\| \le c2^{-j(s_2 - n/p_2)} \left\| f \left| B_{p_2 q_2}^{s_2}(\Omega) \right. \right\|,
\tag{41}
$$

for some c which is independent of j. The desired estimate (29) now follows from (30) (perhaps with cid' instead of id'), (31), (33) and (41).

Step 4 Here we complete the proof of the theorem. We combine (29) and Theorem 3.2.3/2, choosing $k = \frac{1}{4} N_j = 2^{jn-2}$. Let $p_2 \ge p_1$; then (see (29) and (3.3.1/2))

$$
a_k \ge c2^{-j\delta} \alpha_k, \quad k = 2^{jn-2}.
\tag{42}
$$

Suppose that either $0 < p_1 \le p_2 \le 2$ or $2 \le p_1 \le p_2 \le \infty$. Then by Theorem 3.2.3/2(i) and Corollary 3.2.3(iv),

$$
a_k \ge ck^{-\delta/n}, \quad k \in \mathbf{N},
\tag{43}
$$

which is the left-hand inequality of (2). Now let $0 < p_1 \le 2 \le p_2 < \infty$ and take $k = 2^{jn-2}$ in the sense of (42). Then (3.2.3/16) and (3.2.3/12) with $k \sim m = 2^{jn}$ give

$$
\alpha_k \sim 2^{-jn\varepsilon}, \; k = 2^{jn-2}, \text{ where } \varepsilon = \min\left(\frac{1}{2} - \frac{1}{p_2}, \frac{1}{p_1} - \frac{1}{2}\right) \ge 0.
\tag{44}
$$

Together with (42) and $n\lambda = \delta + n\varepsilon$ this gives the left-hand inequalities of (4) and (6). Lastly, suppose that $0 < p_2 \le p_1 \le \infty$. By Theorem 3.2.3/2(ii),

$$
\alpha_k \ge c2^{jn(1/p_2 - 1/p_1)}, \; k = 2^{jn-2}.
\tag{45}
$$

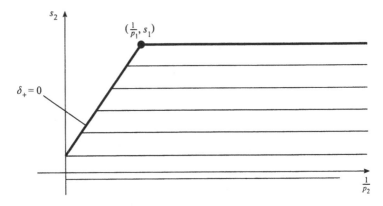

Fig. 3.1. Entropy numbers.

In view of (29) and (45),

$$a_k \geq c2^{-j(s_1-s_2)} = c2^{-j\delta_+}, \ k = 2^{jn-2}, \tag{46}$$

which establishes the left-hand inequality in (2). The proof of the theorem is complete.

Remark 1 The restriction $p_2 < \infty$ in (3) arises from the corresponding restrictions on q in connection with Corollary 3.2.3(ii) and (iii). Because of this, (4) can be extended to the case in which $1 < p_1 \leq 2$ and $p_2 = \infty$. Note also that the different rates of decay in (2) and (4) show that, unlike the entropy numbers (see Theorem 1.3.2), the approximation numbers have no unrestricted interpolation property: there is no common best approximating operator covering all cases.

Remark 2 Figs. 3.2 and 3.3 apply to the approximation numbers, whereas Fig. 3.1 illustrates the simpler situation for the entropy numbers. In these figures level lines are indicated for the exponents ρ in the relations $a_k \sim k^{-\rho}$ and $e_k \sim k^{-\rho}$; see the above theorem and Theorem 3.3.3/2.

3.3.5 Historical remarks

Theorems 3.3.3/2 and 3.3.4, which give upper and lower estimates for the entropy and approximation numbers of embeddings between B and

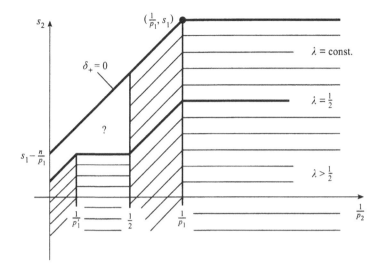

Fig. 3.2. Approximation numbers, $0 < p_1 < 2$.

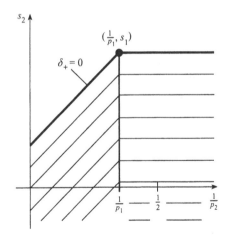

Fig. 3.3. Approximation numbers, $2 \le p_1 < \infty$.

F spaces on bounded domains with smooth boundaries, contain virtually all results of this kind to be found in the literature. Theorem 3.3.3/2, dealing with the entropy numbers, has been known for a long time for classical Besov spaces, that is, when $p_1, p_2 \in (1, \infty)$ and $q_1, q_2 \in [1, \infty]$: see

[Triα], 4.10.3, p. 355 and the references given there. Carl [Carl2] extended this result about classical spaces to $p_1 = 1$ and/or $p_2 = 1$ (where $n = 1$); see also [Kön], 3.c.9, p. 191. These special results cover in particular the estimates which began all this work; that is, if $s \in \mathbf{N}$, $1 < p < q < \infty$, $s - n/p > -n/q$, $W_p^s(\Omega)$ is the usual Sobolev space and e_k is the kth entropy number of the natural embedding of $W_p^s(\Omega)$ in $L_q(\Omega)$, then

$$e_k \sim k^{-s/n} \text{ for all } k \in \mathbf{N}.$$

This was proved first by Birman and Solomyak [BiS1], [BiS2], who devised the method of piecewise-polynomial approximation which has proved to be so useful and versatile, and which has been often used in modified and refined versions. Of these, spline approximations should be particularly emphasised, for when combined with the so-called Ciesielski isomorphism, they enable problems of our type for Besov spaces $B_{pq}^s(\Omega)$ with $p, q \in [1, \infty]$ to be reduced to corresponding problems for diagonal operators acting between ℓ_p sequence spaces. For this approach and other earlier work on classical Besov spaces we refer to the paper of Carl mentioned above, the books of König [Kön] and Pietsch [Pi2], and the survey article by R. Linde [Lin]. The Fourier-analytical approach presented in 3.3 is completely different from these lines of attack, and is based on [ET1], [ET2]. We are unaware of any previous work which covers the possibility that p_1 and/or p_2 may be less than 1.

So far as the approximation numbers a_k are concerned, it should be remarked that in some classical cases, (3.3.4/2) and the upper bound $a_k \leq ck^{-\delta/n}$ have been known for a considerable time: see [Triα], 4.10.2, p. 349, [EEv], pp. 292, 299 and the references given there. For classical spaces it is a familiar fact that the number 2 plays a somewhat sinister role exemplified by Theorem 3.3.4, not only for approximation numbers but also for other width numbers: see König [Kön], 3.c.7, p. 189, Pietsch [Pi2], 6.4.14, p. 252 and especially R. Linde [Lin].

3.4 Limiting embeddings in spaces of Orlicz type

3.4.1 Preliminaries

We shall suppose throughout this section that Ω is a bounded domain in $\mathbf{R}^n$ with C^∞ boundary. In 3.3 we analysed embedding maps between function spaces of B or F type from the standpoint of entropy and approximation numbers, and obtained results which were optimal for entropy numbers and somewhat less satisfactory for approximation numbers, although even there optimal results were obtained in large ranges of the parameter values. Here our objective is different and is motivated by the fact that in certain limiting cases, different target spaces for embeddings arise naturally. This has already been discussed in 2.7.1 and 2.7.3: we remind the reader that if $1 < p < \infty$, then $H_p^{n/p}(\Omega)$ is compactly embedded in $L_q(\Omega)$ for all $q \in [1, \infty)$, is not embedded in $L_\infty(\Omega)$, but is compactly embedded in $L_\infty(\log L)_a(\Omega)$ if, and only if, $a < 1 - \frac{1}{p}$; see Theorem 2.7.3(i). Moreover, we know that if $s > 0$, $s - n/p_1 = -n/p_2$ and $1 < p_1 < p_2 < \infty$, then $H_{p_1}^s(\Omega)$ is continuously but not compactly embedded in $L_{p_2}(\Omega)$; it turns out that if $a < 0$, then $H_{p_1}^s(\Omega)$ is compactly embedded in $L_{p_2}(\log L)_a(\Omega)$.

With this in mind, we focus on the kinds of embeddings mentioned above and on lifted-up versions of them involving logarithmic Sobolev spaces. The restriction to the (fractional) Sobolev spaces $H_p^s(\Omega)$, their logarithmic analogues and $B_{p,p}^s(\Omega)$ is dictated by the applications to be given in Chapter 5; for the same reason we shall concentrate on entropy numbers rather than approximation numbers, although for the case of embeddings in the spaces $L_\infty(\log L)_a(\Omega)$, estimates for the approximation numbers will be given as their exact behaviour has for a long time been the subject of speculation.

3.4.2 Embeddings in $L_\infty(\log L)_{-a}(\Omega)$

As in 2.7.1 we shall write

$$\mathscr{B}_p(\Omega) = B_{pp}^{n/p}(\Omega), \quad \mathscr{H}_p(\Omega) = H_p^{n/p}(\Omega) \tag{1}$$

where now $1 < p < \infty$, and let a_k, e_k be the kth approximation number and entropy number respectively of either of the natural embeddings

$$\text{id} : \mathscr{B}_p(\Omega) \to L_\infty(\log L)_{-a}(\Omega) \tag{2}$$

and

$$\text{id} : \mathscr{H}_p(\Omega) \to L_\infty(\log L)_{-a}(\Omega), \tag{3}$$

where $a > 1/p' = 1 - 1/p$. We know from Theorem 2.7.3 that these embeddings exist and are compact; for convenience we now replace a in part (i) of that theorem by $-a$.

Theorem *Let $1 < p < \infty$, $a > 1/p'$ and suppose that a_k, e_k are as above.*
(i) There are positive constants c_1, c_2 such that for $k = 2, 3, ...,$

$$c_1(\log k)^{1/p'-a} \leq a_k \leq c_2(\log k)^{1/p'-a}. \tag{4}$$

(ii) If in addition $1 + 2/p < a < \infty$, *then there are positive constants* c_1, c_2 *such that for all* $k \in \mathbf{N}$,

$$c_1 k^{-1/p} \le e_k \le c_2 k^{-1/p}. \tag{5}$$

(iii) If in addition to the main hypotheses $1 \le a \le 1 + 2/p$ *and* $\varepsilon > 0$, *then there are two positive numbers* c *and* $c(\varepsilon)$ *such that for all* $k \in \mathbf{N}$,

$$ck^{-1/p} \le e_k \le c(\varepsilon)k^{-\frac{1}{3}(a-1/p')+\varepsilon}. \tag{6}$$

(iv) If in addition to the main hypotheses $1/p' < a \le 1$ *and* $\varepsilon > 0$, *then there are two positive numbers* c *and* $c(\varepsilon)$ *such that for all* $k \in \mathbf{N}$,

$$ck^{-(a-1/p')} \le e_k \le c(\varepsilon)k^{-\frac{1}{3}(a-1/p')+\varepsilon}. \tag{7}$$

Proof Step 1 As in 2.7.2 we denote by $\mathrm{id}_{p,q}$ the embedding corresponding to either the inclusion $\mathscr{B}_p(\Omega) \subset L_q(\Omega)$ or $\mathscr{H}_p(\Omega) \subset L_q(\Omega)$, where $1 < p < \infty$, $1 < q < \infty$ (see 2.3.3 and 2.5.1). Let $e_k(\mathrm{id}_{p,q})$ be the kth entropy number of $\mathrm{id}_{p,q}$.

We shall prove that there is a constant c, independent of q, such that for all $k \in \mathbf{N}$,

$$e_k\left(\mathrm{id}_{p,q}\right) \le cq^{1+2/p}k^{-1/p}. \tag{8}$$

In fact, from Theorem 3.3.2 we know that an estimate of the form

$$e_k\left(\mathrm{id}_{p,q}\right) \le Ck^{-1/p}$$

holds, but the dependence of the constant C upon the various parameters was not clarified. We follow the proof of Theorem 3.3.2, this time paying more attention to the constant involved and making certain necessary modifications. As in that proof we assume that

$$\Omega \subset \{x \in \mathbf{R}^n : |x_j| \le 1 \text{ for } j = 1, ..., n\} = Q,$$

and that $\psi \in C_0^\infty(\mathbf{R}^n)$ is such that $\psi(x) = 1$ if $x \in Q$. As remarked in Step 1 of the proof of Theorem 3.3.2, it is enough to prove (8) with $\mathrm{id}_{p,q}$ replaced by the compact embedding

$$\{f \in \mathscr{B}_p(\mathbf{R}^n) : \mathrm{supp} f \subset Q\} \to L_q(\mathbf{R}^n) \tag{9}$$

and the counterpart with $\mathscr{H}_p(\mathbf{R}^n)$ instead of $\mathscr{B}_p(\mathbf{R}^n)$. For every f of the kind occurring in (9) we have

$$f = \psi \sum_{j=0}^{N}\left(\phi_j\widehat{f}\right)^\vee + \psi \sum_{j=N+1}^{\infty}\left(\phi_j\widehat{f}\right)^\vee = f_N + f^N. \tag{10}$$

By (2.7.2/9) it follows, in analogy to (3.3.2/7), that

$$\left\| f^N \, | L_q(\mathbf{R}^n) \right\| \leq c q^{1-1/p} 2^{-Nn/q} \tag{11}$$

for every f in (9) such that $\| f \, | \mathscr{B}_p(\mathbf{R}^n) \| \leq 1$. Use of (2.7.2/11) gives a similar result for $\mathscr{H}_p(\mathbf{R}^n)$ instead of $\mathscr{B}_p(\mathbf{R}^n)$. In what follows there is no essential difference between the cases $\mathscr{B}_p(\mathbf{R}^n)$ and $\mathscr{H}_p(\mathbf{R}^n)$. We adopt the procedure given in Steps 2 and 3 of the proof of Theorem 3.3.2. With λ as in (3.3.2/10), put $\psi_\lambda(\xi) = \psi(2^\lambda \xi)$; write $N_j = \max\left(N^2 n^2 q^{-2}, 2^{j+2}\sqrt{n}\right)$ and

$$f_N^j(x) = c\psi(x) \sum_{m \in \mathbf{Z}^n, |m| \leq N_j} \left(\phi_j \widehat{f}\right)^\vee (2^{-j}m)\,(\psi - \psi_\lambda)^\vee \left(2^{j+1}x - 2m\right) \tag{12}$$

for $j = 0, 1, ..., N$. Put

$$f_{N,1}(x) = \sum_{j=0}^{N} f_N^j(x), \tag{13}$$

and observe that the same arguments as in Steps 1–3 of the proof of Theorem 3.3.2, now based upon (11), give

$$\|f - f_{N,1} \, | L_q(\mathbf{R}^n)\| \leq c q^{1-1/p} 2^{-Nn/q}. \tag{14}$$

Next, we follow the methods of the proof of Theorem 3.3.2 from Step 4 onwards and use the notation introduced there. With this understanding we choose $K, L \in \mathbf{N}$ such that

$$L/p \sim N/q \quad \text{and} \quad 2^{K+2}\sqrt{n} \sim n^2 N^2/q^2 \tag{15}$$

and introduce the decomposition

$$f_{N,1} = \sum_{j=0}^{K} f_N^j + \sum_{j=K+1}^{L} f_N^j + \sum_{j=L+1}^{N} f_N^j = f_{N,3} + f_{N,4} + f_{N,5}. \tag{16}$$

Now factorise the map $F_j : f \longmapsto f_N^j$ as follows:

$$
\begin{array}{ccc}
\{f \in \mathscr{B}_p(\mathbf{R}^n) : \operatorname{supp} f \subset Q\} & \xrightarrow{\;S_j\;} & \ell_p^{M_j} \\[4pt]
F_j \downarrow & & \downarrow \text{ id} \\[4pt]
L_q(\mathbf{R}^n) & \xleftarrow{\;T_j\;} & \ell_q^{M_j}
\end{array} \tag{17}
$$

where $\mathscr{B}_p(\mathbf{R}^n)$ can also be replaced by $\mathscr{H}_p(\mathbf{R}^n)$. This corresponds to (3.3.2/24), and in it, $M_j = \operatorname{rank} F_j \leq c 2^{jn}$. The maps S_j and T_j are as in 3.3.2, and

$$F_j = T_j \circ \text{id} \circ S_j. \tag{18}$$

The specialised counterparts of (3.3.2/28, 30, 33, 34) are as follows:

$$\|S_j\| \le c, \tag{19}$$

$$\|g_j \,|L_q(\mathbf{R}^n)\| \le c \left(\sum_{\ell \in \mathbf{Z}^n} |g_j(\kappa\ell|^q \right)^{1/q} \tag{20}$$

with $\operatorname{supp}\widehat{g}_j \subset 2Q$ and $\kappa > 0$ chosen suitably,

$$\|T_j\| \le c2^{-jn/q} \tag{21}$$

and

$$e_k^{(j)} \le c2^{-jn/q}\varepsilon_k^{(j)} \text{ for } j \ge K, \tag{22}$$

where $e_k^{(j)}$ is the kth entropy number of F_j and $\varepsilon_k^{(j)}$ is the kth entropy number of $\operatorname{id} : \ell_p^{M_j} \to \ell_q^{M_j}$. It should be emphasised that all the numbers c in (19)–(22) and in the inequality $M_j \le c2^{jn}$ are independent of q. To proceed further, we examine separately the terms $f_{N,4}$, $f_{N,5}$ and $f_{N,3}$, just as we did in Steps 5–7 of the proof of Theorem 3.3.2. With Remark 3.2.2/2 in mind, the counterparts of (3.3.2/39, 40) are

$$\left\| f_{N,4} - \sum_{j=K+1}^{L} G_j \,|L_q(\mathbf{R}^n) \right\| \le c_1 2^{-Nn/q} \le c_2 2^{-Ln/p} \tag{23}$$

and

$$\sum_{j=K+1}^{L} k_j \le c2^{nL}, \tag{24}$$

when c is independent of q and $G_j = G_j(f)$ are optimally chosen elements of $L_q(\mathbf{R}^n)$. This means that we need at most $2^{c2^{nL}}$ elements G_j to establish (23). Next we follow Step 6 of the proof of Theorem 3.3.2 to estimate $f_{N,5}$: this is a crucial step, for it is from here that the dependence upon q in the right-hand side of (8) has its origin. As in (3.3.2/41, 43) we choose

$$k_j = c_1 2^{nL-(j-L)\kappa}, \quad \kappa = c_2/q \text{ and } \mu = c_3/q, \tag{25}$$

for some suitable positive numbers c_1, c_2, c_3 which are independent of q. Then

$$\sum_{j=L+1}^{N} k_j = c_1 2^{nL} \sum_{j=L+1}^{N} 2^{-c_2(j-L)/q} \le c_3 q 2^{nL}, \tag{26}$$

and we also have the following counterparts of (3.3.2/42) and (23):

$$e_{k_j}^{(j)} \le c2^{-Ln/p} 2^{-c(j-L)/q} (j-L)^{1/p-1/q} \tag{27}$$

and

$$\left\| f_{N,5} - \sum_{j=L+1}^{N} G_j \,|L_q(\mathbf{R}^n) \right\| \le \sum_{j=L+1}^{N} e_{k_j}^{(j)}$$

$$\le c_1 2^{-Ln/p} q^{1/p-1/q} \sum_{j=L+1}^{N} 2^{-c_2(j-L)/q} \left(\frac{j-L}{q} \right)^{1/p-1/q}$$

$$\le c_1 2^{-Ln/p} q^{1/p-1/q} \int_0^\infty e^{-c_3 t/q} (t/q)^{1/p-1/q} dt$$

$$\le c_4 q^{1+1/p-1/q} 2^{-Ln/p}. \tag{28}$$

Once more, $2^{cq2^{nL}}$ elements G_j are enough to establish (28). To handle $f_{N,3}$ we follow Step 7 of the proof of Theorem 3.3.2 and, just as in (11) above,

$$\| f_{N,3} \,|L_q(\mathbf{R}^n) \| \le c q^{1-1/p}, \tag{29}$$

when c is independent of q. We cover

$$\{ f_{N,3} : f \in \mathscr{B}_p(\mathbf{R}^n), \| f \,|\mathscr{B}_p(\mathbf{R}^n) \| \le 1, \operatorname{supp} f \subset Q \}, \tag{30}$$

or its counterpart with $\mathscr{H}_p(\mathbf{R}^n)$ instead of $\mathscr{B}_p(\mathbf{R}^n)$, with $L_q(\mathbf{R}^n)$-balls of radius $q 2^{-Ln/p}$. In view of (29) and (12) with $N_j \sim N^2/q^2 \sim cL^2$ we need at most

$$\left(c_1 2^{c_2 L} \right)^{c_3 L^{2n+1}} = c_1 2^{c_4 L^{2n+2}} \tag{31}$$

such balls. To estimate the right-hand side of this we use the generous estimate $c_4 L^{2n+2} \le c_5 2^{nL}$.

Finally we can prove (8), using the method of Step 8 of the proof of Theorem 3.3.2. Cover

$$\{ f_{N,1} : f \in \mathscr{B}_p(\mathbf{R}^n), \| f \,|\mathscr{B}_p(\mathbf{R}^n) \| \le 1, \operatorname{supp} f \subset Q \}, \tag{32}$$

or its counterpart with $\mathscr{H}_p$ instead of $\mathscr{B}_p$, with $L_q(\mathbf{R}^n)$-balls of radius $q^{1+1/p} 2^{-Ln/p}$. By (16), (23) and (24), (26) and (28), together with the remarks above about the covering of (30), it follows that at most $2^{cq2^{nL}}$ balls are needed. From (10), (11) and (14) with $N/q \sim L/p$, and the covering of (32), we see that (32) with f instead of $f_{N,1}$ can also be covered by $2^{cq2^{nL}}$ balls in $L_q(\mathbf{R}^n)$ of radius $q^{1+1/p} 2^{-Ln/p}$. Thus

$$e_{cq2^{nL}} \le q^{1+1/p} \left(2^{-Ln} \right)^{1/p} \tag{33}$$

and (8) follows immediately.

Step 2 Here we prove that

$$a_k \leq c(\log k)^{1/p'-a}, \quad k = 2, 3, ...;$$ (34)

that is, the right-hand estimate of part (i) of the theorem holds. By (14),

$$q^{-a} \|f - f_{N,1} | L_q(\mathbf{R}^n)\| \leq cq^{1/p'-a}2^{-Nn/q}.$$ (35)

The maximum of the right-hand side occurs at some $q = cN$, where c is independent of N. Thus by (2.7.1/6),

$$
\begin{aligned}
\|f - f_{N,1} | L_\infty(\log L)_{-a}(\Omega)\| &\leq c_1 \sup_{j \in \mathbf{N}} j^{-a} \|f - f_{N,1} | L_j(\Omega)\| \\
&\leq c_2 N^{-(a-1/p')}.
\end{aligned}
$$ (36)

By (12) and (13), the rank of the map $f \longmapsto f_{N,1}$ is at most $c2^{nN}$. Together with (36) this shows that

$$a_{c2^{nN}} \leq c_2 N^{-(a-1/p')},$$ (37)

which gives (34).

Step 3 We complete the proof of (i) by showing that

$$a_k \geq c(\log k)^{1/p'-a}, \quad k = 2, 3, ...,$$ (38)

for some $c > 0$.

Subdivide the cube $\{y \in \mathbf{R}^n : y = (y_1, ..., y_n), 0 < y_j < 1\}$ into sub-cubes $\{y \in \mathbf{R}^n : 2^{-\ell}(k_j - 1) < y_j < 2^{-\ell}k_j\}$, where $\ell \in \mathbf{N}$ and $k_j = 1, ..., 2^\ell$; let the centres of these sub-cubes be x^r, $r = 1, ..., 2^{n\ell}$. Define

$$F(x) = \sum_{r=1}^{2^{n\ell}} a_r f\left(2^\ell(x - x^r)\right), \quad a_r \in \mathbf{C},$$

where $f = f_{p\sigma}$ in the sense of Theorem 2.7.1 with δ in (2.7.1/2) chosen so small that supp$f \subset \{y \in \mathbf{R}^n : |y| < 1/4\}$.

By Theorems 2.3.2 and 2.7.1 we have

$$\|F | \mathscr{B}_p(\mathbf{R}^n)\| \sim \left(\sum_{r=1}^{2^{n\ell}} |a_r|^p\right)^{1/p}$$ (39)

together with a corresponding statement about $\mathscr{H}_p$ rather than $\mathscr{B}_p$, and

$$\|F | L_j(\mathbf{R}^n)\| \sim j^{1/p'}2^{-n\ell/j}\left(\sum_{r=1}^{2^{n\ell}} |a_r|^j\right)^{1/j},$$ (40)

where the constants implicit in '$\sim$' are independent of $j \in \mathbf{N}$. Let $\ell_p^{2^{n\ell}}$ be the sequence space normed by the right-hand side of (39). Now consider the sequence of maps

$$\ell_p^{2^{n\ell}} \overset{A}{\to} \mathscr{B}_p(\Omega) \overset{\mathrm{id}_1}{\to} L_\infty(\log L)_{-a}(\Omega) \overset{\mathrm{id}_2}{\to} L_j(\Omega) \overset{B}{\to} \ell_j^{2^{n\ell}}, \tag{41}$$

where id_1 and id_2 are the natural embeddings,

$$A : \{a_r\}_{r=1}^{2^{n\ell}} \longmapsto \sum_{r=1}^{2^{n\ell}} a_r f\left(2^\ell(x - x^r)\right) = F(x) \tag{42}$$

and

$$B : g \longmapsto \left\{ \int_\Omega g(x) f^{j-1}\left(2^\ell(x - x^r)\right) dx \right\}_{r=1}^{2^{n\ell}} \frac{2^{n\ell}}{\|f \,|L_j\|^j}. \tag{43}$$

Here $f = f_{p\sigma}$ as indicated above. We assume tacitly that $\mathrm{supp} F \subset \Omega$; moreover, $\mathscr{B}_p(\Omega)$ can be replaced by $\mathscr{H}_p(\Omega)$ in (41). In view of (39) and its $\mathscr{H}_p$-counterpart, we see that

$$\|A\| \sim 1. \tag{44}$$

Also (43) really does define a map with the claimed properties, for $f^{j-1}(x) \in L_\lambda$ with $1/j + 1/\lambda = 1$ and

$$\left| \int_\Omega g(x) f^{j-1}\left(2^\ell(x - x^r)\right) dx \right|^j \le \int_{Q_r} |g(x)|^j dx \left(\int_\Omega f^j \left(2^\ell(x - x^r)\right) dx \right)^{j-1}, \tag{45}$$

where Q_r is the cube centred at x^r in the above sense. As the final integral in (45) equals $2^{-n(j-1)\ell} \|f \,|L_j\|^{(j-1)j}$, we see from (43) that

$$\begin{aligned}
\left\| Bg \,|\ell_j^{2^{n\ell}} \right\| &\le \|g \,|L_j(\Omega)\| \, \|f \,|L_j\|^{j-1} \, 2^{-n\ell(j-1)/j} 2^{n\ell} \, \|f \,|L_j\|^{-j} \\
&\le c \, \|g \,|L_j(\Omega)\| \, j^{-1/p'} 2^{n\ell/j},
\end{aligned} \tag{46}$$

where we have used (2.7.1/5). Thus

$$\|B\| \le c j^{-1/p'} 2^{n\ell/j} \tag{47}$$

for some c which does not depend upon ℓ and j. In addition, by (2.7.1/6) we see that

$$\|\mathrm{id}_2\| \le c j^u. \tag{48}$$

Identification of g in (43) with F given by (42) shows that

$$B \circ \mathrm{id}_2 \circ \mathrm{id}_1 \circ A = \mathrm{id}_3 \tag{49}$$

is the identity map from $\ell_p^{2n\ell}$ to $\ell_j^{2n\ell}$. Hence

$$
\begin{aligned}
a_k(\mathrm{id}_3) &\leq \|B\| \, a_k(\mathrm{id}_1) \, \|\mathrm{id}_2\| \, \|A\| \\
&\leq c j^{a-1/p'} 2^{n\ell/j} a_k(\mathrm{id}_1);
\end{aligned}
\tag{50}
$$

of course $a_k(\mathrm{id}_1) = a_k$, in our notation. If $1 < p \leq 2$, we choose

$$
\ell = j \geq p', \quad k \sim 2^{2n\ell/p'-2}
\tag{51}
$$

in (50), observe that $a_k(\mathrm{id}_3) \sim 1$ by Corollary 3.2.3(iii), and obtain

$$
a_{k(\ell)} \geq c\ell^{1/p'-a} \text{ with } k(\ell) = \left[2^{2n\ell/p'-2}\right].
\tag{52}
$$

If $2 \leq p < \infty$, then $a_k(\mathrm{id}_3) \sim 1$ if $k = 2^{n\ell-2}$ by Corollary 3.2.3(iv), and the analogue of (52) holds. In both cases (38) holds.

Step 4 We now turn to the estimate from below of the entropy numbers e_k. There are natural embeddings corresponding to the inclusions

$$
\mathscr{B}_p(\Omega) \subset L_\infty(\log L)_{-a}(\Omega) \subset L_q(\Omega)
\tag{53}
$$

for some $q \in (1,\infty)$; the same holds for $\mathscr{H}_p$. Thus by Theorem 3.3.3/1,

$$
e_k \geq ck^{-1/p}, \quad k \in \mathbf{N},
\tag{54}
$$

for some $c > 0$. A second lower estimate follows from the analogues of (49) and (50):

$$
e_k(\mathrm{id}_3) \leq c j^{a-1/p'} 2^{n\ell/j} e_k.
\tag{55}
$$

Moreover, $e_\ell(\mathrm{id}_3) \sim 1$, as can be seen easily; see also (3.2.2/1) and Remark 3.2.2/1. The choice $k = j = \ell$ in (55) gives

$$
e_k \geq ck^{-(a-1/p')}.
$$

Step 5 Here we suppose that $1 + 2/p < a$ and prove that

$$
e_k \leq ck^{-1/p}, \quad k \in \mathbf{N}.
\tag{56}
$$

By Theorem 2.6.2/1 with $p = \infty$ we have

$$
\|f \,|L_\infty(\log L)_{-a}(\Omega)\| \sim \sup_{j\in\mathbf{N}} 2^{-ja} \|f \,|L_{2^j}(\Omega)\|.
\tag{57}
$$

Given $J \in \mathbf{N}$, let

$$
k_j \sim 2^{p(J-j)(a-1-2/p)} \text{ if } j = 1, ..., J,
\tag{58}
$$

where $\sim$ here means that k_j is a nearby natural number. With obvious notation, (8) gives

$$e_{k_j}\left(\mathscr{B}_p \rightarrow 2^{-ja}L_{2^j}\right) \leq c2^{-ja}2^{j(1+2/p)}2^{-(J-j)(a-1-2/p)}$$
$$= c2^{-J(a-1-2/p)} \qquad (59)$$

and a corresponding inequality with $\mathscr{H}_p$ instead of $\mathscr{B}_p$. Cover the unit ball U of $\mathscr{B}_p(\Omega)$ by 2^{k_1} balls in $2^{-a}L_2(\Omega)$ of radius $2e_{k_1}$, each ball having centre in U. Let U_1 be one of these balls, and cover $U \cap U_1$ in $2^{-2a}L_4(\Omega)$ by 2^{k_2} balls of radius $2e_{k_2}$, where we may assume that the centres of these balls are in $U \cap U_1$. By iteration, we obtain a covering of $U \cap U_1 \cap ... \cap U_{J-1}$ in $2^{-Ja}L_{2^J}(\Omega)$ by 2^{k_J} balls of radius $2e_{K_J}$, where the centres of these balls are in $U \cap U_1 \cap ... \cap U_{J-1}$; let these centres be g_ℓ, $\ell = 1, ..., L$, where

$$L = 2^{k_1+\cdots+k_J}, \quad \sum_{j=1}^{J} k_j \sim 2^{pJ(a-1-2/p)}. \qquad (60)$$

Thus when $f \in U$ is given, there is one of these centres g_ℓ such that

$$2^{-ja}\|f - g_\ell |L_{2^j}(\Omega)\| \leq c_1 e_{k_j} \leq c_2 2^{-J(a-1-2/p)} \qquad (61)$$

if $j = 1, ..., J$. Let $j > J$. By Theorem 2.7.2,

$$2^{-ja}\|f - g_\ell |L_{2^j}(\Omega)\| \leq c2^{-j(a-1/p')} \leq c2^{-J(a-1-2/p)}. \qquad (62)$$

With $k = \sum_{j=1}^{J} k_j$, we see that (60)–(62) give

$$e_k \leq c_1 2^{-J(a-1-2/p)} \leq c_2 k^{-1/p}. \qquad (63)$$

In view of the monotonicity of the e_k, (56) follows.

Step 6 To complete the proof, let $1/p' < a \leq 1 + 2/p$ and let $\varepsilon > 0$ be small. Choose $\theta \in (0,1)$ so that

$$a = \frac{1-\theta}{p'} + \left(1 + \frac{2}{p} + \varepsilon\right)\theta. \qquad (64)$$

Then

$$a - \frac{1}{p'} = \theta\left(\varepsilon + \frac{3}{p}\right), \quad \frac{1}{3}\left(a - \frac{1}{p'} - \varepsilon\theta\right) = \frac{\theta}{p} \qquad (65)$$

and

$$j^{-a}\|f |L_j(\Omega)\| = \left(j^{-1/p'}\|f |L_j(\Omega)\|\right)^{1-\theta}\left(j^{-(\varepsilon+1+2/p)}\|f |L_j(\Omega)\|\right)^{\theta}. \qquad (66)$$

Let f belong to the unit ball in $\mathscr{B}_p(\Omega)$ or $\mathscr{H}_p(\Omega)$, and let the g_ℓ be the same elements as in Step 5 with $\varepsilon + 1 + 2/p$ instead of a. Replace f in (66) by $f - g_\ell$ and note that the first factor on the right-hand side can be estimated uniformly with respect to j; see Theorem 2.7.2. A suitable choice of g_ℓ then gives

$$j^{-a} \|f - g_\ell | L_j(\Omega) \| \leq c(\varepsilon) k^{-\theta/p}. \tag{67}$$

The upper estimates in (6) and (7) follow directly from (65) and (67), and the proof of the theorem is complete.

Remark   The first results concerning the approximation numbers of the embedding of $W_p^m(\Omega)$, where $mp = n$, in $L_\infty(\log L)_{-a}$ were obtained in [EM]: there it was proved that

$$a_k = O\left((\log k)^{1/p'-a}\right) \tag{68}$$

if $1/p' < a < 1$. The estimate (68) with e_k instead of a_k is given in [EE], and the main purpose of [EET] was to extend these upper estimates to the embedding of $\mathscr{B}_p(\Omega)$ or $\mathscr{H}_p(\Omega)$ in $L_\infty(\log L)_{-a}(\Omega)$. Until the work of [Tri3], however, nothing was known about the sharpness of these bounds; in particular, it was not even clear that the e_k and the a_k had entirely different behaviour, there being speculation that the rates of decay might be the same and logarithmic.

3.4.3 Interior estimates

The last section was devoted to the study of the limiting embedding of $B_{pp}^{n/p}(\Omega)$ $\left(\text{or } H_p^{n/p}(\Omega)\right)$ in $L_\infty(\log L)_{-a}(\Omega)$ where $1 < p < \infty$ and $a > 1/p'$. Here we study a counterpart of this limiting embedding, namely

$$\mathrm{id} : H_{p_1}^s(\Omega) \to L_{p_2}(\log L)_a(\Omega), \tag{1}$$

where $1 < p_1 < p_2 < \infty$, $s > 0$, $a < 0$ and

$$s - n/p_1 = -n/p_2. \tag{2}$$

This shift from an 'L_∞-situation' to an 'L_p-situation' in which $1 < p < \infty$ is quite natural; our goal is to obtain estimates for the entropy numbers of id which will parallel those obtained in 3.4.2 for the 'L_∞-situation'. We begin by establishing an analogue of (3.4.2/8).

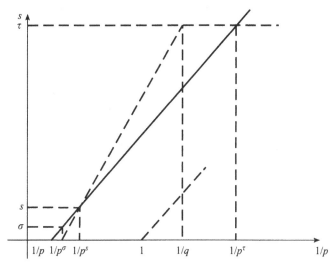

Fig. 3.4.

Proposition 1 *Suppose that* $1 < p < \infty$ *and* $s > 0$, *and define* p^s *by* $1/p^s = 1/p + \frac{s}{n}$. *Then given any* $\varepsilon > 0$, *there is a constant* $c_\varepsilon > 0$ *such that for all* $\sigma > 0$ *with* $p^\sigma > 1$ $(1/p^\sigma = 1/p + \sigma/n)$,

$$e_k \left(\mathrm{id} : H^s_{p^s}(\Omega) \to L_{p^\sigma}(\Omega) \right) \leq c_\varepsilon \sigma^{-(2s/n)-\varepsilon} k^{-s/n}, \quad k \in \mathbf{N}. \tag{3}$$

Proof First we observe that the proof of (3.4.2/8) followed that of Theorem 3.3.2, care being taken to keep control of the constants. We adopt the same principle here, with one difference. Comparison of the critical exponents in (3.4.2/8) and (3) shows that the direct counterpart of $1 + 2/p$ in (3.4.2/8), which is $1 + 2s/n$, is improved in (3) to $\varepsilon + 2s/n$. No corresponding improvement in (3.4.2/8) can be expected. This improvement obliges us to use an interpolation argument in Step 2 which has no counterpart in the L_∞-situation.

Step 1 Let $\tau = s/\sigma$. Then $1/p^\tau = 1/p + \tau/n$ and (see Fig. 3.4)

$$\rho = \frac{1}{p^\tau} - \frac{1}{q} = \frac{\sigma}{n} \cdot \frac{\tau - s}{s} = \frac{1 - \sigma}{n} \sim 1 \tag{4}$$

since we are interested in small positive σ.

We shall prove that for any $\varepsilon > 0$, there is a constant c_ε (independent of σ, τ and k) such that

$$e_k(\text{id} : H_q^\tau(\Omega) \to L_{p^\sigma}(\Omega)) \leq c_3(1/\sigma)^{\tau(\frac{2}{n}+\varepsilon)}k^{-\tau/n}, \quad k \in \mathbf{N}. \tag{5}$$

The proof follows that of Theorem 3.3.2 and Step 1 of Theorem 3.4.2; we use the same notation as in these earlier proofs and concentrate on the changes which are now necessary. As in (3.3.2/6),

$$f = \psi \sum_{k=0}^{N} \left(\phi_j \widehat{f}\right)^\vee + \psi \sum_{j=N+1}^{\infty} \left(\phi_j \widehat{f}\right)^\vee = f_N + f^N. \tag{6}$$

From (4), $\rho \sim 1$; thus the counterpart of (3.3.2/7) is

$$\left\| f^N \, | L_{p^\sigma}(\mathbf{R}^n) \right\| \leq c2^{-N\rho n}. \tag{7}$$

At this stage it becomes important to know how the constants, such as c in (7), depend upon τ; since $1/q \sim \tau \sim 1/\sigma$ their dependence upon σ needs to be known. For the time being, we shall take it for granted that embedding constants for the embedding

$$\text{id} : H_q^\tau(\Omega) \to L_p(\mathbf{R}^n) \tag{8}$$

in the above circumstances can be estimated from above by $c_1 2^{c_2\tau} \sim c_3 2^{c_4\sigma^{-1}}$ and that the same applies to the dependence of c in (3.3.2/12). Discussion of this point will be given below. However, these dependences are unimportant, since the critical constants in our estimates here are of type $c_1\tau^{c_2\tau}$ and all dependences of type $c_1 2^{c_2\tau}$ can be absorbed in the ε in (5). In view of this, we shall not care about these dependences in what follows.

To continue with the proof of (5), note that (3.3.2/13) can be replaced by

$$\left| D^\gamma \overset{\vee}{\psi}(x) \right| \leq c_{\gamma,a} 2^{-|x|^\beta} |x|^{-a} \text{ if } |x| \geq 1, \, 0 < \beta < 1; \tag{9}$$

see [SchmT], 1.2.1 and 1.2.2, p. 16. Then $N^2\delta^2$ in (3.3.2/14, 15) may be replaced by $cN^{1/\beta}\rho^{1/\beta} \sim cN^{1/\beta}$, and (3.3.2/19) has as its counterpart

$$\|f_{N,2} \, | L_{p^\sigma}(\mathbf{R}^n) \| \leq c2^{-N\rho n}. \tag{10}$$

Corresponding to (3.3.2/21) we have

$$2^K \sim N^{1/\beta}\rho^{1/\beta}, \, L\tau \sim \rho N, \tag{11}$$

which gives

$$2^K \sim N^{1/\beta}, \, N \sim s\sigma^{-1}L. \tag{12}$$

This dependence of N upon σ is crucial, having the greatest influence upon the constants which are involved. Bearing in mind the remarks above concerning the dependence on τ of the constants related to embeddings of type (8), we see that Step 4 of the proof of Theorem 3.3.2 causes no trouble, and it remains to deal with $f_{N,3}$, $f_{N,4}$ and $f_{N,5}$ as in Steps 5, 6, 7 of the proof of that theorem. Just as in the proof of Theorem 3.4.2 the important term is $f_{N,5}$, for compared with this the estimates for $f_{N,3}$ and $f_{N,4}$ pale into insignificance; we accordingly deal only with $f_{N,5}$. In this connection, note that κ and μ in (3.3.2/41, 43) here satisfy

$$0 < \kappa < \rho n q \sim \tau^{-1} \sim \sigma, \quad \mu \sim 1. \tag{13}$$

Moreover, as between (3.3.2/43) and (3.3.2/44), the number of balls involved can be estimated from above by 2^M, where

$$M = \sum_{j=L+1}^{N} k_j \le c 2^{nL} \kappa^{-1} \sim 2^{nL} \sigma^{-1}. \tag{14}$$

Also, corresponding to (3.3.2/42) and (3.4.2/28), we now have

$$
\begin{aligned}
\sum_{j=L+1}^{N} e_{k_j}^{(j)} &\le c_1 2^{-L\tau} \sum_{j=L+1}^{N} 2^{-c_2(j-L)} (j-L)^{1/q} \\
&\le c_3 2^{-L\tau} \int_0^\infty e^{-t} t^{\frac{1}{q}-1} dt \\
&\le c_4 2^{-L\tau} \Gamma(1/q) \le c_5 2^{-L\tau} (1/q)^{1/q} \\
&\le c_6 2^{-L\tau} \tau^{\tau/n}.
\end{aligned}
\tag{15}
$$

Here a rough estimate for $\Gamma(1/q)$, based on Stirling's formula, was used; the possibility that the coefficients might depend polynomially upon τ was also borne in mind. Just as in 3.3.2 and 3.4.2 we now obtain

$$e_{c_1 \sigma^{-1} 2^{nL}} \le c_\varepsilon 2^{-L\tau} (1/\sigma)^{\tau\left(\frac{1}{n}+\varepsilon\right)}, \tag{16}$$

where c_1 and c_ε are independent of σ, but may depend upon s, which is assumed to be fixed; e_k here stands for the kth entropy number of id $: H_q^\tau(\Omega) \to L_{p^\sigma}(\Omega)$. With $\sigma^{-1} 2^{nL} \sim k \in \mathbb{N}$ this gives

$$e_k \le c_\varepsilon (\sigma k)^{-\tau/n} \sigma^{-\tau\left(\frac{1}{n}+\varepsilon\right)} \le c_\varepsilon \sigma^{-\tau\left(\frac{2}{n}+\varepsilon\right)} k^{-\tau/n}, \tag{17}$$

and the proof of (5) is complete.

Step 2 We can now prove (3). Let the ϕ_j ($j \in \mathbf{N}_0$) be as in (6), and recall the Paley–Littlewood characterisation

$$\| f \, | H_t^v(\mathbf{R}^n) \| \sim \left\| \left(\sum_{j=0}^{\infty} 2^{2jv} \left| \left(\phi_j \hat{f} \right)^{\vee} (\cdot) \right|^2 \right)^{1/2} \Big| L_t(\mathbf{R}^n) \right\| \tag{18}$$

for $v \in \mathbf{R}$ and $0 < t < \infty$. This is applied to $H_q^\tau(\mathbf{R}^n)$ and $L_{p^\sigma}(\mathbf{R}^n)$ with small positive σ, as in the beginning of Step 1. As p is given and σ is small, (18) holds with L_{p^σ} instead of H_t^v uniformly, in the sense that the corresponding equivalence constants may be assumed to be independent of σ.

From the situation above we have

$$H_{p^s}^s(\mathbf{R}^n) = \left(L_{p^\sigma}(\mathbf{R}^n), H_q^\tau(\mathbf{R}^n) \right)_\sigma \subset \left(L_{p^\sigma}(\mathbf{R}^n), H_q^\tau(\mathbf{R}^n) \right)_{\sigma,\infty}, \tag{19}$$

where $(\cdot, \cdot)_\sigma$ corresponds to the complex interpolation method according to [Triβ], pp. 66–9, whereas $(\cdot, \cdot)_{\sigma,\infty}$ comes from the usual real interpolation method. The equality in (19) is covered by [Triβ], p. 69. The inclusion in (19) with the classical complex interpolation method $[\cdot, \cdot]_\sigma$ instead of $(\cdot, \cdot)_\sigma$ may be found in [Triα], p. 64. Since the same arguments given there apply also to $(\cdot, \cdot)_\sigma$, this justifies the inclusion in (19), which coincides with (1.3.2/4) with $\theta = \sigma$, specialised to the situation above.

Recall that the estimates in Step 1 and in 3.3.2 are based on representations of type (18). We also have (19) with Ω instead of $\mathbf{R}^n$; see [Triβ], 3.3.6, p. 204. Now interpolate the embeddings

$$\left. \begin{array}{l} \text{id} \,:\, H_q^\tau(\Omega) \to L_{p^\sigma}(\Omega) \\[2mm] \text{id} \,:\, L_{p^\sigma}(\Omega) \to L_{p^\sigma}(\Omega); \end{array} \right\} \tag{20}$$

and

then (19) with Ω instead of $\mathbf{R}^n$ shows that Theorem 1.3.2(ii) can be applied. Now this theorem and (5) give

$$\begin{aligned} e_k \left(\text{id} : H_{p^s}^s(\Omega) \to L_{p^\sigma}(\Omega) \right) &\leq c_1 e_k^\sigma \left(\text{id} : H_q^\tau(\Omega) \to L_{p^\sigma}(\Omega) \right) \\ &\leq c_2 (1/\sigma)^{\tau\sigma\left(\frac{2}{n}+\varepsilon\right)} k^{-\sigma\tau/n} \\ &= c_2 (1/\sigma)^{s\left(\frac{2}{n}+\varepsilon\right)} k^{-s/n}. \end{aligned} \tag{21}$$

This completes the proof of (3), subject to the assumptions made about the embedding constants.

Step 3 All that remains is to supply the discussion promised in Step 1 of the embedding constants for embeddings of type (8) under the conditions imposed in the proof above. We refer to Fig. 3.4 in this connection. We may, for that purpose, replace τ at the beginning of Step 1 by $c\tau$ so that, say, $\rho > 1$ in (4). Then

$$H_q^\tau(\mathbf{R}^n) \subset B_{\infty\infty}^1(\mathbf{R}^n) \subset L_\infty(\mathbf{R}^n) \tag{22}$$

since, say, $\tau > \frac{n}{q} + 1$, and the question can be reduced to

$$\sup_{x\in\mathbf{R}^n} \left|\left(\phi_j\widehat{f}\right)^\vee(x)\right| \le c(q)2^{jn/q}\left\|\left(\phi_j\widehat{f}\right)^\vee |L_q(\mathbf{R}^n)\right\|. \tag{23}$$

The dependence on j is simply a scaling argument, and so (23) is reduced to

$$\|g\,|L_\infty(\mathbf{R}^n)\| \le c\,\|g\,|L_q(\mathbf{R}^n)\| \tag{24}$$

if

$$g \in L_q(\mathbf{R}^n) \cap S'(\mathbf{R}^n) \text{ and } \operatorname{supp}\widehat{g} \subset \{y \in \mathbf{R}^n : |y| \le 1\}. \tag{25}$$

But in that case, c may even be chosen independently of q: see [Triβ], 1.3.2, p. 18. As we also used (3.3.2/12), we need to ascertain the behaviour of $c(q)$ in the inequality

$$\left(\sum_{m\in\mathbf{Z}^n} |g(cm)|^q\right)^{1/q} \le c(q)\,\|g\,|L_q(\mathbf{R}^n)\| \tag{26}$$

under the assumptions (25). By [Triβ], 1.3.1 and 1.3.3, pp. 16, 19, we see that

$$c(q) \le c_1 2^{c_2/q} \tag{27}$$

for some positive c_1 and c_2 which are independent of q. This estimate is of the type mentioned after (8).

Remark The proof depends crucially upon (20) and (21), based upon the Paley–Littlewood equivalence (18), applied to L_{p^σ}. This has no counterpart when $p = \infty$, in which case we have Step 1 but not Step 2 and the number 1 in the power of q in (3.4.2/8) cannot be replaced by an arbitrarily small $\varepsilon > 0$; see also the beginning of the above proof. This difference is responsible for the improvement of the L_p-situation given by (3), compared with (3.4.2/8) in the L_∞ situation.

Armed with the proposition above, we can deal with the limiting embedding (1), which we shall now write as

$$\operatorname{id} : H_{p^s}^s(\Omega) \to L_p(\log L)_a(\Omega), \tag{28}$$

where

$$a \le 0, 1 < p < \infty, \ s > 0, \ 1/p^s = 1/p + s/n. \tag{29}$$

We remind the reader again that Ω is supposed to be bounded and to have C^∞ boundary. If $a = 0$, the right-hand side of (28) is $L_p(\Omega)$ and id is continuous but not compact; if $a < 0$, id is also continuous, by (2.6.1/6). Note that it is possible that p^s might be less than 1.

Let e_k be the kth entropy number of id. The following theorem is a counterpart of Theorem 3.4.2, insofar as entropy numbers are concerned, and its proof follows the same general strategy as that of the earlier theorem.

Theorem 1 *Suppose that*

$$1 < p < \infty, \ s > 0 \ and \ a < 0. \tag{30}$$

Then id, given by (28), is compact.

(i) Suppose in addition that $a < -2s/n$. Then

$$c_1 k^{-s/n} \le e_k \le c_2 k^{-s/n}, \quad k \in \mathbf{N}, \tag{31}$$

for some positive constants c_1, c_2 which are independent of k.

(ii) Suppose that in addition to (30),

$$-2s/n \le a < -s/n, \ \varepsilon > 0.$$

Then there are positive constants c, c_ε such that

$$ck^{-s/n} \le e_k \le c_\varepsilon k^{\frac{a}{2}+\varepsilon}, \quad k \in \mathbf{N}. \tag{32}$$

(iii) Suppose that in addition to (30),

$$-s/n \le a < 0 \ and \ \varepsilon > 0.$$

Then there are positive constants c, c_ε such that

$$ck^a \le e_k \le c_\varepsilon k^{\frac{a}{2}+\varepsilon}, \quad k \in \mathbf{N}. \tag{33}$$

Proof Step 1 For shortness we shall omit the symbol 'Ω' in connection with spaces and quasi-norms. Moreover, for technical convenience we introduce the Banach space $G_{p,a}$, defined to be the space of all functions f such that

$$\begin{aligned}
\|f \,|G_{p,a}\| &= \sup_{0 < \sigma < \delta} \sigma^{-a} \|f \,|L_{p^\sigma}\| \sim \sup_{j \ge J} j^a \|f \,|L_{p^{\rho_j}}\| \\
&\sim \sup_{j \ge J'} 2^{ja} \|f \,|L_{p^{\sigma_j}}\|
\end{aligned} \tag{34}$$

is finite. Here $\sigma_j = 2^{-j}$, $\rho_j = 1/j$, δ is small and positive, and J, J' are large: see 2.6.1 and 2.6.2. By Theorem 2.6.2/1, we see that if $\varepsilon \in (0, -a)$, then

$$G_{p,a+\varepsilon} \subset L_p(\log L)_a \subset G_{p,a}. \tag{35}$$

As a convenient shorthand we shall, for the moment, denote the entropy numbers of

$$\text{id} : H_p^s \to G_{p,a} \tag{36}$$

by $e_k(G_{p,a})$. Then by (35), there are two positive numbers c_1 and c_2 with

$$c_1 e_k(G_{p,a}) \le e_k \le c_2 e_k(G_{p,a+\varepsilon}), \quad k \in \mathbf{N}. \tag{37}$$

In view of this, it is enough to prove the theorem, including the lower estimate in (33), for (36) and $e_k(G_{p,a})$.

Step 2 Here we prove the estimates from below. Since plainly

$$G_{p,a} \subset L_{p^\sigma}, \quad \sigma > 0, \tag{38}$$

we see from (3.3.3/1) that

$$e_k(G_{p,a}) \ge ck^{-s/n}. \tag{39}$$

The lower estimate in (33) requires much more effort: we modify the corresponding argument in the proof of Theorem 3.4.2. We use the localisation theorem in 2.3.2, and first introduce the necessary notation. Let f be a non-trivial, non-negative function in $C_0^\infty(Q_d)$, where

$$Q_d = \{x \in \mathbf{R}^n : |x_k| < d \text{ if } k = 1, ..., n\},$$

and assume that d is small and $Q = Q_{1/2} \subset \Omega$. Divide Q in the natural way into $2^{n\ell}$ congruent sub-cubes of side length $2^{-\ell}$ and centres x^r, $r = 1, ..., 2^{n\ell}$. Finally let

$$f^\ell(x) = \sum_{r=1}^{2^{n\ell}} a_r f\left(2^\ell (x - x^r)\right), \quad a_r \in \mathbf{C}. \tag{40}$$

Then by Theorem 2.3.2,

$$\left\| f^\ell \,\middle|\, H_p^s \right\| \sim 2^{-\ell n/p} \left(\sum_{r=1}^{2^{n\ell}} |a_r|^{p^s} \right)^{1/p^s} \tag{41}$$

and

$$\|f^\ell \,|L_{p^\sigma}\| \sim 2^{-\ell n/p^\sigma} \left(\sum_{r=1}^{2^{n\ell}} |a_r|^{p^\sigma}\right)^{1/p^\sigma}, \tag{42}$$

since $s - n/p^s = -n/p$. Note that the constants implicit in '$\sim$' are uniform with respect to ℓ. Now construct the following sequence of maps, similar to (3.4.2/41):

$$\ell_{p^s}^{2^{n\ell}} \overset{A}{\to} H_{p^s}^s \overset{\mathrm{id}_1}{\to} G_{p,a} \overset{\mathrm{id}_2}{\to} L_{p^\sigma} \overset{B}{\to} \ell_{p^\sigma}^{2^{n\ell}}, \tag{43}$$

where id_1 is the embedding id from (36), id_2 is the natural embedding,

$$A : \{a_r\}_{r=1}^{2^{n\ell}} \longmapsto f^\ell \text{ is given by (40),} \tag{44}$$

and

$$B : g \longmapsto \left\{\int_\Omega g(x) \left|f\left(2^\ell (x - x^r)\right)\right|^{p^\sigma - 1} dx\right\}_{r=1}^{2^{n\ell}} 2^{n\ell} \Big/ \|f\,|L_{p^\sigma}\|^{p^\sigma}. \tag{45}$$

With $\mathrm{id}_3 : \ell_{p^s}^{2^{n\ell}} \to \ell_{p^\sigma}^{2^{n\ell}}$ as the identity map, we have

$$\mathrm{id}_3 = B \circ \mathrm{id}_2 \circ \mathrm{id}_1 \circ A \tag{46}$$

and

$$\|A\| \leq c 2^{-n\ell/p}, \tag{47}$$

by (41). To estimate $\|B\|$, observe that

$$\left|\int_\Omega g(x) \left|f\left(2^\ell (x - x^r)\right)\right|^{p^\sigma - 1} dx\right| \leq \left(\int_{\Omega_r} |g(x)|^{p^\sigma} dx\right)^{1/p^\sigma}$$

$$\times \left(\int_{\Omega_r} \left|f\left(2^\ell (x - x^r)\right)\right|^{p^\sigma} dx\right)^{1/(p^\sigma)'}$$

$$\leq \left(\int_{\Omega_r} |g(x)|^{p^\sigma} dx\right)^{1/p^\sigma} c 2^{-\ell n/(p^\sigma)'}, \tag{48}$$

where the Ω_r denote the pairwise disjoint supports of the $f\left(2^\ell (x - x^r)\right)$. Taking the p^σth power and summing over r we obtain

$$\|B\| \leq c 2^{n\ell - n\ell/(p^\sigma)'}. \tag{49}$$

Since

$$\|\mathrm{id}_2\| \leq \sigma^a, \tag{50}$$

by (34), it follows from (46) that

$$e_m(\mathrm{id}_3) \leq c2^{n\ell\left(\frac{1}{p^{\sigma}} - \frac{1}{p}\right)} \sigma^a e_m(\mathrm{id}_1) = c2^{\sigma\ell} \sigma^a e_m(G_{p,a}). \tag{51}$$

It is easy to see that $e_\ell(\mathrm{id}_3) \sim 1$; see also Proposition 3.2.2 and Remark 3.2.2/1. The choice of $m = \ell$ and $\sigma = 1/\ell$ in (51) now gives

$$e_\ell(G_{p,a}) \geq c\ell^a, \tag{52}$$

which by (37) gives the lower estimate in (33).

Step 3 Suppose that $a < -2s/n$. We shall establish the estimate from above in (31). Choose $\varepsilon > 0$ so that $a + \frac{2s}{n} + \varepsilon < 0$, let $L \in \mathbf{N}$, put

$$k_\ell = 2^{-\frac{n}{s}(L-\ell)\left(a + \frac{2s}{n} + \varepsilon\right)}, \quad \ell = 1, ..., L, \tag{53}$$

and assume that $p^{\sigma_\ell} > 1$ (otherwise $\ell = L_0, ..., L$ for some L_0). In this connection we recall that $\sigma_\ell = 2^{-\ell}$. Then by Proposition 1 and (53),

$$e_{k_\ell}\left(\mathrm{id} : H_p^s \to 2^{\ell a} L_{p^{\sigma_\ell}}\right) \leq c_\varepsilon 2^{\ell a} 2^{\ell\left(\frac{2s}{n} + \varepsilon\right)} 2^{(L-\ell)\left(a + \frac{2s}{n} + \varepsilon\right)}$$

$$= c_\varepsilon 2^{L\left(a + \frac{2s}{n} + \varepsilon\right)}. \tag{54}$$

Moreover, if $\ell > L$, then

$$\left\|\mathrm{id} : H_p^s \to 2^{\ell a} L_{p^{\sigma_\ell}}\right\| \leq c2^{\ell a} \leq c2^{La} \leq c2^{L\left(a + \frac{2s}{n} + \varepsilon\right)}. \tag{55}$$

We now reason in much the same way as in Step 5 of the proof of Theorem 3.4.2. Cover the unit ball U in H_p^s by 2^{k_1} balls in $2^a L_{p^{\sigma_1}}$ of radius $2e_{k_1}$; we may and shall assume that these balls have centres in U. Let U_1 be one of these balls; cover $U \cap U_1$ by 2^{k_2} balls in $2^{2a} L_{p^{\sigma_2}}$ of radius $2e_{k_2}$ with centres in $U \cap U_1$. By iteration, we finally arrive at a covering of $U \cap U_1 \cap \cdots \cap U_{L-1}$ by 2^{k_L} balls in $2^{La} L_{p^{\sigma_L}}$ of radius $2e_{k_L}$ and with centres $g_1, ..., g_M$ in $U \cap U_1 \cap \cdots \cap U_{L-1}$. Then

$$M = 2^{k_1 + \cdots + k_L} \tag{56}$$

and the exponent here may be estimated by means of

$$\sum_{\ell=1}^{L} k_\ell \sim c2^{-\frac{n}{s}L\left(a + \frac{2s}{n} + \varepsilon\right)}. \tag{57}$$

Thus given any $f \in U$, there is a centre g_m such that

$$2^{\ell a} \|f - g_m | L_{p^{\sigma_\ell}}\| \leq ce_{k_\ell} \leq c_\varepsilon 2^{L\left(a + \frac{2s}{n} + \varepsilon\right)}, \quad \ell = 1, ..., L, \tag{58}$$

by (54). Together with (55) this means that U can be covered in $G_{p,a}$ by M balls of radius $c2^{L\left(a+\frac{2s}{n}+\varepsilon\right)}$; thus

$$e_k(G_{p,a}) \le ck^{-s/n} \tag{59}$$

and the upper estimate in (31) is established.

Step 4 All that is left is to prove the upper estimates in (32) and (33). Let $a < b < 0$ and $b = \theta a$. By (34) and a scaling argument,

$$\| f \, |G_{p,b} \| \le c \sup_{0<\sigma<\delta} \sigma^{-a\theta} \| f \, |L_{p^{\theta\sigma}} \|. \tag{60}$$

Note that

$$\frac{1-\theta}{p} + \frac{\theta}{p^\sigma} = \frac{1}{p} + \frac{\theta\sigma}{n} = \frac{1}{p^{\theta\sigma}} \tag{61}$$

and that, by Hölder's inequality,

$$\| f \, |G_{p,b} \| \le c \| f \, |L_p \|^{1-\theta} \| f \, |G_{p,a} \|^\theta. \tag{62}$$

Use of the limiting embedding $H_{p^s}^s \to L_p$, plus the interpolation property of entropy numbers in (1.3.2/3), shows that

$$e_k \left(G_{p,b} \right) \le c e_k^\theta \left(G_{p,a} \right). \tag{63}$$

The choice of $a = -\frac{2s}{n} - \varepsilon$, so that $b = -\frac{2s\theta}{n} - \varepsilon\theta$, then gives, by (31) and (63), the estimate

$$e_k \left(G_{p,b} \right) \le ck^{-s\theta/n} = ck^{(b+\varepsilon\theta)/2}, \quad k \in \mathbf{N}, \tag{64}$$

which completes the proof of Theorem 1.

We now look for comparable results when we have an embedding between spaces which are both of the type $H_p^s(\log H)_a(\Omega)$. The first move is to establish the following proposition, which is analogous to Proposition 1.

Proposition 2 *Suppose that*

$$-\infty < s_2 < s_1 < \infty, \ 0 < p_1 < p_2 < \infty, \ s_1 - \frac{n}{p_1} = s_2 - \frac{n}{p_2}. \tag{65}$$

Then given any $\varepsilon > 0$, there is a constant $c_\varepsilon > 0$ such that for all $\sigma > 0$,

$$e_k \left(\mathrm{id} : H_{p_1}^{s_1}(\Omega) \to H_{p_2}^{s_2}(\Omega) \right) \le c_\varepsilon \sigma^{-2(s_1-s_2)n^{-1}-\varepsilon} k^{-(s_1-s_2)/n}, \quad k \in \mathbf{N}. \tag{66}$$

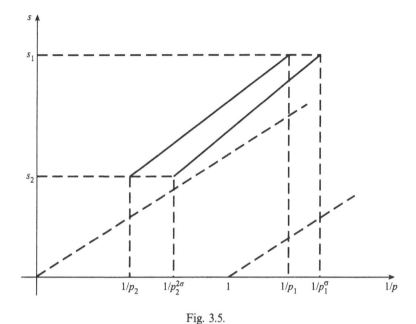

Fig. 3.5.

Proof The proof is the same as that of Proposition 1, with natural technical modifications. See also Fig. 3.5.

With this result we can now give a partial extension of Theorem 1 which focusses on a case in which sharp results are obtained.

Theorem 2 *Suppose that (65) holds and assume that*

$$a_1 \leq 0 \quad and \quad a_2 < a_1 - \frac{2}{n}(s_1 - s_2). \tag{67}$$

Let id $: H_{p_1}^{s_1}(\log H)_{a_1}(\Omega) \to H_{p_2}^{s_2}(\log H)_{a_2}(\Omega)$ *be the natural embedding and let its entropy numbers be denoted by* e_k. *Then*

$$e_k \sim k^{-(s_1-s_2)/n}. \tag{68}$$

Proof We remark that since $a_2 < a_1$, the existence and continuity of id follow immediately from Definition 2.6.3 and the fact that the embedding of $H_{p_1}^{s_1}(\Omega)$ in $H_{p_2}^{s_2}(\Omega)$ is continuous; see (2.3.3/8) and 2.5.1.

As in the proof of Theorem 1, there are obvious counterparts of (34) and (35), now at the levels s_1 and s_2, and of the lower estimate (39). To obtain the estimate from above, all that is necessary is to make a slight technical change in Step 3 of the proof of Theorem 1, corresponding to the use of Proposition 2 rather than Proposition 1.

3.4.4 Duality arguments

Here we show how the entropy numbers of adjoint maps, plus the characterisation of the dual of $H_p^s(\log H)_a(\Omega)$, can be used to obtain results which complement those of Theorem 3.4.3/2. The price to be paid for this is, of course, the restriction $1 < p < \infty$.

Theorem *Suppose that*

$$-\infty < s_2 < s_1 < \infty, \quad 1 < p_1 < p_2 < \infty, \quad s_1 - \frac{n}{p_1} = s_2 - \frac{n}{p_2}, \quad (1)$$

and let

$$a_1 > 0, \quad a_2 < a_1 - \frac{2}{n}(s_1 - s_2). \quad (2)$$

Denote by e_k the entropy numbers of the natural embedding

$$\mathrm{id} : H_{p_1}^{s_1}(\log H)_{a_1}(\Omega) \to H_{p_2}^{s_2}(\log H)_{a_2}(\Omega).$$

Then

$$e_k \sim k^{-(s_1-s_2)/n}. \quad (3)$$

Proof Step 1 First assume that $a_2 \geq 0$. Since $1 < p_1 < p_2 < \infty$, we have the duality assertion (2.6.3/24): $\left[\widetilde{H}_p^s(\log \widetilde{H})_a(\Omega)\right]' = H_{p'}^{-s}(\log H)_{-a}(\Omega)$, $s \geq 0$. We use the results of Bourgain et al [BPST] mentioned in 1.3.1: recall that these imply that if A is a uniformly convex Banach space, B is a Banach space and $T : A \to B$ is a compact linear map with entropy numbers $e_k(T)$ satisfying $e_k(T) = O(k^{-\lambda})$ for some $\lambda > 0$, then $e_k(T^*) = O(k^{-\lambda})$. The spaces $\widetilde{H}_p^s(\log \widetilde{H})_a(\Omega)$ (with $1 < p < \infty$ and $a \leq 0$) can be shown to be uniformly convex; see [CoE] for the kind of arguments needed to establish this via inequalities of Clarkson type; we also refer to [KaT]. Now use Theorem 3.4.3/2, with $\widetilde{H}$ instead of H, together with (2.6.3/24), the lifting of Theorem 2.6.3 and the result of [BPST] mentioned above to obtain the upper estimate in (3). Note that since Ω has the extension property for all the spaces in question (see Remark 2.5.1/1), it does not matter here whether we use the H or $\widetilde{H}$ spaces. The lower estimate in (3) follows from the previous considerations.

Step 2 Here we suppose that $a_2 < 0$, and decompose id as follows:

$$H^{s_1}_{p_1}(\log H)_{a_1}(\Omega) \to H^s_p(\Omega) \to H^{s_2}_{p_2}(\log H)_{a_2}(\Omega), \tag{4}$$

where

$$-\infty < s_2 < s < s_1 < \infty, \quad 1 < p_1 < p < p_2 < \infty, \quad s_1 - \frac{n}{p_1} = s - \frac{n}{p} = s_2 - \frac{n}{p_2} \tag{5}$$

and

$$0 < a_1 - \frac{2}{n}(s_1 - s), \quad a_2 < -\frac{2}{n}(s - s_2). \tag{6}$$

This can be done, since (2) holds. By Step 1 and (3.4.3/68), together with the multiplicative property of entropy numbers, we see that

$$e_{2k-1} \le ck^{-(s_1-s)/n}k^{-(s-s_2)/n} = ck^{-(s_1-s_2)/n},$$

from which the upper estimate implicit in (3) follows. The lower estimate is a consequence of the previous considerations.

3.5 Embeddings in non-smooth domains

Let $\Omega \in MR(n)$ in the sense of Definition 2.5.1/2 and let

$$-\infty < s_2 < s_1 < \infty, \quad p_1, p_2, q_1, q_2 \in (0, \infty] \tag{1}$$

and

$$\delta_+ = s_1 - s_2 - n\left(\frac{1}{p_1} - \frac{1}{p_2}\right)_+ > 0, \tag{2}$$

as in (2.5.1/8), (2.5.1/9) and 3.3.1. If, in addition, Ω has a C^∞ boundary, then we calculated in Theorem 3.3.3/2 and Theorem 3.3.4 the entropy and approximation numbers of the embedding

$$\text{id} : B^{s_1}_{p_1 q_1}(\Omega) \to B^{s_2}_{p_2 q_2}(\Omega) \tag{3}$$

respectively. There we used the fact that there was a bounded linear extension operator from $B^s_{pq}(\Omega)$ to $B^s_{pq}(\mathbf{R}^n)$. For the approximation numbers the existence of such a linear extension operator is crucial for our method, but this is not so for the entropy numbers and we are accordingly able to extend Theorem 3.3.3/2 to non-smooth domains. Let e_k be the kth entropy number of the embedding id given by (3).

Theorem *Let $\Omega \in MR(n)$. Under the above hypotheses there are positive constants c_1 and c_2 such that for all $k \in \mathbf{N}$,*

$$c_1 k^{-(s_1-s_2)/n} \le e_k \le c_2 k^{-(s_1-s_2)/n}. \tag{4}$$

Proof We follow the proof given in [TriW1], 4.2. Let K_1 and K_2 be two open balls in $\mathbf{R}^n$ with $\overline{K}_1 \subset \Omega$ and $\overline{\Omega} \subset K_2$. We may assume that the unit ball $\{f : \|f \,|\, B^{s_1}_{p_1 q_1}(\Omega)\| \leq 1\}$ of $B^{s_1}_{p_1 q_1}(\Omega)$ is the restriction of a subset $\{g : \|g \,|\, B^{s_1}_{p_1 q_1}(K_2)\| \leq c\}$ for some $c > 1$. By Theorem 3.3.3/2 there exist 2^{k-1} elements $g_\ell \in B^{s_2}_{p_2 q_2}(K_2)$ with

$$\min_\ell \|g - g_\ell \,|\, B^{s_2}_{p_2 q_2}(K_2)\| \leq c k^{-(s_1 - s_2)/n}, \quad k \in \mathbf{N}. \tag{5}$$

By restriction one has a similar assertion with Ω instead of K_2, and this proves the right-hand side of (4). The left-hand side of (4) follows from the arguments above with K_1 and Ω instead of Ω and K_2, respectively.

Remark In Chapter 5 we shall study the distribution of eigenvalues of degenerate elliptic operators in bounded domains with C^∞ boundary, basing our arguments on Theorem 3.3.3/2 and Carl's inequality in Corollary 1.3.4. However, it is quite clear that the above theorem combined with the powerful instrument of atomic representations as described in 2.5.3 opens the door for the development of a spectral theory for degenerate elliptic operators in bounded non-smooth domains.

4

Weighted Function Spaces and Entropy Numbers

4.1 Introduction

This chapter deals with weighted function spaces of type $B^s_{pq}(\mathbf{R}^n, w(\cdot))$ and $F^s_{pq}(\mathbf{R}^n, w(\cdot))$ on $\mathbf{R}^n$, where w is a weight function of at most polynomial growth. Preference is given to weights of the form $w(x) = \left(1 + |x|^2\right)^{\alpha/2}$ with $\alpha \in \mathbf{R}$. In particular, the weights w which are admitted have no local singularities. Of course, if $w(x) = 1$, then these spaces coincide with the unweighted spaces $B^s_{pq}(\mathbf{R}^n)$ and $F^s_{pq}(\mathbf{R}^n)$, respectively, which were introduced in 2.2. In 4.2 we develop the theory of the spaces $B^s_{pq}(\mathbf{R}^n, w(\cdot))$ and $F^s_{pq}(\mathbf{R}^n, w(\cdot))$ and in particular clarify under which conditions embeddings between these spaces are compact. Then in analogy to Chapter 3 it makes sense to enquire into the behaviour of the related entropy numbers. This will be done in 4.3, where it will be seen that some new phenomena arise. Corresponding results for approximation numbers are mentioned only very briefly. Applications to the spectral theory of (weighted) degenerate pseudodifferential operators are given in 5.4. In this chapter we mainly follow [HaT1], and partly [HaT2] and [Har2].

4.2 Weighted spaces

4.2.1 Definitions

The basic notation is the same as in 2.2.1 and 2.2.2; in particular, $\langle x \rangle = \left(1 + |x|^2\right)^{1/2}$ if $x \in \mathbf{R}^n$.

Definition 1 *The class W^n of admissible weight functions is the collection of all positive C^∞ functions w on $\mathbf{R}^n$ with the following properties:*

(i) *for all multi-indices γ there exists a positive constant c_γ with*

$$|D^\gamma w(x)| \leq c_\gamma w(x) \quad \text{for all} \ x \in \mathbf{R}^n; \tag{1}$$

(ii) *there exist two constants $c > 0$ and $\alpha \geq 0$ such that*

$$0 < w(x) \leq cw(y)\langle x - y\rangle^\alpha \quad \text{for all} \ x, y \in \mathbf{R}^n. \tag{2}$$

Remark 1 Of course, for appropriate positive constants c_1 and c_2, we have

$$c_1 w(x) \leq w(y) \leq c_2 w(x) \quad \text{for all} \ x, y \in \mathbf{R}^n \ \text{with} \ |x - y| \leq 1. \tag{3}$$

Furthermore, if $w \in W^n$ and $w' \in W^n$, then $w^{-1} \in W^n$ and $ww' \in W^n$.

Remark 2 Let w be a measurable function on $\mathbf{R}^n$ satisfying (2). Let $h \geq 0$ be a C^∞ function on $\mathbf{R}^n$, supported by the unit ball and with, say, $\int h(x)\,dx = 1$. Then

$$(w * h)(x) = \int h(x - y) w(y)\,dy \in W^n. \tag{4}$$

Furthermore, w and $w * h$ are equivalent. This makes it clear that part (i) of the definition above is convenient for us but not necessary in what follows.

Next we define the weighted counterparts of the spaces $B^s_{pq}(\mathbf{R}^n)$ and $F^s_{pq}(\mathbf{R}^n)$ introduced in Definition 2.2.1. All the notation has the same meaning as in 2.2.1; in particular, $\{\varphi_k\}_{k=0}^\infty$ is the dyadic resolution of unity given by (2.2.1/3)–(2.2.1/5) and generated by φ. Recall that $\left(\varphi_k \hat{f}\right)^\vee$ is an entire analytic function for all $f \in \mathscr{S}'(\mathbf{R}^n)$. Furthermore, if $0 < p \leq \infty$ then $L_p(\mathbf{R}^n)$ is the usual Lebesgue space introduced in 2.2.1. Let $w > 0$ be a weight function on $\mathbf{R}^n$ and let $0 < p \leq \infty$; then $L_p(\mathbf{R}^n, w(\cdot))$ stands for the weighted generalisation of $L_p(\mathbf{R}^n)$, quasi-normed by

$$\|f \mid L_p(\mathbf{R}^n, w(\cdot))\| = \|wf \mid L_p(\mathbf{R}^n)\| ; \tag{5}$$

see (2.2.1/1).

Definition 2 Let $w \in W^n$, let $s \in \mathbf{R}, 0 < q \leq \infty$ and let $\{\varphi_k\}$ be the above dyadic resolution of unity.
   (i) Let $0 < p \leq \infty$. The space $B^s_{pq}(\mathbf{R}^n, w(\cdot))$ is the collection of all

$f \in \mathscr{S}'(\mathbf{R}^n)$ such that

$$\left\| f \mid B_{pq}^s \left(\mathbf{R}^n, w\,(\cdot)\right) \right\|_\varphi = \left(\sum_{j=0}^\infty 2^{jsq} \left\| \left(\varphi_j \, \hat{f}\right)^\vee \mid L_p \left(\mathbf{R}^n, w\,(\cdot)\right) \right\|^q \right)^{1/q} \quad (6)$$

(with the usual modification if $q = \infty$) is finite.

(ii) Let $0 < p < \infty$. The space $F_{pq}^s\,(\mathbf{R}^n, w\,(\cdot))$ is the collection of all $f \in \mathscr{S}'(\mathbf{R}^n)$ such that

$$\left\| f \mid F_{pq}^s \left(\mathbf{R}^n, w\,(\cdot)\right) \right\|_\varphi = \left\| \left(\sum_{j=0}^\infty 2^{jsq} \left| \left(\varphi_j \, \hat{f}\right)^\vee (\cdot) \right|^q \right)^{1/q} \mid L_p \left(\mathbf{R}^n, w\,(\cdot)\right) \right\| \quad (7)$$

(with the usual modification if $q = \infty$) is finite.

(iii) Let $w(x) = \langle x \rangle^\alpha$ for some $\alpha \in \mathbf{R}$. Then we put

$$B_{pq}^s \left(\mathbf{R}^n, \alpha\right) = B_{pq}^s \left(\mathbf{R}^n, \langle \cdot \rangle^\alpha\right) \quad (8)$$

and

$$F_{pq}^s \left(\mathbf{R}^n, \alpha\right) = F_{pq}^s \left(\mathbf{R}^n, \langle \cdot \rangle^\alpha\right). \quad (9)$$

(iv) Let $0 < p < \infty$; then we put

$$H_p^s \left(\mathbf{R}^n, w\,(\cdot)\right) = F_{p2}^s \left(\mathbf{R}^n, w\,(\cdot)\right) \quad (10)$$

and write this as $H_p^s(\mathbf{R}^n, \alpha)$ when $w(x) = \langle x \rangle^\alpha$ and $\alpha \in \mathbf{R}$.

Remark 3 Of course, if $w(x) = 1$ then the above spaces coincide with the unweighted spaces $B_{pq}^s\,(\mathbf{R}^n)$ and $F_{pq}^s\,(\mathbf{R}^n)$ introduced in Definition 2.2.1. As mentioned in Remark 2.2.1, the theory of these unweighted spaces has been developed to its full extent in [Triβ] and [Triγ]. It is not difficult to extend this theory to the weighted classes above. Moreover, in [SchmT], 5.1, p. 243 the treatment proceeds in a way parallel to [Triβ] to deal with weighted spaces of this type in the framework of ultra-distributions and for a much larger class of admissible weight functions. The later developments in the theory of the unweighted spaces $B_{pq}^s\,(\mathbf{R}^n)$ and $F_{pq}^s\,(\mathbf{R}^n)$ as presented in [Triγ] also have their more or less obvious counterparts for the weighted spaces under consideration here. In particular, we shall use equivalent quasi-norms in $B_{pq}^s\,(\mathbf{R}^n, w\,(\cdot))$ and $F_{pq}^s\,(\mathbf{R}^n, w\,(\cdot))$ via local means in generalisation of what has been done in [Triγ], 1.8.4, 2.4.6 and 2.5.3, pp. 58, 122, 138, respectively for the unweighted spaces. One has simply to replace $L_p\,(\mathbf{R}^n)$ by $L_p\,(\mathbf{R}^n, w\,(\cdot))$.

Remark 4 As in the unweighted case the two weighted scales $B_{pq}^s(\mathbf{R}^n, w(\cdot))$ and $F_{pq}^s(\mathbf{R}^n, w(\cdot))$ above cover weighted (fractional) Sobolev spaces, weighted Hölder–Zygmund spaces and weighted classical Besov spaces characterised in the usual way via derivatives and differences. We refer to 2.2.2 for the unweighted case and to [SchmT], 5.1, pp. 241–53 and the literature mentioned there for the weighted case.

Notational agreement All spaces in this chapter are defined on $\mathbf{R}^n$, where $n \in \mathbf{N}$ is assumed to be fixed. In view of this we shall omit $\mathbf{R}^n$ in what follows and write $B_{pq}^s(w(\cdot))$ and $B_{pq}^s(\alpha)$ instead of $B_{pq}^s(\mathbf{R}^n, w(\cdot))$ and $B_{pq}^s(\mathbf{R}^n, \alpha)$, etc.

4.2.2 Basic properties

As mentioned above we feel free to use assertions for the unweighted spaces B_{pq}^s and F_{pq}^s on $\mathbf{R}^n$ also for the weighted spaces $B_{pq}^s(w(\cdot))$ and $F_{pq}^s(w(\cdot))$ if their extensions are more or less obvious or if they are covered by the more general spaces treated in [SchmT], 5.1, pp. 241–53.

Theorem *Let $s \in \mathbf{R}$, suppose that $p, q \in (0, \infty]$ (with $p < \infty$ in the F case) and let $w \in W^n$.*

(i) $B_{pq}^s(w(\cdot))$ and $F_{pq}^s(w(\cdot))$ are quasi-Banach spaces (Banach spaces if $p \geq 1$ and $q \geq 1$), and they are independent of the chosen dyadic resolution of unity $\{\varphi_j\}$ (generated by φ).

(ii) The operator $f \longmapsto wf$ is an isomorphic mapping from $B_{pq}^s(w(\cdot))$ onto B_{pq}^s and from $F_{pq}^s(w(\cdot))$ onto F_{pq}^s. In particular,

$$\left\| wf \mid B_{pq}^s \right\| \text{ is an equivalent quasi-norm on } B_{pq}^s(w(\cdot)) \qquad (1)$$

and

$$\left\| wf \mid F_{pq}^s \right\| \text{ is an equivalent quasi-norm on } F_{pq}^s(w(\cdot)). \qquad (2)$$

Remark 1 We take part (i) for granted, as it is either parallel to [Triβ] or covered by [SchmT], 5.1, p. 243. Of course, $B_{pq}^s = B_{pq}^s(\mathbf{R}^n)$ and $F_{pq}^s = F_{pq}^s(\mathbf{R}^n)$ in part (ii); see the notational agreement at the end of 4.2.1. As in Remark 2.2.1 for the quasi-norms on B_{pq}^s and F_{pq}^s we also omit the subscript φ in (4.2.1/6) and (4.2.1/7) in what follows. Moreover, part (ii) is covered by [SchmT], 5.1.3, p. 246, but the proof given there, even when reduced to our simpler situation, is rather tricky because of its use of complex interpolation for quasi-Banach spaces. The earlier proof by J. Franke [Fra1] uses the heavy machinery of paramultiplication. This justifies our insertion here of a short new proof which is based on the one hand on local means and on the other hand on some splitting techniques which later on (in 5.4.2) will be of great service in connection with weighted pseudodifferential operators.

Proof (of part (ii) of the theorem) Step 1 Following [Triγ], 2.5.3, p. 138, we described in Step 2 of the proof of Theorem 2.3.2 the characterisation of B_{pq}^s in terms of local means. By [Triγ], 2.4.6, p. 122, one obtains the corresponding counterpart for F_{pq}^s on replacing (2.3.2/13) by

$$\left\| g \mid F_{pq}^s \right\| \sim \left\| K_0(1,g) \mid L_p \right\| + \left\| \left(\int_0^1 t^{-sq} |K(t,g)(\cdot)|^q \frac{dt}{t} \right)^{1/q} \mid L_p \right\|. \quad (3)$$

All symbols have the same meaning as in Step 2 of the proof of Theorem 2.3.2. Let $s > n/p$ and suppose that $w \in W^n$. By (2.3.2/12), with wf in place of g, together with the Taylor expansion of $w(x+ty)$ with remainder term and (4.2.1/1), we have

$$|K(t,wf)(x)| \le c \sum_{|\alpha| \le m-1} t^{|\alpha|} w(x) \left| \int_{\mathbf{R}^n} y^\alpha K(y) f(x+ty)\, dy \right|$$
$$+ c t^m w(x) \sup_{|x-y| \le 1} |f(y)| \quad (4)$$

with $s < m \in \mathbf{N}$; see also [Triγ], 4.2.2, p. 203, so far as this technique is concerned. Let $w_j \in W^n$ for $j = 1, 2, 3, 4$ and let $w_1(x)w_2(x) = w_3(x)w_4(x)$ for all $x \in \mathbf{R}^n$. Now use (4) with $w_3 f$ and $w_1 w_3^{-1}$ in place of f and w, respectively, and take for granted that (3) holds also for the weighted spaces $F_{pq}^s(w_2(\cdot))$. Of course, one has to replace L_p by $L_p(w_2(\cdot))$. Then we have, by the same arguments as in Step 1 of the proof of Theorem 4.2.2 in [Triγ], pp. 203–04,

$$\left\| \left(\int_0^1 t^{-sq} |K(t,w_1 f)(\cdot)|^q \frac{dt}{t} \right)^{1/q} \mid L_p(w_2(\cdot)) \right\| \le c \left\| w_3 f \mid F_{pq}^s(w_4(\cdot)) \right\| \quad (5)$$

and

$$\left\| w_1 f \mid F_{pq}^s(w_2(\cdot)) \right\| \le c \left\| w_3 f \mid F_{pq}^s(w_4(\cdot)) \right\|. \quad (6)$$

But now (2) follows from (6), and the proof of part (ii) of the theorem for F_{pq}^s-spaces with $s > n/p$ is complete. The proof for the B_{pq}^s spaces with $s > n/p$ is the same.

Step 2 Let $v \in W^n$ and $w \in W^n$. We generalise the lift I_σ, defined in (2.2.2/4), by

$$I_{\sigma,v} : f \longmapsto \left(\langle \xi \rangle^\sigma \widehat{vf} \right)^\vee v^{-1}, \quad \sigma \in \mathbf{R}. \tag{7}$$

In the proposition below we establish the lift property

$$I_{\sigma,v} B_{pq}^s \left(w(\cdot) \right) = B_{pq}^{s-\sigma} \left(w(\cdot) \right) \text{ and } I_{\sigma,v} F_{pq}^s \left(w(\cdot) \right) = F_{pq}^{s-\sigma} \left(w(\cdot) \right), \tag{8}$$

which generalises (2.2.2/5). Taking (8) for granted, (6) can easily be extended to arbitrary real values of s. Let $\sigma \in \mathbf{R}$ be such that $s + \sigma > \frac{n}{p}$, and suppose that $w_3 f \in F_{pq}^s \left(w_4(\cdot) \right)$. By (8) and

$$w_3 I_{-\sigma} f = I_{-\sigma, w_3^{-1}} w_3 f \tag{9}$$

we have

$$\left\| w_3 f \mid F_{pq}^s \left(w_4(\cdot) \right) \right\| \sim \left\| w_3 I_{-\sigma} f \mid F_{pq}^{s+\sigma} \left(w_4(\cdot) \right) \right\|. \tag{10}$$

Application of (6) with $s + \sigma$ in place of s to $I_{-\sigma} f$, together with (10) with w_1 and w_2 in place of w_3 and w_4, yields the desired result. The rest is the same as in the first step.

Proposition *Let v, $w \in W^n$ and let $I_{\sigma,v}$ be given by (7). Then (8) holds.*

Proof Step 1 By our philosophy we may assume that $w(x) = 1$ in (8). We prove that

$$I_{\sigma,v} F_{pq}^s = F_{pq}^{s-\sigma}, \quad s \in \mathbf{R}, \ \sigma \in \mathbf{R}, v \in W^n, \tag{11}$$

by means of the localisation principle for F spaces described in [Triγ], 2.4.7, p. 124. Let ψ be a compactly supported C^∞ function on $\mathbf{R}^n$, let $\psi_k(x) = \psi(x - k)$, $k \in \mathbf{Z}^n$, and suppose that

$$\sum_{k \in \mathbf{Z}^n} \psi_k(x) = 1, \ x \in \mathbf{R}^n. \tag{12}$$

Let $0 < p < \infty$, $0 < q \leq \infty$ and $s \in \mathbf{R}$. Then

$$\left(\sum_{k \in \mathbf{Z}^n} \left\| \psi_k f \mid F_{pq}^s \right\|^p \right)^{1/p} \tag{13}$$

is an equivalent quasi-norm on F_{pq}^s. Let $g = \psi_0 f$ and $l \in \mathbf{Z}^n$. For any multi-index β we have

$$x^\beta I_{\sigma,v} g(x) = \sum_{\gamma + \delta = \beta} c_{\gamma,\delta} v^{-1}(x) \left(D_\xi^\gamma \langle \xi \rangle^\sigma \widehat{y^\delta vg} \right)^\vee (x). \tag{14}$$

Next we use the pointwise multiplier properties in F_{pq}^s (see [Triγ], 4.2.2, p. 203), mapping properties of type (2.2.2/5) and the fact that g is supported near the origin. Then we obtain

$$\left\| \psi_l x^\beta I_{\sigma,v} g \mid F_{pq}^{s-\sigma} \right\| \leq cv^{-1}(l) \left\| vg \mid F_{pq}^s \right\|$$
$$\leq c \langle l \rangle^\alpha \left\| \psi_0 f \mid F_{pq}^s \right\|, \tag{15}$$

in which we used (4.2.1/2) with $v(0)v^{-1}(l) \leq c' \langle l \rangle^\alpha$. In (15) one can replace x^β by $\langle x \rangle^K$ for any $K \in \mathbf{N}$ and ψ_0 by ψ_k with $k \in \mathbf{Z}^n$. The pointwise multiplier property then gives

$$\left\| \psi_l I_{\sigma,v} \psi_k f \mid F_{pq}^{s-\sigma} \right\| \leq c \langle k-l \rangle^{-L} \left\| \psi_k f \mid F_{pq}^s \right\|, \tag{16}$$

where $L > 0$ is at our disposal. Let $\lambda = \min(1,p,q)$. Then (16) and the λ triangle inequality for F_{pq}^s yield

$$\left\| \psi_l I_{\sigma,v} f \mid F_{pq}^{s-\sigma} \right\|^p = \left\| \psi_l I_{\sigma,v} \sum_k \psi_{l+k} f \mid F_{pq}^{s-\sigma} \right\|^p$$
$$\leq \left(\sum_k \left\| \psi_l I_{\sigma,v} \psi_{l+k} f \mid F_{pq}^{s-\sigma} \right\|^\lambda \right)^{p/\lambda}$$
$$\leq c_1 \left(\sum_k \langle k \rangle^{-L\lambda} \left\| \psi_{l+k} f \mid F_{pq}^s \right\|^\lambda \right)^{p/\lambda}$$
$$\leq c_2 \sum_k \langle k \rangle^{-M} \left\| \psi_{l+k} f \mid F_{pq}^s \right\|^p, \tag{17}$$

where $M > 0$ is at our disposal. Summation over all $l \in \mathbf{Z}^n$ and the localisation principle for F_{pq}^s spaces which was mentioned above prove that

$$\left\| I_{\sigma,v} f \mid F_{pq}^{s-\sigma} \right\| \leq c \left\| f \mid F_{pq}^s \right\|. \tag{18}$$

Now (11) follows from (18) and the identity $I_{-\sigma,v} \circ I_{\sigma,v} = \mathrm{id}$.

Step 2 By the same arguments we have

$$\left\| I_{\sigma,v} f \mid B_{\infty\infty}^{s-\sigma} \right\| \leq c \left\| f \mid B_{\infty\infty}^s \right\|. \tag{19}$$

We use the real interpolation formula

$$D_{pq}^s = \left(\Gamma_{pq_0}^{s_0}, \Gamma_{pq_1}^{s_1} \right)_{\theta,q} \tag{20}$$

with $s_0 < s < s_1, s = (1-\theta)s_0 + \theta s_1$ and $p,q,q_0,q_1 \in (0,\infty]$, where $F_{pq_0}^{\cdot} = F_{pq_1}^{\cdot} = B_{\infty\infty}^{\cdot}$ if $p = \infty$; see [Triβ], 2.4.2, p. 64. Now (18), complemented by (19), and the interpolation property prove the B-counterpart of (11).

Remark 2 Again we proved the proposition for the unweighted case. As remarked above, all the basic properties extend to the weighted spaces for the harmless weights $w \in W^n$ under consideration here. This also applies to (20) for a fixed weight in all three spaces. In the proof of the proposition we used the splitting technique developed in [HaT2], 3.1 in connection with mapping properties of pseudodifferential operators in weighted spaces of the above type; see also 5.4.2.

4.2.3 Embeddings: general weights

In 2.3.3 we gave a brief account of continuous embeddings between unweighted spaces B_{pq}^s and F_{pq}^s on $\mathbf{R}^n$. By Theorem 4.2.2(ii) we have immediate counterparts if only one weight function $w \in W^n$ is involved. However, we are interested in embeddings with different weights. By Theorem 4.2.2(ii) and the elementary embeddings for the unweighted spaces B_{pq}^s and F_{pq}^s as described in [Triβ], p. 47, we may restrict ourselves to the F spaces at this moment. Recall that the class of admissible weights was introduced in Definition 4.2.1/1.

Theorem *Let w_1, $w_2 \in W^n$ and suppose that*

$$-\infty < s_2 < s_1 < \infty, \, 0 < p_1 \le p_2 < \infty, \; 0 < q_1 \le \infty \text{ and } 0 < q_2 \le \infty. \tag{1}$$

(i) Then $F_{p_1 q_1}^{s_1}(w_1(\cdot))$ is continuously embedded in $F_{p_2 q_2}^{s_2}(w_2(\cdot))$,

$$F_{p_1 q_1}^{s_1}(w_1(\cdot)) \subset F_{p_2 q_2}^{s_2}(w_2(\cdot)), \tag{2}$$

if, and only if,

$$s_1 - \frac{n}{p_1} \ge s_2 - \frac{n}{p_2} \text{ and } \frac{w_2(x)}{w_1(x)} \le c < \infty \tag{3}$$

for some $c > 0$ and all $x \in \mathbf{R}^n$.

(ii) The embedding (2) is compact if, and only if,

$$s_1 - \frac{n}{p_1} > s_2 - \frac{n}{p_2} \text{ and } \frac{w_2(x)}{w_1(x)} \to 0 \text{ as } |x| \to \infty. \tag{4}$$

Proof Step 1 We prove the "if"-parts. The embedding (2) with $w_1 = w_2$ is an immediate consequence of (4.2.2/2) and (2.3.3/7), complemented by the elementary embeddings from [Triβ], p. 47 to which we have already referred. Then (2) with $w_2(x) \le cw_1(x)$ follows from (4.2.1/7). This proves the "if"-part of (i). Let

$$K_R = \{x \in \mathbf{R}^n : |x| < R\}, \ R > 0, \tag{5}$$

be the ball of radius R centred at the origin and let $F^s_{pq}(K_R)$ be the corresponding spaces introduced in Definition 2.5.1/1. By (1) and the first part of (4) the embedding

$$F^{s_1}_{p_1q_1}(K_R) \subset F^{s_2}_{p_2q_2}(K_R) \tag{6}$$

is compact; see also 3.3.2. Let ψ_R be a C^∞ function on $\mathbf{R}^n$ with, say, $\psi_R(x) = 1$ if $|x| > 2R$ and $\psi_R(x) = 0$ if $|x| < R$. By (4) and the characterisation of the F spaces via local means according to (4.2.2/3) we have

$$\left\| \psi_R f \mid F^{s_2}_{p_2q_2}(w_2(\cdot)) \right\| \le \varepsilon \left\| \psi_R f \mid F^{s_1}_{p_1q_1}(w_1(\cdot)) \right\| \tag{7}$$

where $\varepsilon > 0$ is given and $R = R(\varepsilon)$ is chosen sufficiently large. Now (6) and (7) prove the "if"-part of (ii).

Step 2 To prove the "only if"-parts we claim that there is no continuous embedding from $F^{s_1}_{p_1q_1}(K_R)$ to $F^{s_2}_{p_2q_2}(K_R)$ if $s_2 - \frac{n}{p_2} > s_1 - \frac{n}{p_1}$, and that there is no compact embedding of this type if $s_2 - \frac{n}{p_2} \ge s_1 - \frac{n}{p_1}$ (of course, it is always assumed that (1) holds). Both assertions follow easily from Theorem 2.3.2.

Step 3 It remains to prove that the assertions about the weights in (3) and (4) are necessary. Assume that there exists a sequence $\{x^j\} \subset \mathbf{R}^n$ with $|x^j| \to \infty$ and $w_2(x^j)/w_1(x^j) \to \infty$ as $j \to \infty$. We use the fact that (4.2.2/13) is an equivalent quasi-norm on F^s_{pq}. Then it follows from Theorem 4.2.2(ii) that (2) cannot be true. Similarly, if $w_2(x^j) \ge cw_1(x^j)$ for some $c > 0$ and $|x^j| \to \infty$ then, if there is a continuous embedding (2), it cannot be compact.

Remark Let

$$-\infty < s_2 < s_1 < \infty, \ 0 < p_1 \le p_2 \le \infty, \ 0 < q_1 \le \infty, \ 0 < q_2 \le \infty, \tag{8}$$

where now, compared with (1), $p_1 \le p_2 = \infty$ is admitted. As for the B-counterpart of (2),

$$B^{s_1}_{p_1q_1}(w_1(\cdot)) \subset B^{s_2}_{p_2q_2}(w_2(\cdot)), \tag{9}$$

there are no problems if $s_1 - \frac{n}{p_1} > s_2 - \frac{n}{p_2}$. When $s_1 - \frac{n}{p_1} = s_2 - \frac{n}{p_2}$ one now has the restriction $q_1 \le q_2$; see (2.3.3/5). On the other hand, the embedding (9) is compact if, and only if, (4) holds (under the assumption (8)). Since $p_2 = \infty$ is now admitted, the weighted counterparts $\mathscr{C}^s(w(\cdot)) = B^s_{\infty\infty}(w(\cdot))$ of the Hölder–Zygmund spaces introduced in 2.2.2 are now included.

4.2.4 Embeddings: the weights $\langle x \rangle^\alpha$

Let $w_1, w_2 \in W^n$ (see Definition 4.2.1/1). Then $w_1/w_2 \in W^n$ and by Theorem 4.2.2(ii),

$$\left\| f \mid F^s_{pq}(w_1(\cdot)) \right\| \sim \left\| w_2 f \mid F^s_{pq}\left(w_1 w_2^{-1}(\cdot)\right) \right\| \tag{1}$$

(equivalent quasi-norms). In other words, $f \longmapsto w_2 f$ is an isomorphic mapping from $F^s_{pq}(w_1(\cdot))$ onto $F^s_{pq}(w_1 w_2^{-1}(\cdot))$ for all $w_1 \in W^n$. Of course, we have a corresponding assertion for the B spaces. In other words, if we wish to study continuous or compact embeddings of type (4.2.3/2) we may assume without loss of generality that $w_2 = 1$. Accordingly we put $w_1 = w$ and specify in what follows $w(x) = \langle x \rangle^\alpha$ for some $\alpha > 0$. We are mainly interested in compact embeddings of type (4.2.3/2) in these specialised circumstances. The restriction $p_1 \le p_2$ in Theorem 4.2.3 is convenient. But depending on w_1/w_2 there may also be continuous or compact embeddings of type (4.2.3/2) for some $p_2 < p_1$. In our specialised situation we have rather final answers. For that purpose it is convenient for us to introduce weak unweighted spaces of B^s_{pq} and F^s_{pq} type. Let f^* be the non-increasing rearrangement of a function f on $\mathbf{R}^n$; see (2.6.1/3), now with $\Omega = \mathbf{R}^n$. Let $0 < p < \infty$; then, as usual, the Lorentz space (Marcinkiewicz space, or weak L_p space) $L_{p,\infty} = L_{p,\infty}(\mathbf{R}^n)$ on $\mathbf{R}^n$ with respect to Lebesgue measure is the collection of all measurable functions $f : \mathbf{R}^n \to \mathbf{C}$ such that

$$\| f \mid L_{p,\infty} \| = \sup_{t>0} t^{1/p} f^*(t) < \infty; \tag{2}$$

see also 2.6.1 and the references given there, [BeS], p. 216 and [Triα], 1.18.6, p. 132. Again let $\{\varphi_k\}_{k=0}^\infty$ be the dyadic resolution of unity given by (2.2.1/3)–(2.2.1/5), generated by φ. Recall that $\left(\varphi_k \hat{f}\right)^\vee$ is an entire analytic function for any $f \in \mathscr{S}'(\mathbf{R}^n)$. By our notational agreement at the end of 4.2.1 we omit "$\mathbf{R}^n$" in what follows because all spaces in this chapter are defined on $\mathbf{R}^n$.

Definition *Let* $s \in \mathbf{R}$, $0 < p < \infty$ *and* $0 < q \leq \infty$. *Then* weak-B_{pq}^s *is the collection of all* $f \in \mathscr{S}'$ *such that*

$$\left\| f \mid \text{weak-}B_{pq}^s \right\|_{\varphi} = \left(\sum_{j=0}^{\infty} 2^{jsq} \left\| \left(\varphi_j \hat{f} \right)^{\vee} \mid L_{p,\infty} \right\|^q \right)^{1/q} \tag{3}$$

(with the usual modification if $q = \infty$*) is finite. Similarly,* weak-F_{pq}^s *is the collection of all* $f \in \mathscr{S}'$ *such that*

$$\left\| f \mid \text{weak-}F_{pq}^s \right\|_{\varphi} = \left\| \left(\sum_{j=0}^{\infty} 2^{jsq} \left| \left(\varphi_j \hat{f} \right)^{\vee} (\cdot) \right|^q \right)^{1/q} \mid L_{p,\infty} \right\| \tag{4}$$

(with the usual modification if $q = \infty$*) is finite.*

Remark 1 Of course, one may also replace $L_{p,\infty}$ by the more general Lorentz spaces $L_{p,u}$ with $0 < p \leq \infty$ ($p < \infty$ for the F spaces) and $0 < u \leq \infty$. For $L_{p,u}$ see the end of 2.6.1, especially (2.6.1/11) with $a = 0$ and $\mathbf{R}^n$ in place of Ω. Recall the real interpolation formula

$$\left(L_{p_0,u_0}, L_{p_1,u_1} \right)_{\theta,u} = L_{p,u}, \quad \frac{1}{p} = \frac{1-\theta}{p_0} + \frac{\theta}{p_1}, \quad p_0 \neq p_1, \tag{5}$$

where p_0, p_1, u_0, u_1 and u are positive or infinite, and $0 < \theta < 1$. We refer to [BeL], 5.3, p. 113; [Triα], 1.18.6, Theorem 2 and Remark 5, pp. 134–5; and [BeS], p. 300. Recall that $B_{pq}^s(\alpha)$ $\left(= B_{pq}^s(\mathbf{R}^n, \alpha) \right)$ and $F_{pq}^s(\alpha)$ $\left(= F_{pq}^s(\mathbf{R}^n, \alpha) \right)$ were introduced in Definition 4.2.1/2(iii).

Theorem *(i) Let* $s \in \mathbf{R}$, $0 < p < \infty$ *and* $0 < q \leq \infty$. *Then both* weak-B_{pq}^s *and* weak-F_{pq}^s *are quasi-Banach spaces (Banach spaces if* $p \geq 1$ *and* $q \geq 1$*) and they are independent of the chosen dyadic resolution of unity* $\{\varphi_j\}$, *generated by* φ.

(ii) Let $s \in \mathbf{R}$, $0 < p \leq \infty$ *(with* $p < \infty$ *in the case of the F spaces),* $0 < q \leq \infty$, $\alpha > 0$ *and* $\frac{1}{p_0} = \frac{1}{p} + \frac{\alpha}{n}$. *Then both the continuous embeddings*

$$B_{pq}^s(\alpha) \subset \text{weak-}B_{p_0 q}^s \quad \text{and} \quad F_{pq}^s(\alpha) \subset \text{weak-}F_{p_0 q}^s \tag{6}$$

exist.

Proof Step 1 Replacing L_p in Definition 2.2.1 as the basic space for B_{pq}^s and F_{pq}^s by $L_{p,\infty}$, or more generally by $L_{p,u}$, then the theory for these L_p-based spaces developed in [Triβ] and [Triγ] can be carried over to the related $L_{p,u}$-based spaces. This is a consequence of (5) and the interpolation theory. In other words, we take part (i) of the theorem for granted.

Step 2 By (5), the related interpolation property, and Hölder's inequality

$$L_p \cdot L_{n/\alpha} \subset L_{p_0},\tag{7}$$

we obtain

$$L_p \cdot L_{n/\alpha,\infty} \subset L_{p_0,\infty}.\tag{8}$$

Since $\langle x \rangle^{-\alpha} \in L_{n/\alpha,\infty}$, Definitions 4.2.1/2, (3), (4) and (8) prove (6).

Remark 2 In accordance with Remarks 2.2.1 and 4.2.2/1 we shall omit the subscript φ in (3) and (4) in what follows (having in mind equivalent quasi-norms).

4.2.5 Hölder inequalities

In 2.4 we dealt with Hölder inequalities for the unweighted spaces B_{pq}^s and F_{pq}^s on $\mathbf{R}^n$. These considerations may readily be extended to the weighted spaces $B_{pq}^s(w(\cdot))$ and $F_{pq}^s(w(\cdot))$ introduced in Definition 4.2.1/2. This can be done on the basis of Theorem 4.2.2(ii), but for the sake of simplicity we restrict ourselves to the spaces $H_p^s(w(\cdot))$ defined by (4.2.1/10). As mentioned at the end of 4.2.1 we always omit "$\mathbf{R}^n$" in connection with spaces; otherwise we use the same notation as in 2.4.3 and 2.4.5. In particular, the Hölder inequalities of interest for us are confined to the strip G_2 (see (2.4.3/2) and Fig. 2.1 on p. 52), and r^s has the same meaning as in (2.4.3/3). Now it is quite obvious how to extend Theorem 2.4.5 from the unweighted case to the weighted one.

Theorem *Let r_1, $r_2 \in (1,\infty)$ and suppose that $\frac{1}{r} = \frac{1}{r_1} + \frac{1}{r_2} < 1$. Assume also that $s \in \mathbf{R}$ and $\frac{1}{r^s} = \frac{1}{r_1} + \frac{s}{n} > 0$. Suppose that $w_1, w_2 \in W^n$ and let $w(x) = w_1(x)w_2(x)$. Then*

$$H_{r_1^s}^s(w_1(\cdot))\ H_{r_2^{|s|}}^{|s|}(w_2(\cdot)) \subset H_{r^s}^s(w(\cdot)),\tag{1}$$

where $\frac{1}{r^s} = \frac{1}{r} + \frac{s}{n}$ and r_1^s, $r_2^{|s|}$ are defined analogously.

Proof This assertion is an immediate consequence of Theorem 2.4.5 and Theorem 4.2.2(ii).

Remark We followed [HaT2], 2.2. The theorem will be of great utility later on in connection with the spectral theory of degenerate pseudodifferential operators in $\mathbf{R}^n$; see 5.4. As we have remarked, a generalisation of (1) to weighted B_{pq}^s and F_{pq}^s spaces causes no problems, but we shall not need this later on.

4.3 Entropy numbers

4.3.1 A preparation

By Theorem 4.2.3(ii) and Remark 4.2.3 we know under which conditions the embeddings (4.2.3/2) and (4.2.3/9) are compact. Hence it makes sense to enquire into the behaviour of the corresponding entropy numbers (see Definition 1.3.1/1). We are interested in particular in the special situation related to 4.2.4 where $w_1(x) = \langle x \rangle^\alpha$ for some $\alpha > 0$ and $w_2(x) = 1$. We need a preparatory result which might be considered as a complement to 3.3.3, where we calculated the behaviour of the entropy numbers of compact embeddings between B_{pq}^s and F_{pq}^s spaces on bounded domains. Let

$$-\infty < s_2 < s_1 < \infty, \quad p_1, p_2, q_1, q_2 \in (0, \infty], \tag{1}$$

and

$$s_1 - s_2 > n \left(\frac{1}{p_1} - \frac{1}{p_2} \right)_+. \tag{2}$$

Let

$$K_R = \{x \in \mathbf{R}^n : |x| < R\} \tag{3}$$

be the ball of radius R centred at the origin, and let $B_{pq}^s(K_R)$ be the space on $\Omega = K_R$ introduced in Definition 2.5.1/1. Then the embedding

$$\mathrm{id}_R : B_{p_1 q_1}^{s_1}(K_R) \to B_{p_2 q_2}^{s_2}(K_R) \tag{4}$$

is compact. Let e_k^R be the kth entropy number of id_R; we proved in Theorem 3.3.3/2 that there exist two positive numbers $c_1(R)$ and $c_2(R)$ such that

$$c_1(R) k^{-(s_1-s_2)/n} \le e_k^R \le c_2(R) k^{-(s_1-s_2)/n}, \quad k \in \mathbf{N}. \tag{5}$$

Our interest here lies in the dependence of $c_2(R)$ on R if $R \ge 1$. By (2.3.3/1) we immediately have corresponding assertions for the F spaces. In this sense the following proposition also covers (4) and (5) with F in place of B (where $p_1 < \infty$ and $p_2 < \infty$). Usually we do not care about equivalent quasi-norms, but now the situation is different and one has to fix the underlying quasi-norm. First we remark that we can replace $B^s_{pq}(K_R)$, with $R \geq 1$, by the closed subspace

$$\widetilde{B}^s_{pq}(K_R) = \{f \in B^s_{pq}(\mathbf{R}^n) : \operatorname{supp} f \subset \overline{K_R}\} \tag{6}$$

of $B^s_{pq}(\mathbf{R}^n)$. Actually we did so in the proof of Theorem 3.3.2. Now we assume that all the spaces $B^s_{pq}(\mathbf{R}^n)$ are quasi-normed with respect to a fixed dyadic resolution of unity $\{\varphi_k\}_{k=0}^{\infty}$ according to Definition 2.2.1. Naturally this quasi-norm is also taken for all the spaces in (6).

Proposition *Suppose that (1) and (2) hold together with*

$$n\left(\frac{1}{p_1} - 1\right)_+ < s_1 < \frac{n}{p_1}. \tag{7}$$

Then there exists a constant $c > 0$ such that for all $R \geq 1$,

$$c_2(R) \leq cR^\delta \text{ with } \delta = s_1 - \frac{n}{p_1} - \left(s_2 - \frac{n}{p_2}\right), \tag{8}$$

where $c_2(R)$ is the constant on the right-hand side of (5).

Proof Step 1 Let

$$D_R : f(x) \longmapsto f(Rx), \quad R > 0, \tag{9}$$

be the dilation operator in $\mathbf{R}^n$. Let id^R be given by (4) with $\widetilde{B}$ instead of B (as we remarked above it is sufficient to prove (8) for that case). Then we have

$$\operatorname{id}^R = D_{R^{-1}} \circ \operatorname{id}^1 \circ D_R. \tag{10}$$

Suppose that $R \geq 1$ and

$$-\infty < s_2 < 0 \leq n\left(\frac{1}{p_1} - 1\right)_+ < s_1. \tag{11}$$

Then by Proposition 2.3.1/1 we have

$$\left\| D_R f \mid B^{s_1}_{p_1 q_1}(\mathbf{R}^n) \right\| \leq cR^{s_1 - \frac{n}{p_1}} \left\| f \mid B^{s_1}_{p_1 q_1}(\mathbf{R}^n) \right\|, \tag{12}$$

and Proposition 2.3.1/2 gives

$$\left\| D_{R^{-1}} f \mid B^{s_2}_{p_2 q_2}(\mathbf{R}^n) \right\| \leq cR^{-s_2 + \frac{n}{p_2}} \left\| f \mid B^{s_2}_{p_2 q_2}(\mathbf{R}^n) \right\|. \tag{13}$$

Now (8) follows from (10), (12), (13) and (5) with $R = 1$.

Step 2 Let

$$-\infty < s_2 < s_1, \; s_1 > n\left(\frac{1}{p_1} - 1\right)_+ \quad \text{and } p_1 = p_2. \tag{14}$$

We choose s_3 with $s_3 < \min(s_2, 0)$ and apply the well-known real interpolation formula

$$\left(B^{s_1}_{p_1 q_1}, B^{s_3}_{p_1 q_1}\right)_{\theta, q_2} = B^{s_2}_{p_1 q_2}, \; s_2 = (1 - \theta)s_1 + \theta s_3, \tag{15}$$

for the B spaces on $\mathbf{R}^n$ (now we return to our notational agreement to omit "$\mathbf{R}^n$" for spaces on $\mathbf{R}^n$); see [Triβ], 2.4.2, p. 64. We apply Theorem 1.3.2(i) with

$$A = B_0 = \widetilde{B}^{s_1}_{p_1 q_1}(K_R), \; B_1 = \widetilde{B}^{s_3}_{p_1 q_1}(K_R) \text{ and } B_\theta = \widetilde{B}^{s_2}_{p_1 q_2}(K_R), \tag{16}$$

where (1.3.2/2) follows from (15), perhaps with a constant independent of R. By Step 1 with $p_1 = p_2$ and s_3 in (11) in place of s_2 we obtain (8) for the special case

$$\mathrm{id}^{R,p} : \widetilde{B}^{s_1}_{p_1 q_1}(K_R) \to \widetilde{B}^{s_2}_{p_1 q_2}(K_R). \tag{17}$$

Step 3 Now let s_1, s_2, p_1 and p_2 be restricted by (1), (2) and (7), and let $q \in (0, \infty]$. Then we find numbers $s_3 < 0$ and $p_3 \in (0, \infty]$ such that for some $\theta \in (0, 1)$,

$$s_2 = (1 - \theta)s_1 + \theta s_3, \; \frac{1}{p_2} = \frac{1 - \theta}{p_1} + \frac{\theta}{p_3}, \; s_1 - s_3 > n\left(\frac{1}{p_1} - \frac{1}{p_3}\right)_+; \tag{18}$$

see Fig. 4.1.

By Definition 2.2.1(i) and Hölder's inequality we have

$$\left\| f \mid B^{s_2}_{p_2 q} \right\| \le \left\| f \mid B^{s_1}_{p_1 q} \right\|^{1-\theta} \left\| f \mid B^{s_3}_{p_3 q} \right\|^{\theta}; \tag{19}$$

here we used the fact that the parameter q is fixed. Next we again apply Step 1 with p_3 and s_3 in place of p_2 and s_2 (see (11)), as well as (19) and Theorem 1.3.2(i) and obtain (8) for the special case

$$\mathrm{id}^R_q : \widetilde{B}^{s_1}_{p_1 q}(K_R) \to \widetilde{B}^{s_2}_{p_2 q}(K_R). \tag{20}$$

Step 4 The general case now follows from the special cases (17) and (20), together with the multiplicativity of the entropy numbers

$$e_{k_1 + k_2 - 1}\left(\mathrm{id}^R_q \circ \mathrm{id}^{R,p}\right) \le e_{k_1}\left(\mathrm{id}^R_q\right) e_{k_2}\left(\mathrm{id}^{R,p}\right) \tag{21}$$

if the parameters are chosen in a suitable way. So far as (21) is concerned we refer to Lemma 1.3.1/1(iii).

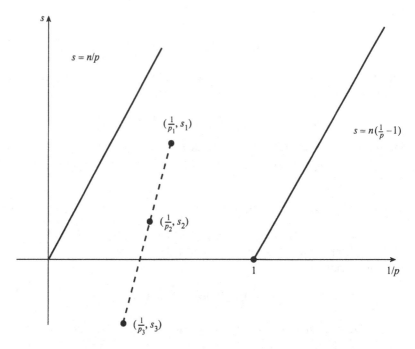

Fig. 4.1.

Remark We proved a little more than explicitly stated in the proposition. The restriction $s_1 < \frac{n}{p_1}$ was needed for the first time in Step 3 in connection with (18) and (19), while in Steps 1 and 2 we used only (11) and (14), respectively. There is hardly any doubt that the proposition remains valid without the technical assumption (7). But this is not important for us. We use the proposition later on in connection with the weighted spaces $B_{pq}^s(\mathbf{R}^n, w(\cdot))$ and $F_{pq}^s(\mathbf{R}^n, w(\cdot))$. Then we have the lifts I_σ and even $I_{\sigma,v}$ according to Proposition 4.2.2 to hand, and to study compact embeddings between these spaces we may always assume, without loss of generality, that (7) holds. As mentioned above, the proposition also applies to the F spaces.

4.3.2 The main theorem

Chapter 3 dealt with entropy and approximation numbers of compact embeddings between spaces of B_{pq}^s and F_{pq}^s type on bounded domains. We are now interested in corresponding assertions for the weighted spaces of this type on $\mathbf{R}^n$ considered in this chapter, and concentrate on entropy numbers. The abstract background is again Chapter 1, and especially 1.3. In particular, the entropy numbers are introduced in Definition 1.3.1/1. The spaces $B_{pq}^s(\alpha)$ and $F_{pq}^s(\alpha)$ with $\alpha \in \mathbf{R}$ have the same meaning as in Definition 4.2.1/2, complemented by the notational agreement at the end of 4.2.1. As above, if $\alpha = 0$ we write simply B_{pq}^s and F_{pq}^s, these being the unweighted spaces on $\mathbf{R}^n$ according to Definition 2.2.1. Our main theme is the study of the entropy numbers of the compact embeddings

$$\mathrm{id}^B : B_{p_1 q_1}^{s_1}(\alpha) \to B_{p_2 q_2}^{s_2}, \ \alpha > 0, \tag{1}$$

and

$$\mathrm{id}^F : F_{p_1 q_1}^{s_1}(\alpha) \to F_{p_2 q_2}^{s_2}, \ \alpha > 0. \tag{2}$$

Theorems 4.2.3 and 4.2.4 give the general basis for what follows. Furthermore, by Theorem 4.2.2(ii) assertions about the entropy numbers of id^B and id^F can be carried over immediately to the embeddings

$$B_{p_1 q_1}^{s_1}\left(w(\cdot)\langle\cdot\rangle^\alpha\right) \to B_{p_2 q_2}^{s_2}(w(\cdot)), \ w \in W^n, \ \alpha > 0, \tag{3}$$

and

$$F_{p_1 q_1}^{s_1}\left(w(\cdot)\langle\cdot\rangle^\alpha\right) \to F_{p_2 q_2}^{s_2}(w(\cdot)), \ w \in W^n, \ \alpha > 0. \tag{4}$$

Of course, the B spaces and F spaces in (1) and (2) can be mixed, either simply by the elementary embedding (2.3.3/1) or by the arguments below. We confine ourselves to (1) and (2). Moreover, with the exception of the interesting limiting case "L" in the notation below, it turns out that the q-indices in (1) and (2) do not play any role, just as in the unweighted case in 3.3. Naturally preference should be given to (1), and (2) can be obtained afterwards via (2.3.3/1) and their obvious counterparts. Let

$$s_1 \in \mathbf{R}, \ p_1, q_1 \in (0, \infty], \ \text{and} \ \alpha > 0. \tag{5}$$

In accordance with Theorem 4.2.4 we introduce p_0 by

$$\frac{1}{p_0} = \frac{1}{p_1} + \frac{\alpha}{n}. \tag{6}$$

As for the target spaces in (1), Theorem 4.2.4 suggests the extension of the restrictions in Theorem 4.2.3 and Remark 4.2.3 by

$$-\infty < s_2 < s_1 < \infty, \ p_0 < p_2 \le \infty, \ 0 < q_2 \le \infty \tag{7}$$

and

$$\delta = s_1 - \frac{n}{p_1} - \left(s_2 - \frac{n}{p_2}\right) > 0. \tag{8}$$

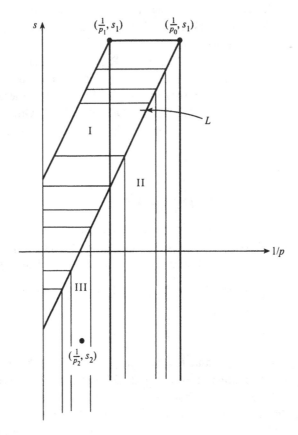

Fig. 4.2. I $: 0 < \delta < \alpha$. II $: \delta > \alpha$, $1/p_1 \leqslant 1/p_2 < 1/p_0$. III $: \delta > \alpha$, $1/p_2 < 1/p_1$. $L : \delta = \alpha, 1/p_2 < 1/p_0$.

In agreement with Theorem 4.2.3(ii), Remark 4.2.3 and Theorem 4.2.4 we divide the region of compact embeddings given by (1) and (2) into four parts; see also Fig. 4.2.

To indicate whether (1) or (2) is considered we let $e_k\left(\mathrm{id}^B\right)$ and $e_k\left(\mathrm{id}^F\right)$ stand for the respective entropy numbers. We write $e_k \sim k^{-\rho}$ for some $\rho > 0$ if there exist two positive numbers c_1 and c_2 such that

$$c_1 k^{-\rho} \le e_k \le c_2 k^{-\rho} \text{ for all } k \in \mathbf{N}. \tag{9}$$

Theorem *Let (5), (6) and (7) be satisfied (in the F case with $p_1 < \infty$ and $p_2 < \infty$).*

(i) *In region I,*

$$e_k \left(\mathrm{id}^B \right) \sim k^{-(s_1-s_2)/n}. \tag{10}$$

(ii) *In region II,*

$$e_k \left(\mathrm{id}^B \right) \sim k^{-\frac{\alpha}{n}+\frac{1}{p_2}-\frac{1}{p_1}}. \tag{11}$$

(iii) *There exist a constant $c > 0$ and, for any $\varepsilon > 0$, a constant $c_\varepsilon > 0$ such that in region III,*

$$ck^{-\frac{\alpha}{n}+\frac{1}{p_2}-\frac{1}{p_1}} \leq e_k \left(\mathrm{id}^B \right)$$
$$\leq c_\varepsilon k^{-\frac{\alpha}{n}+\frac{1}{p_2}-\frac{1}{p_1}} (\log k)^{\varepsilon-\frac{1}{p_2}+\frac{1}{p_1}} \quad \text{for all } k \in \mathbf{N}, k > 1. \tag{12}$$

(iv)$_F$ *There exists a constant $c > 0$ such that on the line L where $\delta = \alpha$ and $\frac{1}{p_2} < \frac{1}{p_0}$,*

$$e_k \left(\mathrm{id}^F \right) \geq ck^{-(s_1-s_2)/n} (\log k)^{\alpha/n} \quad \text{for all } k \in \mathbf{N}, k > 1. \tag{13}$$

(iv)$_B$ *Suppose in addition that either*

$$p_2 = q_2 = \infty \tag{14}$$

or

$$p_2 < \infty \text{ and } q_2 \geq p_2 q_1/p_0. \tag{15}$$

Then there exist constants $c_1 > 0$ and $c_2 > 0$ such that on the line L where $\delta = \alpha$ and $\frac{1}{p_2} < \frac{1}{p_0}$,

$$c_1 k^{-(s_1-s_2)/n} \leq e_k \left(\mathrm{id}^B \right) \leq c_2 \left(\frac{k}{\log k} \right)^{-(s_1-s_2)/n} \quad \text{for all } k \in \mathbf{N}, k > 1. \tag{16}$$

Proof Step 1 In this step and in the following one we establish the estimates from below. In the annuli

$$\mathscr{A}_m = \left\{ x \in \mathbf{R}^n : 2^{m-1} \leq |x| \leq 2^{m+1} \right\}, \quad m \in \mathbf{N}, \tag{17}$$

we have $\langle x \rangle^\alpha \sim 2^{m\alpha}$. We apply Theorem 2.3.2 and modify (2.3.2/2) by

$$f_m^j(x) = \sum\nolimits^{j,m} a_k f \left(2^j x - k \right) \tag{18}$$

where f has the properties required in Theorem 2.3.2 and $\sum^{j,m}$ stands for the sum over all lattice points $k \in \mathbf{Z}^n$ with $2^{-j}k \in \mathscr{A}_m$ which correspond to the "centres" of the shifted functions $f\left(2^j x - k\right)$. Let A_{pq}^s be either B_{pq}^s or F_{pq}^s; then by Theorem 2.3.2 we have

$$\left\| f_m^j \mid A_{p_1 q_1}^{s_1} (\alpha) \right\| \sim 2^{m\alpha} 2^{j\left(s_1 - \frac{n}{p_1}\right)} \left(\sum {}^{j,m} |a_k|^{p_1}\right)^{1/p_1} \tag{19}$$

and

$$\left\| f_m^j \mid A_{p_2 q_2}^{s_2} \right\| \sim 2^{j\left(s_2 - \frac{n}{p_2}\right)} \left(\sum {}^{j,m} |a_k|^{p_2}\right)^{1/p_2}. \tag{20}$$

Put $N_l = 2^{nl}$; then we may assume that the number of admitted lattice points in (19) and (20) is N_{j+m} (neglecting constants). Let span_m^j be the subspace spanned by the functions $f\left(2^j x - k\right)$ for fixed j and m. Let the unit ball U in $A_{p_1 q_1}^{s_1} (\alpha)$ be covered by $2^{N_{j+m}}$ balls in $A_{p_2 q_2}^{s_2}$ of radius $2e_{N_{j+m}}$ (recall the definition of entropy numbers given in Definition 1.3.1/1). Let K be one of these balls, let span_m^j be equipped with a Lebesgue measure and let U_p^N be the unit ball in l_p^N. Then by (20) we have

$$\mathrm{vol}\left(K \cap \mathrm{span}_m^j\right) \le \left[ce_{N_{j+m}} \, 2^{-j\left(s_2 - \frac{n}{p_2}\right)}\right]^{N_{j+m}} \mathrm{vol}\left(U_{p_2}^{N_{j+m}}\right), \tag{21}$$

where c is an appropriate constant. The sum of all these volumes, which can be estimated from above by the right-hand side of (21) times $2^{N_{j+m}}$, must be larger than

$$\mathrm{vol}\left(U \cap \mathrm{span}_m^j\right) \sim \left[2^{-m\alpha} 2^{-j\left(s_1 - \frac{n}{p_1}\right)}\right]^{N_{j+m}} \mathrm{vol}\left(U_{p_1}^{N_{j+m}}\right). \tag{22}$$

Then by (21) and (22) we have

$$e_{N_{j+m}} \ge c 2^{-j\delta} 2^{-m\alpha} \left(\frac{\mathrm{vol}\left(U_{p_1}^{N_{j+m}}\right)}{\mathrm{vol}\left(U_{p_2}^{N_{j+m}}\right)}\right)^{N_{j+m}^{-1}}. \tag{23}$$

By (3.3.3/23) the last factor in (23) is equivalent to $N_{j+m}^{\frac{1}{p_2} - \frac{1}{p_1}}$, and hence is equivalent to $2^{n(j+m)\left(\frac{1}{p_2} - \frac{1}{p_1}\right)}$. Then we obtain

$$e_{2^{n(j+m)}} \ge c 2^{-j(s_1 - s_2)} 2^{-m\left(\alpha - \frac{n}{p_2} + \frac{n}{p_1}\right)}, \quad j \in \mathbb{N}, m \in \mathbb{N}, \tag{24}$$

for some $c > 0$. Choosing $m = 1$ we then obtain the estimates from below in (10) and (16). The choice $j = 1$ gives the estimates from below in (11) and (12).

Step 2 So far as estimates from below are concerned it remains to prove (13). We fix $M \in \mathbb{N}$, and construct in each annulus $\mathscr{A}_m$ the functions

f_m^j given by (18) with $j = M - m$ and respective coefficients a_k, where $m = 1, ..., M$. Then we have in each annulus $N_M = 2^{nM}$ admitted lattice points, and hence altogether $M2^{nM} = N^{(M)}$ lattice points (again neglecting constants). We have the counterpart of (19) and (20) now with F instead of A. Since $j = M - m$ and $\alpha = \delta$ the exponent in (19) is given by

$$M\alpha + (M - m)\left(s_1 - \frac{n}{p_1} - \delta\right) = M\alpha + (M - m)\left(s_2 - \frac{n}{p_2}\right). \qquad (25)$$

In each annulus $\mathscr{A}_m$ we may apply the localisation principle for F spaces described in Step 1 of the proof of Proposition 4.2.2. Then we have, with the modified coefficients $b_k = 2^{(M-m)\left(s_2 - \frac{n}{p_2}\right)} a_k$, the following counterparts of (19) and (20):

$$\left\|\sum_{m=1}^{M} f_m^{M-m} \mid F_{p_1 q_1}^{s_1}(\alpha)\right\| \sim 2^{M\alpha} \left(\sum |b_k|^{p_1}\right)^{1/p_1} \qquad (26)$$

and

$$\left\|\sum_{m=1}^{M} f_m^{M-m} \mid F_{p_2 q_2}^{s_2}\right\| \sim \left(\sum |b_k|^{p_2}\right)^{1/p_2}, \qquad (27)$$

where $\sum$ is taken over $N^{(M)} = M2^{nM}$ summands. Now we are in the same situation as in Step 1 and have the following counterpart of (23):

$$\begin{aligned}
e_{N^{(M)}} &\geq c_1 2^{-M\alpha} \left(\frac{\mathrm{vol}\, U_{p_1}^{N^{(M)}}}{\mathrm{vol}\, U_{p_2}^{N^{(M)}}}\right)^{(N^{(M)})^{-1}} \\
&\geq c_2 2^{-M\delta} M^{\frac{1}{p_2} - \frac{1}{p_1}} 2^{nM\left(\frac{1}{p_2} - \frac{1}{p_1}\right)} \\
&= c_2 2^{-M(s_1 - s_2)} M^{\frac{1}{p_2} - \frac{1}{p_1}}, \qquad (28)
\end{aligned}$$

using the fact that $\alpha = \delta$ and again (3.3.3/23). Let $M2^{nM} \sim 2^K$; then $K \sim M$ and (28) can be rewritten as

$$e_{2^K} \geq c2^{-K(s_1 - s_2)/n} K^{\frac{s_1 - s_2}{n} + \frac{1}{p_2} - \frac{1}{p_1}} = c2^{-K(s_1 - s_2)/n} K^{\delta/n}. \qquad (29)$$

With $\alpha = \delta$ we obtain (13). The proof of part (iv)$_F$ is complete.

Step 3 In the following steps we prove the estimates from above. In particular, in this step and in the following one we establish (10). Hence we have $0 < \delta < \alpha$, and we assume in this step that $\frac{1}{p_2} \leq \frac{1}{p_1}$; compare with Fig. 4.2. Then we have the unweighted embedding in $\mathbf{R}^n$:

$$B_{p_1 q_1}^{s_1} \subset B_{p_2 q_2}^{s_2}. \qquad (30)$$

We complement the annuli $\mathscr{A}_m$ defined in (17) by $\mathscr{A}_0 = \{x \in \mathbf{R}^n : |x| \leq 2\}$ and $\mathscr{A}^m = \{x \in \mathbf{R}^n : |x| > 2^m\}$. Let $B_{pq}^s(\mathscr{A}_j)$ be the B_{pq}^s spaces on $\Omega = \mathscr{A}_j$ according to Definition 2.5.1/1. These spaces should not be confused with the weighted spaces of type $B_{pq}^s(\alpha)$ and $B_{pq}^s(w(\cdot))$ on $\mathbf{R}^n$; see (1) or (3). Let $e_{k,j}$ (with $j = 0, ..., m$ and $k \in \mathbf{N}$) be the entropy numbers of the embeddings

$$\mathrm{id}_j : B_{p_1 q_1}^{s_1}(\mathscr{A}_j) \to B_{p_2 q_2}^{s_2}(\mathscr{A}_j) ; \tag{31}$$

compare with (4.3.1/4). Let $\lambda = \min(1, p_2, q_2)$; then $B_{p_2 q_2}^{s_2}$ is a λ Banach space and we obtain by Lemma 1.3.1/1(iii) and the usual technique of the resolution of unity based on $\{\mathscr{A}_j\}_{j=0}^m$ and $\mathscr{A}^m$,

$$e_k^\lambda(\mathrm{id}^B) \leq c \sum_{j=0}^m 2^{-j\alpha\lambda} e_{k_j,j}^\lambda + c2^{-m\alpha\lambda}, \quad k = \sum_{j=0}^m k_j, \tag{32}$$

where the last term comes from (30). Let I_σ be the lift defined by (2.2.2/4). Then I_σ is also a lift in $B_{pq}^s(\alpha)$; see Proposition 4.2.2. In particular, we may assume without loss of generality that s_1 is restricted by (4.3.1/7) if $p_1 < \infty$, and $s_1 > 0$ if $p_1 = \infty$. Then by Proposition 4.3.1, complemented by Remark 4.3.1 if $p_1 = \infty$, (4.3.1/5) and (32),

$$e_k^\lambda(\mathrm{id}^B) \leq c \sum_{j=0}^m 2^{j(\delta-\alpha)\lambda} k_j^{-\lambda(s_1-s_2)/n} + c2^{-m\alpha\lambda}. \tag{33}$$

Choose $k_j = 2^{b(m-j)}$ for some $b > 0$. Then we have $k \sim 2^{bm}$ and since $\delta < \alpha$,

$$\begin{aligned} e_k^\lambda(\mathrm{id}^B) &\leq c_1 2^{-\lambda bm(s_1-s_2)/n} \sum_{j=0}^m 2^{j\left(\delta-\alpha+b\frac{s_1-s_2}{n}\right)\lambda} + c_1 2^{-m\alpha\lambda} \\ &\leq c_2 2^{-\lambda bm(s_1-s_2)/n} + c_2 2^{-m\alpha\lambda} \\ &\leq c_3 2^{-\lambda bm(s_1-s_2)/n} \end{aligned} \tag{34}$$

if $b > 0$ is chosen sufficiently small. Since $k \sim 2^{bm}$ we obtain

$$e_k(\mathrm{id}^B) \leq ck^{-(s_1-s_2)/n}, \quad k \in \mathbf{N}. \tag{35}$$

This completes the proof of (10), provided that $\frac{1}{p_2} \leq \frac{1}{p_1}$.

Step 4 Again let $0 < \delta < \alpha$ and suppose in this step that $\frac{1}{p_1} < \frac{1}{p_2} < \frac{1}{p_0}$; compare with Fig. 4.2. We prove (35). By the same arguments as in the proof of Theorem 4.2.4 we have

$$B_{p_1 q_1}^{s_1}(\alpha) \subset B_{p\infty}^{s_1} \quad \text{if} \quad \frac{1}{p_1} < \frac{1}{p} < \frac{1}{p_0}, \tag{36}$$

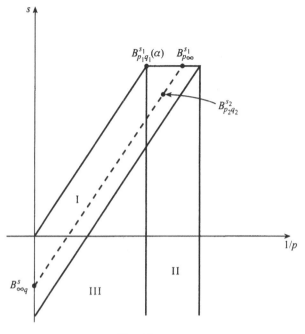

Fig. 4.3.

where the spaces on the right-hand side of (36) correspond to the upper line of region I in Fig. 4.2. Since $\left(\frac{1}{p_2}, s_2\right)$ also belongs to region I we find numbers θ, p, q and s with

$$0 < \theta < 1, \quad \frac{1}{p_1} < \frac{1}{p} < \frac{1}{p_0}, \quad 0 < q \le \infty, \quad s \in \mathbf{R}, \tag{37}$$

$$s_2 = (1 - \theta)s + \theta s_1, \quad \frac{1}{p_2} = \frac{\theta}{p}, \quad \frac{1}{q_2} = \frac{1 - \theta}{q}, \tag{38}$$

and

$$\delta = s_2 - \frac{n}{p_2} = s = s_1 - \frac{n}{p}, \tag{39}$$

which corresponds to the broken line in Fig. 4.3.

By Definition 2.2.1(i) and Hölder's inequality we have

$$\left\| f \mid B_{p_2 q_2}^{s_2} \right\| \le \left\| f \mid B_{\infty q}^{s} \right\|^{1-\theta} \left\| f \mid B_{p\infty}^{s_1} \right\|^{\theta} \tag{40}$$

along the broken line in Fig. 4.3 between the indicated end-points $B_{\infty q}^{s}$

and $B^{s_1}_{p\infty}$. By (35) we have

$$e_k \left(\mathrm{id}^B : B^{s_1}_{p_1 q_1} (\alpha) \to B^s_{\infty q} \right) \leq ck^{-(s_1-s)/n}, \ k \in \mathbf{N}. \tag{41}$$

From Theorem 1.3.2(i), (40), (36) and (41) we obtain

$$e_k \left(\mathrm{id}^B : B^{s_1}_{p_1 q_1} (\alpha) \to B^{s_2}_{p_2 q_2} \right) \leq ce_k \left(\mathrm{id}^B : B^{s_1}_{p_1 q_1} (\alpha) \to B^s_{\infty q} \right)^{1-\theta}$$
$$\leq c'k^{-(s_1-s_2)/n}, \ k \in \mathbf{N}. \tag{42}$$

The proof of (10) is complete.

Step 5 In this step and in the following one we prove the estimate from above in (11). First we deal with the special case $p_1 = p_2$ which corresponds to that part of the vertical line in Fig. 4.2 with $\left(\frac{1}{p_1}, s_1 \right)$ as endpoint, where

$$\delta = s_1 - s_2 > \alpha. \tag{43}$$

Let $0 < \varepsilon < \delta - \alpha$ and $k_j = 2^{mn(\alpha+\varepsilon)/\delta} 2^{jn(\delta-\alpha-\varepsilon)/\delta}$. Then by (32) and (33) we find that $k \sim 2^{nm}$ and

$$e^\lambda_k \left(\mathrm{id}^B \right) \leq c_1 \sum_{j=0}^m 2^{j(\delta-\alpha)\lambda} 2^{-\lambda m(\alpha+\varepsilon)} 2^{-\lambda j(\delta-\alpha-\varepsilon)} + c_1 2^{-\lambda m \alpha}$$
$$\leq c_2 2^{-\lambda m \alpha}. \tag{44}$$

Hence we have the desired estimate

$$e_k \left(\mathrm{id}^B \right) \leq ck^{-\alpha/n}, \ k \in \mathbf{N}. \tag{45}$$

Step 6 Now we prove the upper estimate in (11) for the interior of region II, which means that $\delta > \alpha$, $\frac{1}{p_1} < \frac{1}{p_2} < \frac{1}{p_0}$. By Theorem 4.2.4 we have

$$B^{s_1}_{p_1 q_1} (\alpha) \subset \text{weak-}B^s_{p_0 \infty} \text{ with } s < s_1. \tag{46}$$

Recall the real interpolation formula (4.2.4/5) for the Lorentz–Marcinkiewicz spaces. We are looking for the counterpart of (40) along the line $\alpha < \delta = $ const. with the spaces $B^{s_2}_{p_1 q}$ treated in Step 5 and the weak-$B^s_{p_0 \infty}$ spaces as end points. By Definitions 2.2.1(i) and 4.2.4, the interpolation formula mentioned above and Hölder's inequality we obtain the analogue of (40). From (45) and the analogue of (42) we have

$$e_k \left(\mathrm{id}^B \right) \leq ck^{-\frac{\alpha}{n} + \frac{1}{p_2} - \frac{1}{p_1}}, \ k \in \mathbf{N}. \tag{47}$$

Now (11) follows from (47) and the corresponding estimate from below proved in the first step.

Step 7 In this step and in the following one we prove (iv)$_B$. First we deal with (14), which means

$$\mathrm{id}^B : B^{s_1}_{p_1 q_1}(\alpha) \to \mathscr{C}^{s_2} = B^{s_2}_{\infty\infty}, \quad \alpha = \delta = s_1 - s_2 - \frac{n}{p_1}. \tag{48}$$

As in Step 3 we may assume, without loss of generality, that s_1 is restricted by (4.3.1/7) if $p_1 < \infty$ and $s_1 > 0$ if $p_1 = \infty$, in such a way that Proposition 4.3.1, complemented by Remark 4.3.1 if $p_1 = \infty$, can be applied. The localisation principle described in Step 1 of the proof of Proposition 4.2.2 can also be applied to $\mathscr{C}^{s_2}$; see [Triγ], 2.4.7, p. 124. Then the counterpart of (32) with $k_j = k$ is given by

$$e_{mk}\left(\mathrm{id}^B\right) \le c \sup_{j=0,\ldots,m} 2^{-j\alpha} e_{k,j} + c2^{-m\alpha}. \tag{49}$$

By Proposition 4.3.1 (and Remark 4.3.1 if $p_1 = \infty$) we have $e_{k,j} \le c2^{j\delta}k^{-(s_1-s_2)/n}$, and hence, since $\delta = \alpha$,

$$e_{mk}\left(\mathrm{id}^B\right) \le ck^{-(s_1-s_2)/n} + c2^{-m\alpha}. \tag{50}$$

Choose $k^{(s_1-s_2)/n} \sim 2^{m\alpha}$, so that $m \sim \frac{s_1-s_2}{n\alpha}\log k$. Then we have

$$e_{ck\log k}\left(\mathrm{id}^B\right) \le ck^{-(s_1-s_2)/n}, \quad k = 2, 3, \ldots, \tag{51}$$

which finally results in

$$e_k\left(\mathrm{id}^B\right) \le c\left(\frac{k}{\log k}\right)^{-(s_1-s_2)/n}, \quad k \in \mathbf{N}, k > 1. \tag{52}$$

The proof of (16) under the condition (14) is complete.

Step 8 We prove (16) under the condition (15). For that purpose we again use interpolation. One endpoint is given by (48) with (52). The other endpoint comes from Theorem 4.2.4,

$$\mathrm{id}^B : B^{s_1}_{p_1 q_1}(\alpha) \to \text{weak-}B^{s_1}_{p_0 q_1}, \tag{53}$$

where p_0 is given by (6). In other words, we interpolate between the end points of the line L in Fig. 4.2. Now the situation is similar to that in Step 6. We again apply Theorem 1.3.2(i) and obtain

$$e_k\left(\mathrm{id}^B\right) \le c\left(\frac{k}{\log k}\right)^{-(s_1-s_2)/n}, \quad k \in \mathbf{N}, k > 1, \tag{54}$$

for the embedding

$$\mathrm{id}^B : B^{s_1}_{p_1 q_1}(\alpha) \to B^{s_2}_{p_2 q_2} \tag{55}$$

if

$$\frac{1}{p_2} = \frac{\theta}{p_0} \text{ and } \frac{1}{q_2} = \frac{\theta}{q_1} \text{ for some } \theta \text{ with } 0 < \theta < 1. \tag{56}$$

But this is just the limiting case in (15). Larger values of q_2 can be incorporated afterwards by use of the monotonicity of the B_{pq}^s spaces with respect to the q-index.

Step 9 It remains to prove the right-hand side of (12). Let $\varepsilon > 0$ be given and small. We choose s_3 such that

$$0 < \varepsilon = s_1 - \frac{n}{p_1} - \left(s_3 - \frac{n}{p_2}\right), \tag{57}$$

and temporarily write

$$\mathrm{id}_\varepsilon^B : B_{p_1q_1}^{s_1}(\alpha) \to B_{p_2\infty}^{s_3}(\alpha - \varepsilon) \tag{58}$$

for the natural embedding. By the lift property from (4.2.2/8) and Step 8 we obtain the counterpart of (54), namely

$$e_k\left(\mathrm{id}_\varepsilon^B\right) \le c\left(\frac{k}{\log k}\right)^{-\frac{\varepsilon}{n} - \frac{1}{p_1} + \frac{1}{p_2}}, \ k \in \mathbf{N}, k > 1. \tag{59}$$

Furthermore, for the moment write

$$\mathrm{id}_{\alpha-\varepsilon}^B : B_{p_2\infty}^{s_3}(\alpha - \varepsilon) \to B_{p_2q_2}^{s_2} \tag{60}$$

for the natural embedding. Then we have

$$s_3 - s_2 > \alpha - \varepsilon \tag{61}$$

and by Step 5,

$$e_k\left(\mathrm{id}_{\alpha-\varepsilon}^B\right) \le ck^{-(\alpha-\varepsilon)/n}, \ k \in \mathbf{N}. \tag{62}$$

By Lemma 1.3.1/1(ii) we have

$$e_{2k}\left(\mathrm{id}^B\right) \le e_k\left(\mathrm{id}_{\alpha-\varepsilon}^B\right) e_k\left(\mathrm{id}_\varepsilon^B\right), \ k \in \mathbf{N}. \tag{63}$$

Now the right-hand side of (12) follows from (59), (62) and (63). The proof is complete.

Remark 1 Recall that $e_k\left(\mathrm{id}^B\right)$ in (10), (11) and (12) can be replaced by $e_k\left(\mathrm{id}^F\right)$. This follows simply from (2.3.3/5) and its weighted counterpart. In all cases we indicated in Fig. 4.2 the level lines of the exponents in (10), (11) and (12). As for (12), it is not so clear whether the left-hand side of (12) or the right-hand side with $\varepsilon = 0$ gives the correct behaviour (or something in between).

Remark 2 In contrast to parts (i), (ii) and (iii) of the theorem we have in (iv) separate assertions for the B spaces and the F spaces. To prove (13) we used in Step 2 the localisation principle for F spaces. There is no counterpart for the B spaces with the exception of B_{pp}^s with $0 < p \leq \infty$. Note especially that Step 2 can also be applied if either $p_1 = q_1 = \infty$ and/or $p_2 = q_2 = \infty$. In particular, if we again put

$$B_{\infty\infty}^{s_1}(\alpha) = \mathscr{C}^{s_1}(\alpha), \quad B_{\infty\infty}^{s_2} = \mathscr{C}^{s_2} \tag{64}$$

(see (48)) and

$$\mathrm{id}^{\mathscr{C}} : \mathscr{C}^{s_1}(\alpha) \to \mathscr{C}^{s_2}, \quad \alpha = s_1 - s_2, \tag{65}$$

then we can apply both (13) and (14), (16) to obtain

$$e_k\left(\mathrm{id}^{\mathscr{C}}\right) \sim \left(\frac{k}{\log k}\right)^{-(s_1-s_2)/n}, \quad k \in \mathbf{N}, k > 1, \; \alpha = s_1 - s_2 > 0. \tag{66}$$

Remark 3 The most interesting case of the theorem is the line L, characterised by $\delta = \alpha$. By (13) and the "if"-part of the elementary embedding (2.3.3/5) we have on this line

$$e_k\left(\mathrm{id}^B\right) \geq ck^{-(s_1-s_2)/n}(\log k)^{\alpha/n}, k \in \mathbf{N}, k > 1, \tag{67}$$

if

$$q_1 \geq p_1 \text{ and } q_2 \leq p_2. \tag{68}$$

There are no parameters such that (15) and (68) hold simultaneously. It is somewhat surprising that one can even use $(\mathrm{iv})_F$ and $(\mathrm{iv})_B$ of the theorem to prove the "only if"- part of (2.3.3/5); see [HaT1], 4.3, pp. 143–4. This rather sophisticated assertion has been proved recently; see 2.3.3 for details and references. Accepting the above theorem, the proof in [HaT1] seems to be much simpler.

Remark 4 Recently some remarkable progress concerning estimates and equivalences for the entropy numbers related to the line L has been made by Haroske in her thesis [Har1]. In particular, in some of these limiting cases there is a genuine dependence of $e_k(\mathrm{id}^B)$ in (1) on the q-indices.

4.3.3 Approximation numbers

Recall that Theorem 4.2.3(ii) and Remark 4.2.3 clarify under which conditions the embeddings (4.2.3/2) and (4.2.3/9) are compact. In 4.3.2 we gave a detailed and, with the exception of some limiting cases, rather final description of the behaviour of the entropy numbers e_k of these compact embeddings. In addition to the entropy numbers one can also use the approximation numbers a_k (see Definition 1.3.1/2) to measure compactness. In 3.3.4 we calculated the behaviour of the approximation numbers of compact embeddings between B^s_{pq} and F^s_{pq} spaces on bounded domains. The situation is more complicated than for the entropy numbers; see Theorems 3.3.2 and 3.3.3/1 together with Figs. 3.2 and 3.3 on p. 127. For the weighted spaces now under consideration we have, for the entropy numbers, the situation indicated in Fig. 4.2 on p. 170. Roughly speaking, for the corresponding approximation numbers one has to combine Figs. 3.2 and 3.3 with Fig. 4.2. This gives a strong hint that the outcome is somewhat complicated. A detailed study clarifying many but not all cases has been given by D. Haroske in [Har2]. We describe briefly some of her results and refer for details and proofs to [Har2]. Once more let

$$\mathrm{id}^B : B^{s_1}_{p_1 q_1}(\alpha) \to B^{s_2}_{p_2 q_2}, \quad \alpha > 0, \tag{1}$$

be the embedding operator introduced in (4.3.2/1). Since the indices q_1 and q_2 do not play any role in what follows all assertions also hold for id^F given by (4.3.2/2). Just as in 4.3.2 we assume that

$$-\infty < s_1 < \infty, 0 < p_1 < \infty, 0 < q_1 \le \infty \text{ and } \alpha > 0, \tag{2}$$

$$\frac{1}{p_0} = \frac{1}{p_1} + \frac{\alpha}{n}, \tag{3}$$

$$-\infty < s_2 < s_1 < \infty, \ p_0 < p_2 < \infty, \ 0 < q_2 \le \infty \tag{4}$$

and

$$\delta = s_1 - \frac{n}{p_1} - s_2 + \frac{n}{p_2} > 0, \tag{5}$$

now excluding some limiting cases. As remarked above we have to combine the difficulties reflected by Fig. 4.2 with those portrayed in Figs. 3.2 and 3.3. This means, as is shown in [Har2], that one now has nine regions instead of the three in Fig. 4.2 (not to speak about L). Here we describe only a few typical cases using the same numbers for the regions as in [Har2]. Let p' be given by $\frac{1}{p} + \frac{1}{p'} = 1$ if $1 \le p \le \infty$ and by $p' = \infty$

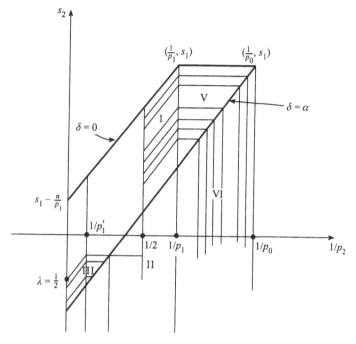

Fig. 4.4. $0 < p_1 < 2, \alpha > \frac{n}{p_1} > 0.$

if $0 < p < 1$. Let the various regions be defined by

I   : $0 < p_1 \le p_2 \le 2$ or $2 \le p_1 \le p_2 < \infty$, $0 < \delta < \alpha$,

II   : $0 < p_1 \le p_2 \le 2$ or $2 \le p_1 \le p_2 < \infty$, $\delta > \alpha$,

III : $0 < p_1 < 2 < p_2 < \infty$, $0 < \delta < \alpha$,

$$\lambda = \frac{s_1 - s_2}{n} - \max \left(\frac{1}{2} - \frac{1}{p_2}, \frac{1}{p_1} - \frac{1}{2} \right) > \frac{1}{2},$$

V   : $p_0 < p_2 \le p_1$, $0 < \delta < \alpha$,

VI   : $p_0 < p_2 \le p_1$, $\delta > \alpha$.

(6)

Figs. 4.4 and 4.5 display two typical cases.
Of course, Fig. 4.5, where $p_1 \ge 2$, is the counterpart of Fig. 4.2. Furthermore, one can imagine that Fig. 4.4, where $p_1 < 2$, is the combination
of Fig. 4.2 and Fig. 3.2 on p. 127. The number λ given by (6) is the

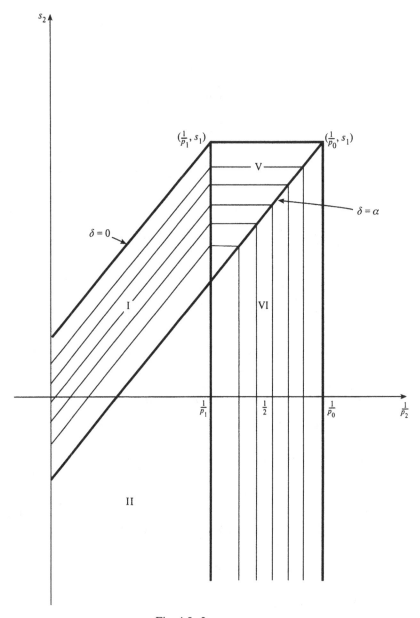

Fig. 4.5. $2 \leq p_1 < \infty$.

same as in (3.3.4/3). We also indicated in Fig. 4.4 the other regions without describing them in detail; see [Har2]. Again we write $a_k \sim k^{-\rho}$ in accordance with what was said just before Theorem 4.3.2 and where the a_k are the approximation numbers of id^B given by (1). As in Fig. 4.2 we have also indicated in Figs. 4.4 and 4.5 the level lines of the respective exponents ρ.

Theorem *Let (2), (3) and (4) be satisfied.*
  (i) In region I,

$$a_k \sim k^{-\delta/n}. \tag{7}$$

  (ii) In region II,

$$a_k \sim k^{-\alpha/n}. \tag{8}$$

  (iii) In region III,

$$a_k \sim k^{-\lambda}, \tag{9}$$

where λ is given by (6).
  (iv) In region V,

$$a_k \sim k^{-(s_1 - s_2)/n}. \tag{10}$$

  (v) In region VI,

$$a_k \sim k^{-\frac{\alpha}{n} + \frac{1}{p_2} - \frac{1}{p_1}}. \tag{11}$$

Remark 1 Proofs and further details may be found in [Har2]. When $p_1 \geq 2$ the results are of a rather final character, save for some limiting cases. The situation is different when $p_1 < 2$; in [Har2] further estimates may be found, but there is no answer comparable with the case $p_1 \geq 2$.

Remark 2 The theorem above should also be compared with Theorem 4.3.2. The behaviour of the entropy numbers e_k and the approximation numbers a_k is the same if $p_0 < p_2 \leq p_1$. But if $p_1 < p_2$ then the behaviour of these two numbers is different. This also sheds new light upon Theorem 1.3.3 which is applicable in the cases treated above. It shows that the non-linear procedure connected with entropy numbers may be much better than the linear procedures related to approximation numbers if the (quasi-)Banach space structure of the spaces involved is different. Even more striking is the different behaviour of entropy and approximation numbers of the limiting embedding in spaces of Orlicz type discussed in 3.4.

5

Elliptic Operators

5.1 Introduction

This chapter deals with the distribution of eigenvalues of degenerate elliptic operators in domains and on $\mathbf{R}^n$. It is based on the results of the previous chapters and demonstrates the symbiotic relationship between the diverse ingredients treated so far:

(i) spectral theory in quasi-Banach spaces, especially the connection between entropy numbers and eigenvalues obtained in 1.3.4;

(ii) some new results in the theory of function spaces, especially the assertions about Hölder inequalities in 2.4;

(iii) sharp estimates of the behaviour of entropy numbers of compact embeddings between function spaces on bounded domains obtained in Chapter 3;

(iv) corresponding assertions for weighted spaces on $\mathbf{R}^n$ described in Chapter 4.

The combination of these ingredients is the basis for the study of the distribution of eigenvalues of degenerate elliptic operators. In 5.2 we concentrate on elliptic operators in bounded smooth domains in non-limiting situations. As a by-product we obtain some results, based on the Birman–Schwinger principle, about the problem of the "negative spectrum" of self-adjoint operators. But we shall be very brief here and defer a detailed study of this topic until 5.4, when we deal with corresponding problems on $\mathbf{R}^n$, which are more natural for problems of the "negative spectrum". In 5.3 we complement the results of 5.2 by the study of limiting situations, again on bounded smooth domains. Finally, 5.4 deals with corresponding problems on $\mathbf{R}^n$, including a more detailed study of the "negative spectrum" of some self-adjoint elliptic operators in $L_2(\mathbf{R}^n)$.

This chapter is largely based on [ET3], [ET4] and [HaT2].

5.2 Elliptic operators in domains: non-limiting cases

5.2.1 Introduction; the Birman–Schwinger principle

As explained in 5.1, our goal here is to study the eigenvalue distribution for elliptic operators in bounded smooth domains. We accomplish this by using the sharp estimates for entropy numbers of embeddings obtained in Chapter 3, together with mapping properties of (pseudo)differential operators and the function space material of Chapter 2. Information about eigenvalue distributions then follows in a strikingly simple manner merely by travelling around $\left(\frac{1}{p}, s\right)$-diagrams for the relevant function spaces and combining the tools mentioned above. This illustrates both the power and the simplicity of our approach to such problems.

The operators handled here typically have the structure

$$B = b_2 C b_1, \tag{1}$$

where b_1 and b_2 are singular functions belonging to some space H_p^s, and C may be the inverse of a general regular elliptic differential operator or a fractional power of it, or an (exotic) pseudodifferential operator. The operator may be non-symmetric and may act in a (quasi-)Banach space rather than a Hilbert space, but despite this our method gives results, often sharp results, for eigenvalue distributions with very little effort: it is simply a matter of putting together in an appropriate manner the results of the earlier chapters.

Some comparison with other methods seems desirable at this point. In the symmetric case in which $b_1 = b_2 = b$ and C is symmetric with respect to, say, the L_2 inner product, then if b and C fit naturally into the Hilbert space techniques used, deep and impressive results have been obtained about the distribution of eigenvalues, now counted with respect to their geometric multiplicity; these results extend the classical theory (see [CoH]) for smooth, self-adjoint elliptic differential operators. For this, we refer especially to Birman and Solomyak [BiS2], [BiS3], [BiS4], [BiS5], Rosenbljum [Ro1], [Ro2] and Tashkian [Tas]. The mapping properties of operators of type $b(X)a(D)$, including the distribution of eigenvalues, have been examined in great detail in a Hilbert space setting, mostly $L_2(\mathbf{R}^n)$: see Birman, Karadzov and Solomyak [BKS], Lieb [Lie] and Simon [Sim1], [Sim2] and the references contained in these works. As would be expected, in these special situations the deep Hilbert space techniques used by the authors listed above often give better results than those obtained by our simple arguments which are not confined to symmetric operators or Hilbert spaces.

As our methods are not especially sensitive to the type of space used, we present the results for the most part in the context of (quasi-)Banach function spaces. The exception to this is the material dealing with the negative spectrum and using an entropy number version of the Birman–Schwinger principle. We now turn to this.

Let A be a self-adjoint operator acting in a Hilbert space $\mathcal{H}$ and suppose that A is positive; that is, $(Af, f) > 0$ for all f in the domain $\mathrm{dom}(A)$ of A, $f \neq 0$. Let V be a closable operator acting in $\mathcal{H}$, suppose that $K : \mathcal{H} \to \mathcal{H}$ is a compact linear operator such that

$$Ku = VA^{-1}V^*u \text{ for all } u \in \mathrm{dom}\left(VA^{-1}V^*\right) \tag{2}$$

and assume that $\mathrm{dom}(A) \cap \mathrm{dom}(V^*V)$ is dense in $\mathcal{H}$. Let $e_k = e_k(K)$ be the kth entropy number of K; if M is a finite set, denote by $\#M$ the number of elements of M. The Birman–Schwinger principle (see Theorem 5.3, p. 193 of [Sche]; see also [Sim2], p. 87. [BiS5], [BiS6] and, of course, the original papers of Birman [Bi] and Schwinger [Schw]) states the following:

Theorem *Under the above conditions, $A - V^*V$ has a self-adjoint extension H, with spectrum $\sigma(H)$, such that*

$$\#\{\sigma(H) \cap (-\infty, 0]\} \leq \#\{\sigma(K) \cap [1, \|K\|]\}. \tag{3}$$

Of course, eigenvalues are counted according to their multiplicities; thus $\sigma(H) \cap (-\infty, 0]$ consists solely of a finite number of eigenvalues of finite multiplicity (if there are any). As an immediate consequence of this and Carl's inequality (1.3.4/11), we have the following entropy version of the principle which is particularly useful for our purposes, the e_k essentially coming from compact embeddings between function spaces.

Corollary *Under the above assumptions,*

$$\#\{\sigma(H) \cap (-\infty, 0]\} \leq \#\left\{k \in \mathbf{N} : \sqrt{2}e_k \geq 1\right\}. \tag{4}$$

5.2.2 Elliptic differential operators: mapping properties

Here Ω will be a bounded domain in $\mathbf{R}^n$ with C^∞ boundary. We shall collect the results originating in the work of Agmon, Douglis and Nirenberg [ADN] and Agmon [Ag1], [Ag2], and follow the presentation given in [Triα], 4.9.1, p. 334 and Chapter 5, p. 361, referring for details also to [Triβ], Chapter 4, p. 230.

Let A be a properly elliptic operator,

$$Af = \sum_{|\alpha| \leq 2m} a_\alpha(x) D^\alpha f, \quad \text{where each } a_\alpha \in C^\infty\left(\overline{\Omega}\right), \tag{1}$$

and suppose there are boundary operators

$$B_j f = \sum_{|\alpha| \leq l_j} b_{j,\alpha}(x) D^\alpha f, \quad \text{where each } b_{j,\alpha} \in C^\infty(\partial\Omega), \tag{2}$$

with $j = 1, ..., m$ and $0 \leq l_1 < ... < l_m \leq 2m - 1$, which form a normal system satisfying the complementing condition. Then $\{A; B_1, ..., B_m\}$ is called a regular elliptic system; see [Triβ], 4.1.2, p. 213 for details. As in [Triβ], p. 231, we assume that the problem

$$\left.\begin{array}{rl} Af &= 0 \text{ in } \Omega, \\[2mm] B_j f &= 0 \text{ on } \partial\Omega, \ j = 1, ..., m, \end{array}\right\} \tag{3}$$

has only the trivial C^∞ solution. The basic result that we need is the following:

Theorem *Suppose that*

$$\left(\frac{1}{p}, s\right) \in G_3 := \left\{ \left(\frac{1}{p_1}, s_1\right) : 0 < p_1 < \infty, \ s_1 \geq n\left(\frac{1}{p_1} - 1\right)_+ \right\}. \tag{4}$$

Then under the above assumptions, A maps

$$\left\{ f \in H_p^{s+2m}(\Omega) : B_j f = 0 \text{ on } \partial\Omega \text{ for } j = 1, ..., m \right\} \tag{5}$$

isomorphically onto $H_p^s(\Omega)$.

Remark 1 Let $A_{s,p}$ be the operator A with domain of definition (5). Then (3) ensures that 0 does not belong to the spectrum $\sigma\left(A_{s,p}\right)$ of $A_{s,p}$, and consequently $\sigma\left(A_{s,p}\right)$ consists of isolated eigenvalues of finite algebraic multiplicity. In this generality the theorem is due to Franke and Runst [FraR]. For its classical part, that is, when $1 < p < \infty$, see Agmon [Ag1]. A partial extension from $1 < p < \infty$ to $p \leq 1$ may also be found in

[Triβ], Chapter 4, p. 233. We also refer to the recent paper by Johnsen [Jo].

Duality Let $1 < p < \infty$. Then in the sense of the dual pairing $(L_p, L_{p'})$, or simply duality theory in L_2 for unbounded operators, the dual $(A_p)'$ of $A_p = A_{0,p}$, denoted by $A'_{p'}$, is again generated by a regular elliptic system $\{A'; C_1, ..., C_m\}$ with

$$A'f = \sum_{|\alpha| \leq 2m} (-1)^{|\alpha|} D^\alpha (\bar{a}_\alpha f);$$

see [Triα], p. 381 for a short description and [LiM] for details.

Fractional powers These will be needed later on in the course of splitting arguments for some more complicated operators, especially in the limiting situations to be discussed in 5.3. As in the case of duality, we shall confine ourselves to the ground level $s = 0$ and $1 < p < \infty$. The basic theory of fractional powers is given in [Triα], 1.15, p. 98, and its application to A_p and $A'_{p'}$ is based on [Triα], 4.9.1, p. 334. Then the fractional powers A_p^κ can be constructed for every $\kappa \in \mathbf{R}$ as shown in [Triα], 1.15: for this construction one needs only integer powers of A_p and of the related resolvents, the latter being expressed in the sense of Agmon, Douglis and Nirenberg via Poisson kernels which have nothing to do with the chosen p. Hence the related formulae for A_p^κ are also independent of p. In view of these observations, the fact that $0 \notin \sigma(A_p)$, and standard Banach space arguments, we have

$$\left(A_p^\kappa\right)' = \left(A'_{p'}\right)^\kappa \text{ for } \kappa \in \mathbf{R}, \ 1 < p < \infty. \tag{6}$$

We may restrict ourselves to $|\kappa| \leq 1$. Note that if $-1 \leq \kappa < 0$, then A_p^κ is compact in L_p.

Proposition *There is a constant $c > 0$ such that for all $k \in \mathbf{N}$ and every $\kappa \in [-1, 0)$,*

$$e_k\left(A_p^\kappa\right) \leq ck^{2m\kappa/n}. \tag{7}$$

Proof The domain of A_p is given by (5) with $s = 0$. Thus by the theorem above and Theorem 3.3.3/2,

$$e_k\left(A_p^{-1}\right) \sim k^{-2m/n}; \tag{8}$$

we recall that statements of the form $b_k \sim d_k$ mean that $c_1 b_k \leq d_k \leq c_2 b_k$, $k \in \mathbf{N}$, where c_1, c_2 are positive constants independent of k. Now (7) follows from the interpolation properties of dom $\left(A_p^{|\kappa|} \right)$ (see [Triα], 1.15, especially 1.15.3, p. 103) and of the entropy numbers (see Theorem 1.3.2).

Remark 2 The upper estimate (7) will be sufficient for our purposes, but in fact it can be improved to

$$e_k \left(A_p^\kappa \right) \sim k^{2m\kappa/n}, \ k \in \mathbf{N}, -1 \leq \kappa < 0. \tag{9}$$

We outline the proof of this. By [Triα], 1.15.3, p. 103 and [Triα], 4.3.3, p. 320,

$$\operatorname{dom}\left(A_p^{|\kappa|} \right) \subset H_p^{2m|\kappa|}(\Omega), \ |\kappa| \leq 1, \tag{10}$$

where with the exception of those κ such that $2m|\kappa| - 1/p = l_j$ for some $j \in \{1, ..., m\}$, dom $\left(A_p^{|\kappa|} \right)$ is even a closed subspace of $H_p^{2m|\kappa|}(\Omega)$. Thus the above proof gives (9) with the possible exception of these special, isolated values of κ. Now assume that

$$e_{k_j} \left(A_p^{\kappa_0} \right) k_j^{-2m\kappa_0/n} \to 0 \tag{11}$$

for some sequence $\{k_j\}$ of natural numbers tending to ∞ and one of these exceptional values of κ_0, $\kappa_0 \neq 0$. Apply once more the interpolation argument for entropy numbers: hence the analogue of (11) holds with κ_0 replaced by any κ with $0 < |\kappa| \leq |\kappa_0|$, and so we have a contradiction. This proves (9). Note the interesting conclusion that

$$e_k \left(A_p^\kappa \right) \sim e_k \left(\left(A_p^\kappa \right)' \right) = e_k \left(A_{p'}^{'\kappa} \right) \sim k^{2m\kappa/n}, \ -1 \leq \kappa < 0. \tag{12}$$

As we know from 1.3.1, general assertions of this type are known for Hilbert spaces but not in general Banach spaces; see also Remark 1.3.1/5.

5.2.3 Pseudodifferential operators: mapping properties

We first recall the basic definitions of these operators. Let $\kappa \in \mathbf{R}$ and $\gamma \in [0, 1]$; then the Hörmander class $S_{1,\gamma}^\kappa$ consists of all complex-valued C^∞ functions $(x, \xi) \longmapsto a(x, \xi)$ on $\mathbf{R}^n \times \mathbf{R}^n$ such that for all multi-indices α, β there is a positive constant $c_{\alpha,\beta}$ with

$$\left| D_\xi^\alpha D_x^\beta a(x, \xi) \right| \leq c_{\alpha,\beta} \langle \xi \rangle^{\kappa - |\alpha| + \gamma|\beta|} \tag{1}$$

for all $x, \xi \in \mathbf{R}^n$. Here D_ξ^α and D_x^β refer to derivatives with respect to ξ and x respectively; also $\langle \xi \rangle = \left(1 + |\xi|^2 \right)^{1/2}$. Given any such function (or symbol) a, the corresponding pseudodifferential operator $a(x,D)$ is defined by

$$a(x,D)f(x) = \int_{\mathbf{R}^n} e^{ix\cdot\xi} a(x,\xi) \hat{f}(\xi)\,d\xi$$

$$= (2\pi)^{-n/2} \int_{\mathbf{R}^{2n}} e^{i(x-y)\cdot\xi} a(x,\xi) f(y)\,dy\,d\xi$$

$$= \int_{\mathbf{R}^n} \mathscr{K}(x,y) f(y)\,dy, \quad f \in \mathscr{S}', \tag{2}$$

where the final inequality is the kernel representation in the sense of Schwartz, with $\mathscr{K} \in \mathscr{D}'(\mathbf{R}^n \times \mathbf{R}^n)$. The class of all such pseudodifferential operators will be denoted by $\Psi_{1,\gamma}^{\kappa}$; $\Psi_{1,1}^0$ is called the exotic class. For the necessary background material we refer to [Tay1], [Tay2] or [Hör]. We shall need the following

Theorem *(i) Let* $\gamma \in [0,1)$, $0 < p < \infty$ *and* $\kappa, s \in \mathbf{R}$. *Then if* $a \in S_{1,\gamma}^{\kappa}$,

$$a(\cdot,D) : H_p^s(\mathbf{R}^n) \to H_p^{s-\kappa}(\mathbf{R}^n) \tag{3}$$

is a continuous linear map.

(ii) Let $0 < p < \infty$ *and* $\kappa, s \in \mathbf{R}$ *with* $s - \kappa > n\left(\dfrac{1}{p} - 1\right)_+$. *Then if* $a \in S_{1,1}^{\kappa}$, *the conclusion of (i) holds.*

Remark 1 For (i) we refer to Päivärinta [Päi] and Triebel [Triγ], 6.2.2, p. 258. Part (ii) with $\kappa = 0$, that is, $a \in \Psi_{1,1}^0$, is due to Runst [Run]; different proofs were given by Torres [Torr1], [Torr2]. Then (ii) follows since $(x,\xi) \longmapsto a(x,\xi)\langle\xi\rangle^{-\kappa} \in S_{1,1}^0$, in view of the fact that $I_\kappa : f \longmapsto (\langle\xi\rangle^\kappa \hat{f})^\vee$ is an isomorphic map of $H_p^s(\mathbf{R}^n)$ onto $H_p^{s-\kappa}(\mathbf{R}^n)$ and the representation $a(x,D) = a(x,D) \circ I_{-\kappa} \circ I_\kappa$.

Now let Ω be a bounded domain in $\mathbf{R}^n$ with C^∞ boundary. To handle the restrictions of operators of type (2) from $\mathbf{R}^n$ to Ω, we must first extend $f \in H_p^s(\Omega)$ from Ω to $\mathbf{R}^n$. There are bounded, linear extension operators ext from $H_p^s(\Omega)$ to $H_p^s(\mathbf{R}^n)$: see [Triγ], 4.5, p. 225 and [Triβ], 3.3.4, p. 201. Let re be the natural restriction operator from $H_p^s(\mathbf{R}^n)$ to $H_p^s(\Omega)$; then

$$\text{re} \circ a(\cdot,D) \circ \text{ext} : H_p^s(\Omega) \to H_p^{s-\kappa}(\Omega) \tag{4}$$

if $a \in S_{1,\gamma}^{\kappa}$ with $0 \leq \gamma \leq 1$, and where $s - \kappa > n(\frac{1}{p} - 1)_+$ in the exotic case $\gamma = 1$. However, this result depends upon the way in which ext is constructed. To avoid any possible ambiguity we restrict ourselves

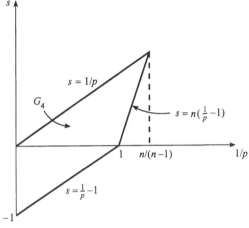

Fig. 5.1.

to those spaces $H_p^s(\Omega)$ for which the characteristic function of Ω is a pointwise multiplier in $H_p^s(\Omega)$ or, equivalently, for which

$$\mathrm{ext}^\Omega f(x) = \left\{ \begin{array}{l} f(x),\ x \in \Omega, \\ \\ 0,\ x \in \mathbf{R}^n \backslash \Omega, \end{array} \right\} \tag{5}$$

is an extension operator. Let

$$G_4 = \left\{ \left(\frac{1}{p}, s \right) : 0 < \frac{1}{p} < n/(n-1), \max\left(\frac{1}{p} - 1, n\left(\frac{1}{p} - 1 \right) \right) < s < \frac{1}{p} \right\};$$
$$\tag{6}$$

see Fig. 5.1. We then have the following

Proposition *The map* ext^Ω *is an extension operator from* $H_p^s(\Omega)$ *to* $H_p^s(\mathbf{R}^n)$ *if (resp. only if)* $\left(\frac{1}{p}, s \right) \in G_4$ *(resp.* $\left(\frac{1}{p}, s \right) \in \overline{G}_4$*).*

Remark 2 For this result we refer to [Triβ], 2.8.7, 3.3.2, pp. 158, 197, respectively, and Franke [Fra2].

It follows that if $a \in S_{1,\eta}^\kappa$, then

$$a^\Omega(\cdot, D) = \mathrm{re} \circ a(\cdot, D) \circ \mathrm{ext}^\Omega : H_p^s(\Omega) \to H_p^{s-\kappa}(\Omega), \tag{7}$$

with $s - \kappa > n(\frac{1}{p} - 1)_+$ in the exotic case $\gamma = 1$, makes sense and is intrinsic in Ω.

5.2.4 Elliptic operators: spectral properties

As before, Ω will stand for a bounded domain in $\mathbf{R}^n$ with C^∞ boundary, and as all spaces in this subsection will be taken with respect to Ω, we shall write L_p for $L_p(\Omega)$, etc.

Let $A_{s,p}$ be the regular elliptic differential operator introduced in Remark 5.2.2/1, with inverse $A_{s,p}^{-1}$ which we shall simply write as A^{-1} as it will be clear from the context between which spaces A^{-1} acts, compactly or isomorphically. We shall study the map B, where

$$Bf = b_2 A^{-1} b_1 f, \tag{1}$$

and b_1, b_2 belong to some spaces L_{r_j} or $H_{r_j}^s$ (see 2.4.3) and are such that B is a compact operator acting in L_p or H_p^s. For simplicity we shall for the moment limit ourselves to the 'ground level' $s = 0$, so that the basic spaces are the L_p spaces. However, it will be plain that any other space with parameters p and s satisfying $0 < p < \infty$, $n(\frac{1}{p} - 1) < s < \frac{n}{p}$ will do as a basic space, and we shall consider this more general situation in 5.2.5. There it will be shown that sometimes there are advantages in starting from other basic spaces (see the example given in (ii) in 5.2.5) or from the use of spaces with $p < 1$ (see the example given in (iv) in 5.2.5).

As usual, we shall denote by $\{\mu_k\}$ the sequence of eigenvalues of the compact map B, counted with respect to their algebraic multiplicity and ordered by decreasing modulus, so that $|\mu_1| \geq |\mu_2| \geq \dots.$

Proposition *Suppose that*

$$p, r_1, r_2 \in [1, \infty], \tag{2}$$

and that

$$\frac{1}{p} + \frac{1}{r_1} < 1, \quad \sigma := n\left(\frac{1}{p} - \frac{1}{r_2}\right) > 0, \quad \delta := \frac{2m}{n} - \left(\frac{1}{r_1} + \frac{1}{r_2}\right) > 0. \tag{3}$$

Let

$$b_1 \in L_{r_1}, \quad b_2 \in L_{r_2}. \tag{4}$$

Then $B : L_p \to L_p$ is compact and

$$|\mu_k| \leq c \, \|b_1 \mid L_{r_1}\| \, \|b_2 \mid L_{r_2}\| \, k^{-2m/n}, \quad k \in \mathbf{N}, \tag{5}$$

where c is independent of b_1, b_2 and k.

Proof The idea is to use the factorisation $B = b_2 \circ \mathrm{id} \circ A^{-1} \circ b_1$ of B with (see Fig. 5.2)

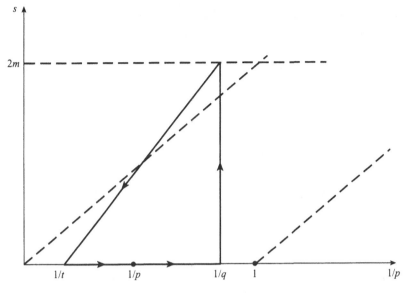

Fig. 5.2.

$$
\left.
\begin{aligned}
b_1 &: L_p \to L_q, \quad \text{with } \tfrac{1}{q} = \tfrac{1}{r_1} + \tfrac{1}{p}, \\[2mm]
A^{-1} &: L_q \to H_q^{2m}, \\[2mm]
\text{id} &: H_q^{2m} \to L_t, \quad \text{with } \tfrac{1}{t} = \tfrac{1}{p} - \tfrac{1}{r_2}, \\[2mm]
b_2 &: L_t \to L_p.
\end{aligned}
\right\}
\tag{6}
$$

Since $\delta > 0$ and $\tfrac{1}{q} - \tfrac{1}{t} = \tfrac{1}{r_1} + \tfrac{1}{r_2}$, it follows that id is compact and by Theorem 3.3.3/2

$$
e_k(\text{id}) \sim k^{-2m/n}.
\tag{7}
$$

Together with Carl's inequality (1.3.4/11) this gives (5).

Remark 1 As will be seen, the method of proof is extremely simple, involving just the information about mapping properties and entropy numbers derived earlier, and a journey around the $(\tfrac{1}{p}, s)$–diagram.

Remark 2 Suppose that $b_1(x) \neq 0$ and $b_2(x) \neq 0$ a.e. Then B is invertible in L_p and, at least formally, $D := B^{-1}$,

$$
D = b_1^{-1} A b_2^{-1},
\tag{8}
$$

is a degenerate elliptic differential operator, considered as an unbounded operator in L_p; see 1.2 for the spectral-theoretic background. Let $\{\lambda_k\}$ be the sequence of eigenvalues of D, counted according to algebraic multiplicity and ordered by increasing modulus. Since $\lambda_k = \mu_k^{-1}$ it follows from (5) that for some $c > 0$,

$$|\lambda_k| \geq c \, \|b_1 \mid L_{r_1}\|^{-1} \|b_2 \mid L_{r_2}\|^{-1} k^{2m/n}, \quad k \in \mathbf{N}. \tag{9}$$

For elliptic operators of order $2m$ and with smooth coefficients it is well known that the kth eigenvalue behaves like a multiple of $k^{2m/n}$, so that the right-hand side of (9) gives the correct behaviour. When $p = 2$ and D is symmetric, then D is self-adjoint in L_2 and one even has asymptotic formulae for λ_k and the counting function $N(\lambda) := \sum_{\lambda_k < \lambda} 1$; see Birman and Solomyak [BiS2], [BiS3] and also Rosenbljum [Ro1], [Ro2], who deals with the interesting limiting case $\delta = 0$ (δ is defined in (3)).

More sophisticated applications can plainly be given. For example, if b_1 and b_2 belong to some spaces $H_r^s s$, $\frac{1}{r^s} = \frac{1}{r} + \frac{s}{n}$, $s > 0$, then one may expect to have a smoothness theory for degenerate elliptic differential operators which includes assertions about the smoothness of root vectors (alias associated eigenfunctions). We shall not try to present the most general cases, but restrict ourselves to the case of maximal smoothness which can be obtained in this way. This case occurs when

$$b_2 \in H_p^\sigma = H_{r_2^\sigma}^\sigma, \tag{10}$$

where σ has the same meaning as in (3); as we shall see, less smoothness will do for b_1. The restriction imposed by (10) is natural, as B is considered to be acting in L_p.

Theorem 1 *Suppose that p, r_1 and r_2 satisfy (2) and (3), with $r_1, r_2 < \infty$, and put*

$$\kappa = \left(\frac{n}{p} + \frac{n}{r_1} - 2m \right)_+. \tag{11}$$

Assume that

$$b_1 \in H_{r_1}^\kappa, \quad b_2 \in H_p^\sigma = H_{r_2}^\sigma. \tag{12}$$

Then $B : L_p \to L_p$ is compact and

$$|\mu_k| \leq c \left\| b_1 \mid H_{r_1}^\kappa \right\| \left\| b_2 \mid H_p^\sigma \right\| k^{-2m/n}, \quad k \in \mathbf{N}, \tag{13}$$

where c is independent of b_1, b_2 and k. Furthermore, we have the following

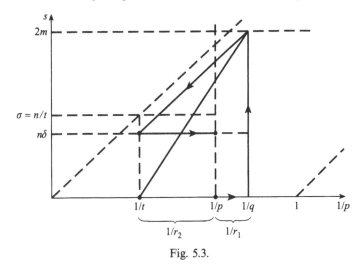

Fig. 5.3.

(i) Im $B \subset H_p^{\sigma-\varepsilon}$ and Im $B \subset L_{r_2-\varepsilon}$ *for all small enough* $\varepsilon > 0$ ($\varepsilon < \sigma$, $\varepsilon < r_2 - 1$).

(ii) Let $\frac{1}{r_2} < \frac{1}{u} < 1 - \frac{1}{r_1}$ *(in the sense of (3) with an arbitrary u instead of the given p). Then B is compact in* L_u *and the corresponding root spaces coincide for all those values of u.*

Proof The inequality (13) follows from the proposition above and the fact that $H_{r_1^*}^\kappa \subset L_{r_1}$ and $H_p^\sigma \subset L_{r_2}$; see (2.3.3/8). To prove (i) and (ii) we travel around the $(\frac{1}{p}, s)$-diagram and use (12) rather than the weaker condition (4). There are two typical situations, depicted in Fig. 5.2 and Fig. 5.3.

Of course, we always have (6), which in the sense of Fig. 5.3 is complemented by

$$
\left.
\begin{array}{ll}
b_1 : L_p \to L_q, & \text{since } H_{r_1^*}^\kappa \subset L_{r_1}, \\[2ex]
A^{-1} : L_q \to H_q^{2m}, & \\[2ex]
\text{id} : H_q^{2m} \to H_t^{n\delta} & \text{along the indicated line of slope } n, \\[2ex]
b_2 : H_t^{n\delta} \to H_p^{n\delta} & \text{since } H_{r_2}^\sigma \subset H_{r_2}^{n\delta}.
\end{array}
\right\}
\tag{14}
$$

In the last two steps we used embeddings along lines of slope n (see (2.3.3/8)) and Theorem 2.4.3, complemented by (2.4.3/17) since we are

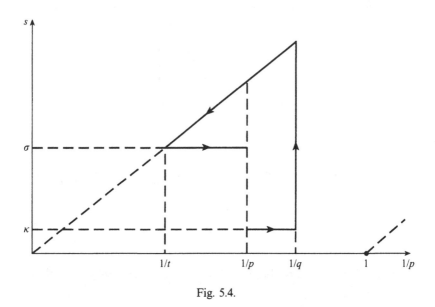

Fig. 5.4.

inside G_1 (see (2.4.3/1)). In the case of Fig. 5.2 we come immediately via id to the space H_t^σ instead of $H_t^{n\delta}$, and in the last line of (14) from $H_t^{\sigma-\varepsilon}$ to $H_p^{\sigma-\varepsilon}$ for any small $\varepsilon > 0$. In the case of Fig. 5.3 we arrive at least at the situation in which

$$\operatorname{Im}B \subset H_p^{n\delta}.$$

Instead of starting with L_p in (14) we can now begin with any space H_p^ρ with $0 < \rho \le n\delta$. Then any smoothness parameter s in the spaces in (14) is lifted up to $s + \rho$ as long as Theorem 2.4.3 and (2.4.3/17) can be applied, and also Theorem 5.2.2. After finitely many steps of this kind we arrive at the situation portrayed in Fig. 5.4 (if we did not have from the very beginning the situation in Fig. 5.2), and now (i) follows from Theorem 2.4.3 and (2.4.3/17).

Moreover, (ii) follows from (i) by standard arguments, in particular by embedding along lines of slope n and using Hölder's inequality. The proof is complete.

Remark 3 Suppose that as in Remark 2, B is invertible with inverse D. Then (13) gives immediately the obvious counterpart of (9). What is perhaps more interesting still is that under the hypotheses of the theorem,

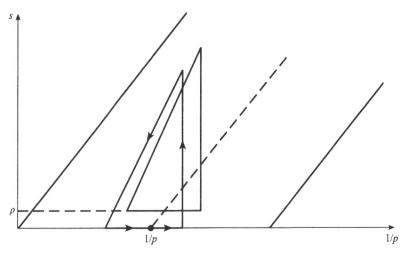

Fig. 5.5.

the root spaces of D are independent of the admissible values of u in the sense of (ii) of the theorem. This extends what is known in the classical case: see Agmon [Ag1], [Ag2].

We have dealt with the smoothness question in a traditional way: given p, the operator is considered in L_p and smoothness is treated in terms of H_p^σ. Plainly L_p can be replaced by any space H_p^s with $\left(\frac{1}{p}, s\right) \in G_1$ (see (2.4.3/1)), and in the proof of the above theorem this is precisely what we did. Further use of this procedure will be found in 5.2.5. The matter of smoothness may also be approached in a somewhat different way, one which is suited to the Hölder inequalities of 2.4.3 and in which one proceeds along lines of slope n in the $\left(\frac{1}{p}, s\right)$-diagram instead of being faced with a fixed p. We indicate briefly how this works. Since Ω is bounded, H_p^ρ in the above proof (with $\rho > 0$) may be replaced by $H_{p^\rho}^\rho$.

The triangles are then pushed along the line with slope n and foot point $1/p$ as far as is allowed by the smoothness of b_1, b_2 prescribed by (12); see Fig. 5.5. Ignoring the unsymmetrical role played by b_1 and b_2 we arrive with no effort at the following

Corollary *Suppose that* p, r_1 *and* r_2 *satisfy* (2) *and* (3), *with* $r_1, r_2 < \infty$. *Let* $v > 0$ *and assume that*

$$b_1 \in H^v_{r_1}, \quad b_2 \in H^v_{r_2}. \tag{15}$$

Then B *is compact in* L_p,

$$|\mu_k| \leq c \left\| b_1 \mid H^v_{r_1} \right\| \left\| b_2 \mid H^v_{r_2} \right\| k^{-2m/n}, \quad k \in \mathbf{N}, \tag{16}$$

where c *is independent of* b_1, b_2 *and* k, *and*

$$\mathrm{Im}\, B \subset H^v_{p^v}. \tag{17}$$

Remark 4 Naturally the conditions on b_1 can be weakened in the sense of (12). It is equally plain that the two approaches to smoothness can be combined for a given p along the lines $\left(\frac{1}{p}, s\right)$ and $\left(\frac{1}{p^s}, s\right)$.

Fractional powers Let A^κ_p denote a fractional power of the regular elliptic differential operator A_p of 5.2.2. As in the earlier part of this section we shall write A^κ instead of A^κ_p as it will be clear from the context which p is intended. We shall also assume that $|\kappa| \leq 1$. The operator B in (1) is now replaced by

$$Bf = b_2 A^{-\kappa} b_1 f, \quad 0 < \kappa \leq 1. \tag{18}$$

In view of the mapping and lifting properties described in 5.2.2 there is no difficulty in extending what has just been done for the operator in (1) to that given by (18), except for the corollary, which now has to be limited to those values with $q^v > 1$, where $\frac{1}{q} = \frac{1}{p} + \frac{1}{r_1}$ and $\frac{1}{q^v} = \frac{1}{q} + \frac{v}{n}$. As we are simply concerned with the illustration of the procedure, we restrict discussion to the following counterpart of the proposition above.

Theorem 2 *Let*

$$r_1, r_2, p \in [1, \infty], \quad \kappa \in (0, 1] \tag{19}$$

where

$$\frac{1}{p} + \frac{1}{r_1} < 1, \quad \sigma = n\left(\frac{1}{p} - \frac{1}{r_2}\right) > 0, \quad \delta = \frac{2m\kappa}{n} - \left(\frac{1}{r_1} + \frac{1}{r_2}\right) > 0. \tag{20}$$

Suppose that

$$b_1 \in L_{r_1}, \quad b_2 \in L_{r_2}. \tag{21}$$

Then the map B given by (18) is compact in L_p and

$$|\mu_k| \le c \, \|b_1 \mid L_{r_1}\| \, \|b_2 \mid L_{r_2}\| \, k^{-2m\kappa/n}, \quad k \in \mathbf{N}, \tag{22}$$

where c is independent of b_1, b_2 and k.

Proof This is the same as for the proposition at the beginning of this section, but now based upon Proposition 5.2.2 and Remark 5.2.2/2.

Remark 5 Despite the minor difficulty with $\mathrm{dom}(A_p^\kappa)$ for exceptional values of κ which is discussed in Remark 5.2.2/2, all the figures in this section can be taken over, with κm instead of m. This gives a full counterpart of Theorem 1 and a partial one, with the restriction given above, of the corollary.

5.2.5 Elliptic operators: generalisations

In the last section we have typically been concerned with operators B of the form

$$Bf = b_2 A^{-1} b_1 f, \tag{1}$$

where, for example, b_1 and b_2 are pointwise multipliers in some space H_p^s and

$$A^{-1} : L_p \to H_p^{2m} \tag{2}$$

is the inverse of a regular elliptic differential operator A. Here we mention a few natural generalisations of this basic situation.

(i) The regular elliptic differential operator A may be replaced by a degenerate elliptic differential operator of order $2m$, again denoted by A, with possible degeneracies at the boundary. Assuming that the inverse of A exists, the normal situation would be that A^{-1} maps L_p into a weighted space $H_p^{2m}(a)$, normed by $\sum_{|\alpha| \le 2m} \|a D^\alpha f \mid L_p\|$, so that

$$\|A^{-1} f \mid H_p^{2m}(a)\| \le c \, \|f \mid L_p\|, \quad a^{-1} \in L_u$$

for some $u \in (1, \infty)$. Fig. 5.6 shows a typical piece of travelling around in the $\left(\frac{1}{p}, s\right)$-diagram which generalises that of Fig. 5.2 on p. 193.
By analogy with (5.2.4/6) and (5.2.4/14) we use the following factorisation

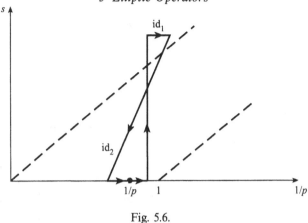

Fig. 5.6.

of the map $B = b_2 A^{-1} b_1$:

$$
\left.
\begin{array}{ll}
b_1 & : \ L_p \to L_q, \ \text{with} \ \frac{1}{q} = \frac{1}{p} + \frac{1}{r_1}, \\[2mm]
A^{-1} & : \ L_q \to H_q^{2m}(a), \\[2mm]
\mathrm{id}_1 & : \ H_q^{2m}(a) \to H_v^{2m}, \ \frac{1}{v} = \frac{1}{q} + \frac{1}{u}, \\[2mm]
\mathrm{id}_2 & : \ H_v^{2m} \to L_t, \frac{1}{p} = \frac{1}{t} + \frac{1}{r_2}, \\[2mm]
b_2 & : \ L_t \to L_p.
\end{array}
\right\} \tag{3}
$$

With the necessary restrictions on p this gives

$$
|\mu_k| \leq c \, \|b_1 \mid L_{r_1}\| \, \|a^{-1} \mid L_u\| \, \|b_2 \mid L_{r_2}\| \, k^{-2m/n} \tag{4}
$$

where

$$
0 \leq \frac{1}{r_1} + \frac{1}{r_2} + \frac{1}{u} < \frac{2m}{n}. \tag{5}
$$

When travelling around in the $\left(\frac{1}{p}, s\right)$-diagram it is not necessary to remain in the strip G_1 defined in (2.4.3/1). Suppose $H_p^{2m}(a)$ can also be normed in the sense of [Triγ], Theorems 3.5.3 and 5.2.2, pp. 194, 245, respectively, with dx replaced by $a^p(x)dx$, so that an equivalent norm is

$$
\|af \mid L_p(\Omega)\| + \left\| \left(\int_0^1 t^{-4m} \left(d_t^M f \right)^2 (\cdot) \frac{dt}{t} \right)^{1/2} a \mid L_p(\Omega) \right\|, \quad M > 2m.
$$

Then it is is possible to have $v \leq 1$, since Hölder's inequality applied in connection with id_1 then always works, even in fractional cases. The interest in this generalisation comes from Hilbert space techniques in which A, or rather $\sqrt{A}$, is generated by quadratic forms of the type

$$\int_\Omega \sum_{j,k=1}^n a_{jk}(x) \frac{\partial f}{\partial x_j} \frac{\partial \overline{f}}{\partial x_k} dx, \tag{6}$$

where $\left(a_{jk}(x)\right)_{j,k=1}^n$ is a positive, symmetric, possibly degenerate matrix. In this setting problems of this type have been treated by Birman and Solomyak [BiS2], [BiS3], who obtained similar results. Moreover, in this weighted case crucial improvements in the sense of Rosenbljum have been obtained by Tashkian [Tas]. There is naturally also a smoothness theory for the operator B above.

(ii) Instead of modifying the operator A and leaving b_1, b_2 essentially unchanged, we can take the obvious alternative. Thus let $A^{-\kappa}$ be a fractional power of a regular elliptic differential operator A of order $2m$, with $0 < \kappa \leq 1$, and take

$$B = \left(\sum_{|\beta| \leq l_2} b_{2,\beta}(x) D^\beta \right) \circ A^{-\kappa} \circ \left(\sum_{|\beta| \leq l_1} b_{1,\beta}(x) D^\beta \right) = B_2 \circ A^{-\kappa} \circ B_1 \tag{7}$$

with $l_1 + l_2 < 2m\kappa$. Now one can travel around in the $\left(\frac{1}{p}, s\right)$-diagram, perhaps as indicated in Fig. 5.7 with $H_p^{l_1}$ as the basic space and, as shown, using the decomposition

$$B = B_2 \circ \mathrm{id} \circ A^{-\kappa} \circ B_1$$

where $b_{1,\beta}$ and $b_{2,\beta}$ belong to some appropriate L_u space. When Theorem 2.4.3 is applied we must be in G_1 or G_2: see (2.4.3/1) or (2.4.3/2) respectively. This imposes conditions on the parameters which we leave to the reader to formulate.

(iii) Here we take B to be of the form (7), and outline a modified line of approach derived from the Hölder inequality of Theorem 2.4.3 and G_2 (see (2.4.3/2)). Suppose, for example, that

$$Bf = b_2(x) A^{-\kappa} \left(\sum_{j=1}^n b_{1,j}(x) \frac{\partial f}{\partial x_j} \right), \tag{8}$$

where

$$b_{1,j} \in H_{r_1^1}^1, \; j = 1, ..., n, \; b_2 \in L_{r_2} \tag{9}$$

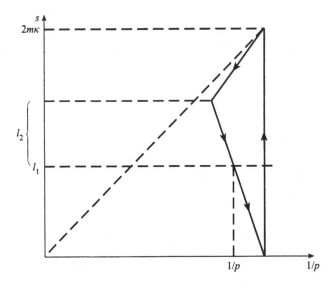

Fig. 5.7.

and

$$\frac{1}{r_1^1} = \frac{1}{r_1} + \frac{1}{n}.$$

Then the counterpart of (5.2.4/6) is, see also Fig. 5.8,

$$\left.\begin{array}{lll} \sum_{j=1}^{n} b_{1,j}\frac{\partial}{\partial x_j} & : L_p \to H_q^{-1}, \ \frac{1}{q} = \frac{1}{p} + \frac{1}{r_1}, \\[2mm] A^{-\kappa} & : H_q^{-1} \to H_q^{2m\kappa-1}, \\[2mm] \text{id} & : H_q^{2m\kappa-1} \to L_t, \ \frac{1}{t} = \frac{1}{p} - \frac{1}{r_2}, \\[2mm] b_2 & : L_t \to L_p. \end{array}\right\} \qquad (10)$$

To remain within G_2 (see (2.4.3/2)) when Theorem 2.4.5 is applied (with $s = -1$), the restriction

$$\frac{1}{p'} > \frac{1}{r_1} + \frac{1}{n} = \frac{1}{r_1^1}$$

is needed in addition to (5.2.4/2). Hence

$$\frac{1}{r_1} + \frac{1}{p} < 1 - \frac{1}{n} \ \text{and} \ \frac{1}{p} > \frac{1}{r_2}, \ \frac{1}{r_1} + \frac{1}{r_2} < \frac{2m\kappa - 1}{n} ; \qquad (11)$$

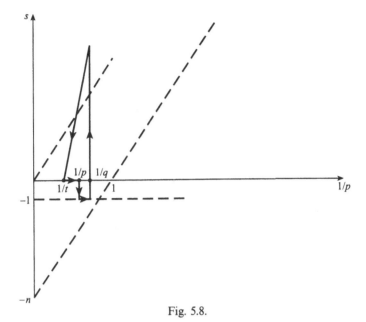

Fig. 5.8.

see (5.2.4/3). Under these conditions B is compact in L_p and

$$|\mu_k| \leq c \left(\sum_{j=1}^{n} \left\| b_{1,j} \mid H^1_{r^1_1} \right\| \right) \| b_2 \mid L_{r_2} \| \, k^{-(2m\kappa-1)/n}, \ k \in \mathbf{N}. \tag{12}$$

A smoothness theory, in the sense of Theorem 5.2.4/1 and Corollary 5.2.4, can also be developed without difficulty.

(iv) As remarked just before Proposition 5.2.4 we have concentrated, mostly for the sake of simplicity, on the 'ground level' $s = 0$. The exceptions to this arose in a natural way from the smoothness theory of Theorem 5.2.4/1 and Corollary 5.2.4 and in (ii) above. Routine extensions based on the tools supplied can naturally be given. However, it may happen that the full use of the strip G_1, and especially of its parts with $p < 1$, gives more satisfactory results than those obtained so far. To illustrate this, we describe how the *positive* ε in Theorem 5.2.4/1(i) can be replaced by $\varepsilon = 0$ under slightly modified circumstances: in view of (5.2.4/12) this is sharp. To do this we shift (and specialise) Fig. 5.3 within G_1 as shown in Fig. 5.9.

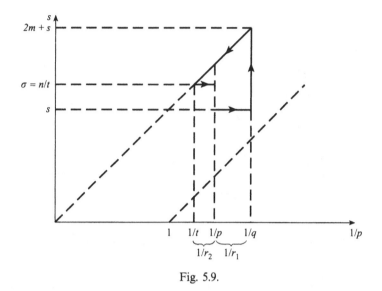

Fig. 5.9.

Compared with Fig. 5.2 and Fig. 5.3 we have replaced the basic space L_p by H_p^s with $\left(\frac{1}{p}, s\right) \in G_1$. Also

$$1 + \frac{1}{r_2} \le \frac{1}{p} < 1 - \frac{1}{r_1} + \frac{s}{n},$$

with

$$r_1, r_2 \in (1, \infty), \quad 2m + s = n/q,$$

expresses analytically what is shown in Fig. 5.9. Let $b_1 \in H_{r_1^s}^s$ and $b_2 \in H_{r_2^\sigma}^\sigma = H_p^\sigma$: see (5.2.4/12) and (5.2.4/3). Put

$$B = b_2 A^{-1} b_1;$$

B is now considered as a compact operator acting in H_p^s, and there is an obvious counterpart of (5.2.4/13). However, under the conditions indicated above, part (i) of Theorem 5.2.4/1 can be decisively improved.

The counterpart of (5.2.4/14) is

$$b_1 \quad : \quad H_p^s \to H_q^s \text{ since } b_1 \in H_{r_1}^s \text{ and Theorem 2.4.5 holds,}$$

$$A^{-1} \quad : \quad H_q^s \to H_q^{s+2m} = H_q^{n/q} \text{ by Theorem 5.2.2, with } q < 1,$$

$$\text{id} \quad : \quad H_q^{n/q} \to H_t^{n/t} \text{ along the indicated line of slope } n, (2.3.3/8),$$

$$b_2 \quad : \quad H_t^{n/t} \to H_p^\sigma \text{ by Remark 2.4.4/1 since } t \le 1, \ \sigma = n/t.$$

$$(13)$$

Compared with 5.2.4 the crucial difference is the last line, which does not hold when $t > 1$: see Remark 2.4.4/1. Thus in the case discussed, Theorem 5.2.4/1(i) can be improved to

$$\text{Im} B \subset H_p^\sigma \subset L_{r_2},$$

which is optimal.

5.2.6 Pseudodifferential operators: spectral properties

Estimates of the eigenvalues of these operators can now be obtained in an extremely simple way by following the same procedure as in 5.2.4. Let $a^\Omega(x, D)$ be the pseudodifferential operator introduced in (5.2.3/7), with symbol

$$a \in S_{1,\gamma}^{-\kappa}, \ \kappa > 0, \ 0 \le \gamma \le 1, \tag{1}$$

and put

$$B = b_2 a^\Omega(\cdot, D) b_1. \tag{2}$$

It is now just a matter of repetition of what was done in 5.2.4, with the additional restrictions caused by G_4 (see (5.2.3/6)) in Proposition 5.2.3. Everything else, including Proposition 5.2.4, is unaltered.

Theorem *Suppose that*

$$r_1, r_2, p \in [1, \infty], \ \kappa > 0, \ \gamma \in [0, 1], \tag{3}$$

where

$$\frac{1}{p} + \frac{1}{r_1} < 1, \ \sigma = n\left(\frac{1}{p} - \frac{1}{r_2}\right) > 0, \ \delta = \frac{\kappa}{n} - \left(\frac{1}{r_1} + \frac{1}{r_2}\right) > 0. \tag{4}$$

Let

$$b_1 \in L_{r_1}, \ b_2 \in L_{r_2}. \tag{5}$$

Then B is compact in L_p and its eigenvalues μ_k satisfy

$$|\mu_k| \le c\,\|b_1 \mid L_{r_1}\|\,\|b_2 \mid L_{r_2}\|\,k^{-\kappa/n}, \quad k \in \mathbf{N}, \tag{6}$$

where c is independent of b_1, b_2 and k.

Proof This is the same as that of Proposition 5.2.4. No restrictions arise from G_4 and Proposition 5.2.3, nor in the exotic case (see (5.2.3/6) and afterwards).

Remark It is plain that partial counterparts of Theorem 5.2.4/1 and Corollary 5.2.4 can be proved, taking into account the restrictions imposed by G_4 and Proposition 5.2.3 when $a^\Omega(x, D)$ is applied.

5.2.7 The negative spectrum

Here we apply the Birman–Schwinger principle described in 5.2.1 to the operators treated in 5.2.2 and 5.2.3. Of course, we have to limit ourselves to the symmetric case with $L_2(\Omega)$ as the basic space. Let G be a positive, self-adjoint operator acting in $L_2(\Omega)$ and given by

$$Gf = gA^\kappa g, \ \ g(x) > 0 \text{ a.e. in } \Omega, \ 0 < \kappa \le 1, \tag{1}$$

where A is a regular elliptic, positive, self-adjoint operator of order $2m$. We perturb G by a multiplication operator $f \longmapsto Vf$, giving

$$H_\alpha f = Gf - \alpha Vf, \ \ V(x) > 0 \text{ a.e. in } \Omega, \ \alpha > 0, \tag{2}$$

and ask for the behaviour of $\#\{\sigma(H_\alpha) \cap (-\infty, 0]\}$ as $\alpha \to \infty$. This is the usual question in this field of research, but it is usually studied in the context of $\mathbf{R}^n$ rather than Ω because of its connection with quantum mechanics. In view of this we shall merely give a simple example to illustrate the procedure, and shall return to the same problem on $\mathbf{R}^n$ in 5.4, when a more thorough examination of the question will be made.

Our idea here is to apply Theorem 5.2.4/2, and with this in mind we assume that $p = 2$, $r_1 = r_2 = r > 2$ and $m\kappa r > n$ in that theorem. Application of the Birman–Schwinger principle in the version given by Corollary 5.2.1 requires that we should estimate the entropy numbers $e_k(K_\alpha)$ of the map K_α given by

$$K_\alpha f = \alpha V^{1/2} g^{-1} A^{-\kappa} g^{-1} V^{1/2} f; \tag{3}$$

see (5.2.1/2). Of course, $e_k(K_\alpha) = \alpha\,e_k(K_1)$.

Theorem *Suppose that $r > 2$, $m\kappa r > n$ and that $V^{1/2}g^{-1} \in L_r$. Then*

$$\#\{\sigma(H_\alpha) \cap (-\infty, 0]\} \le c \left(\alpha \left\| V^{1/2}g^{-1} \mid L_r \right\|^2 \right)^{\frac{n}{2m\kappa}} \tag{4}$$

for some $c > 0$ which is independent of α.

Proof In (5.2.4/19–21) we have $r_1 = r_2 = r > 2$, $p = 2$, $b_1 = b_2 = V^{1/2}g^{-1}$. Then by Corollary 5.2.1 we have to count the $k \in \mathbf{N}$ such that

$$1 \le \sqrt{2}\alpha e_k(K_1) \le c\alpha \left\| V^{1/2}g^{-1} \mid L_r \right\|^2 k^{-2m\kappa/n}, \tag{5}$$

since $|\mu_k|$ in (5.2.4/22) can be replaced by $e_k(K_1)$. This gives (4) immediately.

5.3 Elliptic operators in domains: limiting cases

5.3.1 Introduction

Throughout 5.3, Ω will stand for a bounded domain in $\mathbf{R}^n$ with C^∞ boundary. Our objective is to extend the material of 5.2 concerning eigenvalue distributions to limiting situations. These are of two types, corresponding to the natural occurrence of Orlicz spaces of the form $L_\infty(\log L)_{-a}(\Omega)$, and of logarithmic Sobolev spaces. The methods used to obtain the eigenvalue estimates are, of course, the same as in 5.2: we simply use the mapping properties of the operators together with appropriate factorisations and knowledge of the behaviour of the entropy numbers of embedding maps in limiting situations. We shall thus be relying on the estimates for entropy numbers obtained in 3.4.

5.3.2 Orlicz spaces

Our first aim is to extend Theorem 5.2.4/2 to limiting situations. Let B be defined by

$$Bf = b_2 A^{-\kappa} b_1 f, \quad 0 < \kappa \le 1, \tag{1}$$

where, as in 5.2.4, $A^{-\kappa}$ is a fractional power of a regular elliptic differential operator A of order $2m$; b_1 and b_2 will be required to belong to certain spaces.

Theorem 1 *Let*

$$r_1, r_2 \in [1, \infty], \quad 0 < \kappa \le 1, \tag{2}$$

with

$$1 \geq \frac{1}{r_1} + \frac{1}{r_2} = \frac{2m\kappa}{n}. \tag{3}$$

(i) Suppose also that

$$r_2 < \infty, \quad \frac{1}{r_1} + \frac{1}{r_2} < 1, \quad \frac{1}{r_2} \leq \frac{1}{p} < \frac{1}{r_1'}, \quad an > n + 4m\kappa, \tag{4}$$

and

$$b_1 \in L_{r_1}, \quad b_2 \in L_{r_2} (\log L)_a. \tag{5}$$

Then B is a compact operator from L_p to itself, and

$$|\mu_k| \leq c \, \|b_1 \mid L_{r_1}\| \, \|b_2 \mid L_{r_2} (\log L)_a\| \, k^{-2m\kappa/n}, \quad k \in \mathbf{N}, \tag{6}$$

where c is independent of b_1, b_2 and k.

(ii) Suppose that in addition to the general hypotheses,

$$r_1 < \infty, \quad \frac{1}{r_1} + \frac{1}{r_2} < 1, \quad \frac{1}{r_2} < \frac{1}{p} \leq \frac{1}{r_1'}, \quad an > n + 4m\kappa, \tag{7}$$

and

$$b_1 \in L_{r_1} (\log L)_a, \quad b_2 \in L_{r_2}. \tag{8}$$

Then B is a compact operator from L_p to itself, and there exists $c > 0$, as before, such that

$$|\mu_k| \leq c \, \|b_2 \mid L_{r_2}\| \, \|b_1 \mid L_{r_1} (\log L)_a\| \, k^{-2m\kappa/n}, \quad k \in \mathbf{N}, \tag{9}$$

(iii) In addition to the general hypotheses suppose that

$$1 < r_1 < \infty, \quad \frac{1}{r_1} + \frac{1}{r_2} = 1, \quad a_j > 1 + \frac{2}{r_j} \quad j = 1, 2, \tag{10}$$

$$b_1 \in L_{r_1} (\log L)_{a_1}, \quad b_2 \in L_{r_2} (\log L)_{a_2}. \tag{11}$$

Then B is a compact operator in L_{r_2} and there exists $c > 0$, as before, such that

$$|\mu_k| \leq c \, \|b_1 \mid L_{r_1} (\log L)_{a_1}\| \, \|b_2 \mid L_{r_2} (\log L)_{a_2}\| \, k^{-1}, \quad k \in \mathbf{N}. \tag{12}$$

Proof Step 1 First we prove (i) when $p = r_2$: see Fig. 5.10 and also Fig. 5.2 on p. 193.

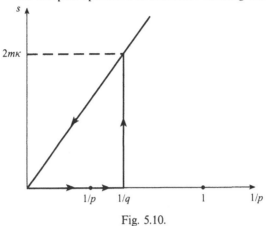

Fig. 5.10.

Corresponding to (5.2.4/6) we have

$$
\left.
\begin{aligned}
b_1 \quad &: \quad L_p \to L_q \text{ with } \tfrac{1}{q} = \tfrac{1}{p} + \tfrac{1}{r_1}, \\[1.5ex]
A^{-\kappa} \quad &: \quad L_q \to H_q^{2m\kappa} = H_q^{n/q}, \\[1.5ex]
\text{id} \quad &: \quad H_q^{n/q} \to L_\infty (\log L)_{-a}, \\[1.5ex]
b_2 \quad &: \quad L_\infty (\log L)_{-a} \to L_p.
\end{aligned}
\right\}
\tag{13}
$$

By Theorem 3.4.2, id is a compact embedding with

$$
e_k (\text{id}) \sim k^{-2m\kappa/n}, \quad k \in \mathbf{N}.
\tag{14}
$$

Note that if $f \in L_\infty (\log L)_{-a}$, then $f^p \in L_\infty (\log L)_{-ap}$; see (2.6.1/2). Recall that $L_1 (\log L)_{ap}$ and $L_\infty (\log L)_{-ap}$ are associated spaces; see Proposition 2.6.1/2 and its proof. Since $b_2^p \in L_1 (\log L)_{ap}$ (see Definition 2.6.1) we have

$$
\begin{aligned}
\|f b_2 \mid L_p\|^p \;&=\; \int_\Omega |f|^p \, |b_2|^p \, dx \\[1ex]
&\leq\; \big\| f^p \mid L_\infty (\log L)_{-ap} \big\| \, \big\| b_2^p \mid L_1 (\log L)_{ap} \big\| \\[1ex]
&\leq\; c \, \|f \mid L_\infty (\log L)_{-a}\|^p \, \|b_2 \mid L_{r_2} (\log L)_a\|^p,
\end{aligned}
\tag{15}
$$

since $p = r_2$. Now (14) and (15) justify (13), and (6) with $p = r_2$ follows from (14) and Carl's inequality (1.3.4/11).

Step 2 The adjoint of B (acting in $L_{p'}$), with $p = r_2$, is

$$B' = b_1 \left(A^{-\kappa}\right)' b_2. \tag{16}$$

This is a compact map of $L_{p'}$ to itself, with $p' = r_2'$. From the discussion in 5.2.2 we see that $A^{-\kappa} = \left((A')^{-\kappa}\right)'$. Interchange of b_1 and b_2 together with this observation, Step 1 and the Riesz–Schauder theory now give (ii) when $p = r_1'$. This also gives (ii) for all $p \in [r_1', r_2)$, in view of the monotonicity of the L_p spaces (since Ω is bounded) and the compactness of B, for which we refer to Remark 1 below. The remaining cases of (i) now follow by duality.

Step 3 To prove (iii) we use the decomposition

$$B = B_2 \circ B_1, \text{ with } B_1 = A^{-\kappa/r_1} b_1, \ B_2 = b_2 A^{-\kappa/r_2}, \tag{17}$$

apply (i) with ∞ instead of r_1 and $p = r_2$ to B_2 and (ii) with ∞ instead of r_2 and $p = r_1' = r_2$ to B_1, and simply note that

$$e_{2k} \left(B_2 \circ B_1\right) \leq e_k \left(B_2\right) e_k \left(B_1\right).$$

The proof is complete.

Remark 1 As for the compactness of B used in Step 2, we may assume that $p \in \left(r_1', r_2\right)$. Then in analogy to (13) we even have

$$\|B\| \leq c \|b_1 \mid L_{r_1}\| \, \|b_2 \mid L_{r_2}\|.$$

For $j = 1, 2$ and $N \in \mathbf{N}$ define b_j^N by $b_j^N(x) = b_j(x)$ if $|b_j(x)| \leq N$, $b_j^N(x) = 0$ otherwise; let B^N be the corresponding operator. Plainly B^N is compact in L_p. The compactness of B now follows from

$$\left\|B - B^N\right\| \leq c \left\|b_1 - b_1^N \mid L_{r_1}\right\| \, \|b_2 \mid L_{r_2}\| + c \|b_1 \mid L_{r_1}\| \left\|b_2 - b_2^N \mid L_{r_2}\right\|$$

$$\to 0 \text{ as } N \to \infty.$$

Remark 2 The proof makes it plain that we use to the full extent duality and fractional powers of regular elliptic operators as given in 5.2.2. Note that the unnatural restrictions on a arise from the entropy number estimates of Theorem 3.4.2. If we use the other cases covered by that theorem, then of course we can obtain estimates of type (6), (9) and (12) with smaller powers of k^{-1} than $2m\kappa/n$, but these estimates are presumably not sharp. Furthermore, the proof is based upon consideration of the end point cases $p = r_2$ and $p = r_1'$ first, the rest following by monotonicity and duality. It is reasonable to expect improvements so far as the restrictions on a in (i) and (ii) are concerned, if $p \in (r'_1, r_2)$. This turns out to be the case, and we shall return to this point in Theorem 5.3.3/1 and Remark 5.3.3/1. In the symmetric case, when $b_1 = b_2$, $p = 2$ and $\kappa = 1$, there are again better results: see Rosenbljum [Ro1], [Ro2] in connection with (i) and (ii), and Solomyak [Sol] for (iii).

Remark 3 At least partly, the operator B in (1) can be replaced by the pseudodifferential map B from (5.2.6/2). The exotic case $\gamma = 1$ must be excluded when duality arguments are included. We leave the detailed implementation of this to the reader.

Examples Some concrete illustrations of the results given above may help to shed some light on the situation. Suppose that

$$\Omega = \{ y \in \mathbf{R}^n : |y| < 1/2 \}, \tag{18}$$

let $r \in (1, \infty)$ and suppose that for some $\lambda \in \mathbf{R}$,

$$b(x) = |x|^{-n/r} |\log |x||^{-\lambda/r}, \quad x \in \Omega, x \neq 0. \tag{19}$$

With Theorem 1 in mind, we look for those $\lambda > 0$ such that

$$b \in L_r (\log L)_a. \tag{20}$$

By (2.6.1/1) this is equivalent to

$$\int_{|x|<1/2} |x|^{-n} |\log |x||^{-\lambda} |\log |x||^{ra} dx < \infty, \tag{21}$$

and hence to

$$\int_0^{1/2} t^{-1} |\log t|^{ra-\lambda} dt < \infty, \text{ and } \lambda > 1 + ar. \tag{22}$$

Suppose that b_1, b_2 are given by (19) with r_1, λ_1 and r_2, λ_2 respectively in such a way that the counterparts of (22) hold. Let B be given by (1), so that

$$Bf(x) = |x|^{-n/r_2} |\log |x||^{-\lambda_2/r_2} A^{-\kappa} |x|^{-n/r_1} |\log |x||^{-\lambda_1/r_1} f(x). \tag{23}$$

Let $2m\kappa = n$ and suppose that (10) holds. Then (11) and (12) also hold if $\lambda_j > 3 + r_j$ and we have

$$|\mu_k| \leq ck^{-1}, \quad k \in \mathbf{N}, \tag{24}$$

which is the expected classical order.

In contrast to this, if $\lambda_1 = \lambda_2 = 0$, then a homogeneity argument shows that B is not even compact, at least if $\kappa = 1$ so that $2m = n$. To verify this in detail, we consider the inverse D of the operator,

$$Df = |x|^{n/r_1} A |x|^{n/r_2} f; \qquad (25)$$

recall that $\frac{1}{r_1} + \frac{1}{r_2} = 1$. Let $\omega = \left\{ y \in \mathbf{R}^n : \frac{1}{4} < |y| < \frac{1}{2} \right\}$, let $\psi \in C_0^\infty (\omega), \psi \neq 0$, and for each $j \in \mathbf{N}$ define $\psi_j (x) = \psi \left(2^j x \right)$; the ψ_j have mutually disjoint supports. Since $2m = n$ and hence $H_p^{2m} = W_p^n$, the usual Sobolev space, we have

$$\left\| \sum_{j=1}^\infty a_j \psi_j \mid L_p \right\| \sim \left(\sum_{j=1}^\infty 2^{-jn} |a_j|^p \right)^{1/p}, \qquad (26)$$

$$\left\| A \sum_{j=1}^\infty a_j \psi_j \mid L_p \right\| \sim \left\| \sum_{j=1}^\infty a_j \psi_j \mid W_p^n \right\| \sim \left(\sum_{j=1}^\infty 2^{jn(p-1)} |a_j|^p \right)^{1/p} \qquad (27)$$

and hence

$$\left\| D \sum_{j=1}^\infty a_j \psi_j \mid L_p \right\| \sim \left(\sum_{j=1}^\infty 2^{-jn} |a_j|^p \right)^{1/p}. \qquad (28)$$

Since the right-hand sides of (26) and (28) coincide, we see that

$$f \longmapsto |x|^{-n/r_2} A^{-1} |x|^{-n/r_1} f \qquad (29)$$

cannot be compact in L_p. This illustrates the big difference between the operators in (23) and (29). In a similar way we can construct examples and counterexamples related to parts (i) and (ii) of Theorem 1, as well as for the non-limiting cases presented in 5.2.4–5.2.6.

To conclude this section, we observe that Theorem 5.2.7 concerning the negative spectrum can immediately extended to a limiting situation.

Theorem 2 *Let H_α be given by (5.2.7/2), let $2m\kappa = n$ and suppose that $Vg^{-2} \in L_1 (\log L)_{2a}$ with $a > 2$. Then*

$$\# \left\{ \sigma (H_\alpha) \cap (-\infty, 0] \right\} \leq c\alpha \left\| Vg^{-2} \mid L_1 (\log L)_{2a} \right\| \qquad (30)$$

for some $c > 0$ which is independent of V, g and α.

The proof follows precisely the same lines as that of Theorem 5.2.7, now based on part (iii) of Theorem 1 above.

Remark 4 Similar results can plainly be obtained for pseudodifferential operators. Indeed, since the inverse of a non-exotic elliptic pseudodifferential operator, if it exists, is again an elliptic pseudodifferential operator, the method of proof can be applied to perturbations of elliptic operators by pseudodifferential operators of lower order using the composition theorem for pseudodifferential operators. We shall not pursue this here as it seems more interesting on $\mathbf{R}^n$ than on bounded domains, and is discussed in 5.4.7.

Remark 5 By way of an example, suppose that Ω is given by (18) and let

$$Gf = |x|^{n/2} A^\kappa |x|^{n/2} f, \quad 2m\kappa = n; \tag{31}$$

write

$$H_\alpha f = Gf - \alpha \, |\log |x||^{-\gamma} f. \tag{32}$$

By (26) and (28), with $p = 2$, G is not an operator with pure point spectrum and G^{-1} is not compact, at least for $\kappa = 1$. To apply (30) we must check whether or not

$$|\log |x||^{-\gamma} |x|^{-n} \in L_1 (\log L)_{2a} \text{ with } a > 2. \tag{33}$$

By (22), this holds if $\gamma > 5$. In this case we have (30).

5.3.3 Logarithmic Sobolev spaces

Once more we seek to extend Theorem 5.2.4/2 (and indeed also Theorem 5.2.6) to limiting situations, this time involving logarithmic Sobolev spaces rather than those Orlicz spaces discussed in 5.3.2 and which are typically related to end point cases as described in Remark 5.3.2/2. Let B be defined by

$$Bf = b_2 A^{-\kappa} b_1 f, \quad 0 < \kappa \le 1, \tag{1}$$

where $A^{-\kappa}$ is either a fractional power of a regular elliptic differential operator of order $2m$ as in 5.2.4 or a pseudodifferential operator $A^{-\kappa} = a^\Omega (\cdot, D) \in \Psi_{1,\gamma}^{-2m\kappa}$ (see Remark 5.2.3/2), where again Ω will be omitted; b_1 and b_2 belong to suitable spaces. We begin at the ground level, as far as smoothness is concerned. Recall that $L_\infty (\log L)_0 - L_\infty$.

Theorem 1 *Let*

$$r_1, r_2 \in (1, \infty], \quad \kappa \in (0, 1], \quad m \in \mathbf{N} \tag{2}$$

with

$$1 > \frac{1}{r_1} + \frac{1}{r_2} = \frac{2m\kappa}{n}, \tag{3}$$

and let $a_1, a_2 \in \mathbf{R}$ be such that

$$\frac{1}{r_2} < \frac{1}{p} < \frac{1}{r_1} \quad \text{and} \quad a_1 + a_2 > \frac{4m\kappa}{n} \tag{4}$$

(with $a_1 \le 0$ if $r_1 = \infty$ and $a_2 \le 0$ if $r_2 = \infty$). Suppose that

$$b_1 \in L_{r_1}(\log L)_{a_1} \quad \text{and} \quad b_2 \in L_{r_2}(\log L)_{a_2}, \tag{5}$$

and let $A^{-\kappa}$ be either a fractional power of a regular elliptic differential operator A of order $2m$ or an element of $\Psi_{1,\gamma}^{-2m\kappa}$, with $0 \le \gamma \le 1$. Then the map B given by (1) is a compact map from L_p to itself, and there exists $c > 0$ such that

$$|\mu_k| \le c \left\| b_1 \mid L_{r_1}(\log L)_{a_1} \right\| \left\| b_2 \mid L_{r_2}(\log L)_{a_2} \right\| k^{-2m\kappa/n}, \quad k \in \mathbf{N}. \tag{6}$$

Proof As in 5.3.2 we use the decomposition (see Fig. 5.11)

$$\left.\begin{array}{rl}
b_1 & : L_p \to L_q(\log L)_{a_1-\varepsilon}, \ \frac{1}{q} = \frac{1}{p} + \frac{1}{r_1}, \ \varepsilon > 0 \text{ small}, \\[2mm]
A^{-\kappa} & : L_q(\log L)_{a_1-\varepsilon} \to H_q^{2m\kappa}(\log H)_{a_1-\varepsilon}, \\[2mm]
\text{id} & : H_q^{2m\kappa}(\log H)_{a_1-\varepsilon} \to L_t(\log L)_{-a_2+\varepsilon}, \ \frac{1}{t} = \frac{1}{p} - \frac{1}{r_2}, \\[2mm]
b_2 & : L_t(\log L)_{-a_2+\varepsilon} \to L_p.
\end{array}\right\} \tag{7}$$

The first and last embeddings follow from Theorems 2.6.2/1, 2 and Hölder's inequality, where $\varepsilon > 0$ plays the same role as in (3.4.3/34, 35); see also the arguments in the proof of Theorem 3 below (in some cases $\varepsilon = 0$ can be chosen). The second embedding is covered by either (5.2.2/10) or (5.2.3/4) and Definition 2.6.3. Finally, id is compact, since

$$2m\kappa - \frac{n}{q} = -\frac{n}{t} \quad \text{and} \quad a_1 + a_2 - 2\varepsilon > \frac{4m\kappa}{n} \tag{8}$$

if $\varepsilon > 0$ is chosen small enough, and it follows either by Theorem 3.4.3/2 or by Theorem 3.4.4 that

$$e_k(\text{id}) \sim k^{-2m\kappa/n}. \tag{9}$$

Carl's inequality (1.3.4/11) now gives (6). The proof is complete.

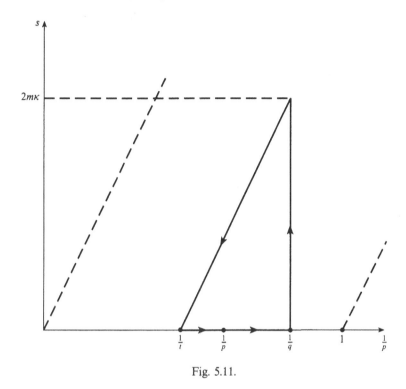

Fig. 5.11.

Remark 1 This theorem should be compared with Theorem 5.3.2/1, where we started from an L_∞-situation and used duality and compositions to obtain results of this general type. However, a typical restriction obtained in that way would be $a_1 + a_2 > 1 + \frac{4m\kappa}{n}$ in (4). Moreover, the treatment above also enables us to consider limiting cases with the same degree of flexibility as the non-limiting cases in 5.2.4: we can now proceed as in 5.2, using smoothness theory, coincidence of root spaces starting from different basic spaces, etc.

Examples We discuss the same examples as in 5.3.2. Let $\Omega = \{y \in \mathbf{R}^n : |y| < 1/2\}$, $r \in (1, \infty)$ and $\lambda \in \mathbf{R}$. Then

$$b(x) = |x|^{-n/r} \|\log |x|\|^{-\lambda} \in L_r (\log L)_a (\Omega) \qquad (10)$$

if, and only if,

$$\lambda > \frac{1}{r} + a, \qquad (11)$$

as shown in 5.3.2. Our concern is with the eigenvalues μ_k of the operator

$$B = b_2 A^{-\kappa} b_1, \quad 0 < \kappa \le 1, \tag{12}$$

where $A^{-\kappa}$ is either a fractional power of the regular elliptic differential operator A of order $2m$ or a pseudodifferential operator in $\Psi_{1,\gamma}^{-2m\kappa}(\Omega)$ with $0 \le \gamma \le 1$, and

$$b_j(x) = |x|^{-n/r_j} \, \big| \log |x| \big|^{-\lambda_j}, \quad j = 1, 2. \tag{13}$$

Theorem 2 *Let Ω be the above domain, let*

$$r_1, r_2 \in (1, \infty), \quad \kappa \in (0, 1], \quad m \in \mathbf{N} \tag{14}$$

with

$$1 > \frac{1}{r_1} + \frac{1}{r_2} = \frac{2m\kappa}{n}, \tag{15}$$

and let $\lambda_1, \lambda_2 \in \mathbf{R}$ be such that

$$\frac{1}{r_2} < \frac{1}{p} < \frac{1}{r_1}, \quad \lambda_1 + \lambda_2 > \frac{6m\kappa}{n}. \tag{16}$$

Then the map B defined by (12), with b_1 and b_2 given by (13), is compact in L_p and there is a constant $c > 0$ such that

$$|\mu_k| \le c k^{-2m\kappa/n}, \quad k \in \mathbf{N}. \tag{17}$$

Proof The result follows immediately from Theorem 1 together with (10), (11) and (15).

Remark 2 This example was discussed in 5.3.2 in certain limiting situations. Here we have gained more flexibility in the non-limiting situations, and it is particularly remarkable that, for example, λ_1 might be negative. Of course, this must be compensated by λ_2 so that (16) holds. We shall discuss later further results of this nature, in which the b_j may belong to spaces of the form $H_p^s (\log H)_a$.

So far in this section we have focussed on the 'ground level' $s = 0$ and on p restricted to $(1, \infty)$. The restriction stems from the fact that the theory of fractional powers of a regular elliptic differential operator A is available only on an L_p-basis with $1 < p < \infty$, in contrast to the corresponding theory for A itself described in 5.2.2. For pseudodifferential operators we have the restrictions imposed by Proposition 5.2.3 and Remark 5.2.3/2. As was pointed out in Remark 1 we now have the same flexibility in

limiting cases as for non-limiting ones. We shall lift the theory up to the
setting of the spaces $H_p^s (\log H)_a$, but rather than giving an exhaustive
discussion of this we choose to give, in the next two theorems, examples
of what can be done.

Theorem 3 *Let*

$$r_1, r_2 \in (1, \infty), \quad m \in \mathbf{N}, \tag{18}$$

where

$$1 > \frac{1}{r_1} + \frac{1}{r_2} = \frac{2m}{n}. \tag{19}$$

Suppose that

$$\frac{1}{r_2} < \frac{1}{p} < 1 - \frac{1}{r_1} \text{ and } a > \frac{4m}{n}, \tag{20}$$

that $s > 0$ and also

$$b_1 \in H_{r_1^s}^s \text{ and } b_2 \in H_{r_2^s}^s (\log H)_a, \tag{21}$$

*where $\frac{1}{r_j^s} = \frac{1}{r_j} + \frac{s}{n}$. Let A^{-1} be the inverse of a regular elliptic differential
operator A of order $2m$. Then the map*

$$B = b_2 A^{-1} b_1 \tag{22}$$

is compact in $H_{p^s}^s$ and its eigenvalues μ_k satisfy

$$|\mu_k| \le c \left\| b_1 \mid H_{r_1^s}^s \right\| \left\| b_2 \mid H_{r_2^s}^s (\log H)_a \right\| k^{-2m/n}, \quad k \in \mathbf{N}, \tag{23}$$

where $c > 0$ is a constant independent of b_1, b_2 and k.

Proof Factorise B in the form (see Fig. 5.12)

$$B = b_2 \circ \text{id} \circ A^{-1} \circ b_1, \tag{24}$$

where

$$\left.\begin{array}{lll}
b_1 & : H_{p^s}^s \to H_{q^s}^s, & \frac{1}{q} = \frac{1}{p} + \frac{1}{r_1}, \\[2mm]
A^{-1} & : H_{q^s}^s \to H_{q^s}^{s+2m}, & \\[2mm]
\text{id} & : H_{q^s}^{s+2m} \to H_{t^s}^s (\log H)_{-a+\varepsilon}, & a - \varepsilon > 4m/n, \frac{1}{t} = \frac{1}{p} - \frac{1}{r_2}, \\[2mm]
b_2 & : H_{t^s}^s (\log H)_{-a+\varepsilon} \to H_{p^s}^s.
\end{array}\right\} \tag{25}$$

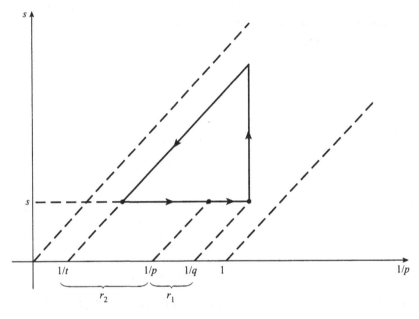

Fig. 5.12.

The first mapping is a consequence of (2.4.3/17), while the second follows from 5.2.2. For id, Theorem 3.4.3/2 tells us that

$$e_k \,(\mathrm{id}) \sim k^{-2m/n}, \ k \in \mathbf{N}. \tag{26}$$

As for b_2, we temporarily extend the definition of $G_{p,a}$ given in (3.4.3/34). Let $p \in (0,\infty)$, $s \in \mathbf{R}$ and $a \leq 0$ (respectively, $a > 0$). Then $G_{p,a}^s$ is the set of all $f \in \mathscr{D}'(\Omega)$ such that

$$\sup_{j \geq J} 2^{ja} \left\| f \mid H_{p^{\sigma_j}}^s \right\| < \infty, \tag{27}$$

$$\text{(respectively, } \sup_{j \geq J} 2^{ja} \left\| g_j \mid H_{p^{\lambda_j}}^s \right\| < \infty). \tag{28}$$

These replace (2.6.3/3) (respectively, (2.6.3/5)). Comparison with Definition 2.6.3 shows that for all $\varepsilon > 0$,

$$H_u^s (\log H)_v \subset G_{u,v}^s \subset H_u^s (\log H)_{v-\varepsilon} \,; \tag{29}$$

and from (2.4.3/17) it now follows easily that

$$H^s_{t^s}(\log H)_{-a+\varepsilon} H^s_{r^s_2}(\log H)_a \subset G^s_{t^s,-a+\varepsilon} G^s_{r^s_2,a} \subset H^s_{p^s}, \tag{30}$$

which establishes the continuity of the last embedding in (25). Carl's inequality (1.3.4/11) now completes the proof of the theorem.

Remark 3 The embeddings (30) may be thought of as an extension of (2.4.3/17) from (fractional) Sobolev(–Hardy) spaces to logarithmic (fractional) Sobolev(–Hardy) spaces, although it would have been preferable, from the standpoint of a generalisation, not to have replaced l_p in (2.6.3/3) and (2.6.3/5) by l_∞ as we did. Since we are not concerned to avoid the ε in (30), this more refined approach was not necessary.

It is natural to seek to replace the functions b_1 and b_2 in Theorem 3 by (pseudo)differential operators of lower order than A. We now give an example which illustrates the kind of results thus obtainable; see also 5.2.5, especially (5.2.5/11).

Theorem 4 *Let*

$$p, r_1, r_2 \in (1,\infty), \quad \kappa \in (0,1], \quad m \in \mathbf{N}, \tag{31}$$

where

$$1 > \frac{1}{r_1} + \frac{1}{r_2} = \frac{2m\kappa - 1}{n}, \tag{32}$$

$$\frac{1}{p} > \frac{1}{r_2}, \quad \frac{1}{p} + \frac{1}{r_1} < 1 - \frac{1}{n}. \tag{33}$$

Let $a_1, a_2 \in \mathbf{R}$ be such that

$$a_1 + a_2 > \frac{2}{n}(2m\kappa - 1), \tag{34}$$

suppose that

$$b_{1,j} \in L_{r_1}(\log L)_{a_1}, \quad j = 1,...,n, \quad b_2 \in H^1_{r'_2}(\log H)_{a_2}, \tag{35}$$

where $\frac{1}{r'_2} = \frac{1}{r_2} + \frac{1}{n}$, and let $A^{-\kappa}$ be either a fractional power of a regular elliptic differential operator A of order $2m$ or an element of $\Psi^{-2m\kappa}_{1,\gamma}$ with

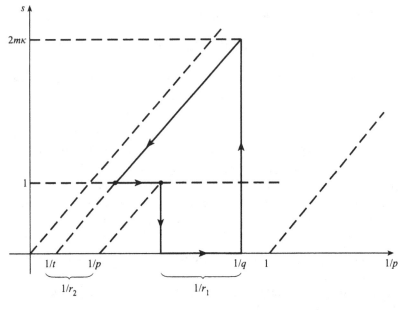

Fig. 5.13.

$0 \le \gamma \le 1$. *Then the map B given by*

$$B = b_2 A^{-\kappa} \sum_{j=1}^{n} b_{1,j} \frac{\partial}{\partial x_j} \tag{36}$$

is compact in $H_{p^1}^1$ and there is a constant $c > 0$ such that its eigenvalues μ_k satisfy

$$|\mu_k| \le c \sum_{j=1}^{n} \left\| b_{1,j} \mid L_{r_1} (\log L)_{a_1} \right\| \left\| b_2 \mid H_{r_2}^1 (\log H)_{a_2} \right\| k^{-(2m\kappa-1)/n}, \quad k \in \mathbf{N}. \tag{37}$$

Proof This time we factorise B (see Fig. 5.13) as

$$B = b_2 \circ \mathrm{id} \circ A^{-\kappa} \circ \sum_{j=1}^{n} b_{1,j} \frac{\partial}{\partial x_j}, \tag{38}$$

where

$$
\sum_{j=1}^{n} b_{1,j} \frac{\partial}{\partial x_j} \; : \; H_{p^1}^1 \to L_q (\log L)_{a_1-\varepsilon}, \; \frac{1}{q} = \frac{1}{r_1} + \frac{1}{p} + \frac{1}{n} = \frac{1}{r_1} + \frac{1}{p^1},
$$

$$
A^{-\kappa} \; : \; L_q (\log L)_{a_1-\varepsilon} \to H_q^{2m\kappa} (\log H)_{a_1-\varepsilon},
$$

$$
\text{id} \quad : \; H_q^{2m\kappa} (\log H)_{a_1-\varepsilon} \to H_{t^1}^1 (\log H)_{-a_2+\varepsilon}, \; a_1 + a_2 - 2\varepsilon > \frac{2}{n} (2m\kappa - 1),
$$

$$
b_2 \quad : \; H_{t^1}^1 (\log H)_{-a_2+\varepsilon} \to H_{p^1}^1.
$$

$$(39)$$

The first mapping is continuous since $b_{1,j}$ maps L_{p^1} to $L_q (\log L)_{a_1-\varepsilon}$ ($\varepsilon > 0$, as in (7)), while we have used the second several times already. Theorems 3.4.3/2 and 3.4.4 show that

$$
e_k (\text{id}) \sim k^{-(2m\kappa-1)/n}, \; k \in \mathbf{N}, \tag{40}
$$

and the final mapping is covered by (30). The proof is now completed in the familiar way.

To conclude this section we give a counterpart of Theorem 5.2.7 on the negative spectrum.

Theorem 5 *Let H_α be given by (5.2.7/2), let $2 < r < \infty$, $rm\kappa = n$, and suppose that*

$$
V^{1/2} g^{-1} \in L_r (\log L)_a \; \text{with} \; ra > 2. \tag{41}
$$

Then

$$
\# \{\sigma (H_\alpha) \cap (-\infty, 0]\} \leq c\alpha^{r/2} \left\| V^{1/2} g^{-1} \mid L_r (\log L)_a \right\|^r. \tag{42}
$$

Proof This is just the same as the proof of Theorem 5.2.7, but with use being made of Theorem 1, with $r_1 = r_2 = r$ and $a_1 = a_2 = a$

5.4 Elliptic operators in $\mathbf{R}^n$

5.4.1 Introduction; the Birman–Schwinger principle revisited

In the preceding sections of this chapter we dealt with degenerate elliptic operators in bounded domains in $\mathbf{R}^n$ and with their spectral properties. These considerations were based on the (unweighted) spaces $B_{pq}^s (\Omega)$, $F_{pq}^s (\Omega)$ and their special cases $H_p^s (\Omega)$, the entropy numbers of related compact embeddings, together with their connections with eigenvalues, and the Birman–Schwinger principle. This principle was the basis for estimates of the negative spectrum. In bounded domains Ω we dealt with problems of this type more or less as a by-product. Now we follow this programme with the whole of $\mathbf{R}^n$ as the underlying domain, and rely on the theory of the weighted spaces developed in Chapter 4. First we are interested in mapping and spectral properties of pseudodifferential operators of type $\Psi^\kappa_{1,\gamma}$, with $\kappa \in \mathbf{R}$ and $0 \le \gamma \le 1$, in the weighted spaces $B^s_{pq}(\mathbf{R}^n, w(\cdot))$ and $F^s_{pq}(\mathbf{R}^n, w(\cdot))$, where $w \in W^n$; see 4.2.1. This will be done in 5.4.2 and 5.4.3. In the following three subsections 5.4.4–5.4.6 we study the distribution of eigenvalues and the behaviour of corresponding root spaces for degenerate pseudodifferential operators, preferably of type $b_2(x)b(x,D)b_1(x)$, where b_1 and b_2 are appropriate functions and $b(x,D) \in \Psi^\kappa_{1,\gamma}$. We use the same techniques as in the preceding sections. The "negative spectrum" (bound states) of related symmetric operators in L_2 will then be considered in greater detail. This is justified by the quantum-mechanical background of problems of this type which are usually connected with the whole of $\mathbf{R}^n$ (and not with bounded domains). This will be done in the subsections 5.4.7–5.4.9. Throughout 5.4 we follow [HaT2] where additional material can also be found.

Since the Birman–Schwinger principle now plays a greater role than before we give a new formulation which is better adapted to our needs. The abstract version has been given in 5.2.1. We follow [HaT2], 2.4. Let $A = a(x,D)$ be a positive-definite self-adjoint pseudodifferential operator in $L_2(\mathbf{R}^n)$, typically of positive order, and let

$$V = a(x)\, v(x,D)\, a(x) \qquad (1)$$

be a degenerate pseudodifferential operator, symmetric in $L_2(\mathbf{R}^n)$, where $v(x,D)$ is a symmetric pseudodifferential operator of lower order than A and a is a real function typically belonging to $L_p(\mathbf{R}^n)$ or $H^s_p(\mathbf{R}^n, w(\cdot))$ according to (4.2.1/10). More precise conditions will be given at the beginning of 5.4.7 which is the first of three subsections dealing with the "negative spectrum". In any case all operators in these subsections may be assumed to be defined at least on $\mathscr{S}(\mathbf{R}^n)$, the rest being always a matter of completion. We are thus not greatly concerned in the formulations below about the respective domains of defintion. We assume that

$$VA^{-1} \text{ is compact (in } L_2(\mathbf{R}^n)). \qquad (2)$$

Then, after completion, $A^{-1}V$ is also compact. Furthermore, $A - V$ is self-adjoint on $\mathrm{dom}(A)$ and the essential spectra of A and $A-V$ coincide. Our interest is in the number of eigenvalues of $A - V$ which are smaller than or equal to 0. Recall that M is a finite set, we denote by $\#M$ the number of elements of M.

Proposition (Birman–Schwinger principle revisited; see 5.2.1) *Let the above conditions be satisfied and let* $\sigma(A - V)$ *be the spectrum of* $A - V$. *Then*

$$\#\{\sigma(A - V) \cap (-\infty, 0]\} \leq \#\left\{k \in \mathbf{N} : \sqrt{2}e_k\left(VA^{-1}\right) \geq 1\right\} \qquad (3)$$

and

$$\#\{\sigma(A - V) \cap (-\infty, 0]\} \leq \#\left\{k \in \mathbf{N} : \sqrt{2}e_k\left(A^{-1}V\right) \geq 1\right\}. \qquad (4)$$

Remark 1 We refer to 5.2.1 where one finds the usual abstract version of the Birman–Schwinger principle in Hilbert spaces which we used in this form in 5.2.7. The concrete versions above were more convenient for us in [HaT2]. The necessary explanations will be given later on, see 5.4.7. Of course, eigenvalues are counted according to their multiplicities. By (3) and (4), $\sigma(A - V) \cap (-\infty, 0]$ consists solely of a finite number of eigenvalues of finite multiplicity (if there are any). Furthermore, $e_k\left(VA^{-1}\right)$ and $e_k\left(A^{-1}V\right)$ are the entropy numbers of the compact operators VA^{-1} and $A^{-1}V$, respectively. Compared with the usual formulation there are two modifications. Firstly, one has usually the eigenvalues of VA^{-1} and $A^{-1}V$ in (3) and (4) instead of $\sqrt{2}e_k$. We used this replacement also in 5.2.1 with a reference to Carl's inequality (1.3.4/11). Now the entropy version above seems to be more natural, since in contrast to the operator K in 5.2.1, its replacements VA^{-1} and $A^{-1}V$ need not be symmetric. Secondly, as we just mentioned, compared with the usual formulation given in 5.2.1, the above formulation is somewhat unsymmetrical in A and V. However, without going into details, Simon's beautiful proof in [Sim2], pp. 86–7, also covers the above version.

Remark 2 Let $v(x, D) = $ id in (1): this is the case in which we are mostly interested. Then V is a multiplication operator and the proposition above can be complemented by

$$\#\{\sigma(A - V) \cap (0, \infty]\} \leq \#\left\{k \in \mathbf{N} : \sqrt{2}e_k\left(aA^{-1}a\right) \geq 1\right\}. \qquad (5)$$

This is nearer to the formulation given in 5.2.1 and in the references given there. In [HaT2] we used the version of the Birman–Schwinger principle given by the above proposition. In 5.4.7 to 5.4.9 we do not consider the most general situation treated in [HaT2], and so (5) will be sufficient for most of the cases studied here.

5.4.2 Pseudodifferential operators: mapping properties

Somewhat in contrast to the preceding sections 5.2, 5.3 we are now exclusively interested in (degenerate) pseudodifferential operators on $\mathbf{R}^n$. We recall that the basic definitions of these were given in 5.2.3, where the Hörmander class $S^{\kappa}_{1,\gamma}$ and the related class $\Psi^{\kappa}_{1,\gamma}$ of pseudodifferential operators were introduced. To study mapping or spectral properties of pseudodifferential operators we may assume $\kappa = 0$, since there is a one-to-one relation between $\Psi^{\kappa}_{1,\gamma}$ and $\Psi^{0}_{1,\gamma}$ given by

$$a(x,D) = a^0(x,D)(\mathrm{id} - \Delta)^{\frac{\kappa}{2}}, \quad \kappa \in \mathbf{R}, \tag{1}$$

where Δ stands for the Laplacian in $\mathbf{R}^n$, $a(x,D) \in \Psi^{\kappa}_{1,\gamma}$ and $a^0(x,D) \in \Psi^{0}_{1,\gamma}$. This can easily be seen from (5.2.3/1) and (5.2.3/2) since $\langle \xi \rangle^{\kappa}$ is the symbol of $(\mathrm{id} - \Delta)^{\frac{\kappa}{2}}$. Recall that $\Psi^{0}_{1,1}$ is called the exotic class. We use the notation introduced in 4.2.1, needing the class W^n and the spaces $B^s_{pq}(w(\cdot))$ and $F^s_{pq}(w(\cdot))$ with $w \in W^n$ where, according to the notational agreement at the end of 4.2.1, we omit "$\mathbf{R}^n$" in the respective notation: all spaces, classes of functions etc. are defined on $\mathbf{R}^n$ in what follows. Finally we use the standard abbreviations $\sigma_p = n\left(\frac{1}{p} - 1\right)_+$ and $\sigma_{pq} = n\left(\frac{1}{\min(p,q)} - 1\right)_+$, see (2.2.3/17).

Theorem *Let $0 \leq \gamma \leq 1$, $a(x,D) \in \Psi^{0}_{1,\gamma}$, $0 < q \leq \infty$ and $w \in W^n$.*

(i) Let $0 < p \leq \infty$, $s \in \mathbf{R}$ (with $s > \sigma_p$ in the exotic case $\gamma = 1$). Then

$$a(x,D) \text{ is a bounded linear map from } B^s_{pq}(w(\cdot)) \text{ into itself.} \tag{2}$$

(ii) Let $0 < p < \infty$, $s \in \mathbf{R}$ (with $s > \sigma_{pq}$ in the exotic case $\gamma = 1$). Then

$$a(x,D) \text{ is a bounded linear map from } F^s_{pq}(w(\cdot)) \text{ into itself.} \tag{3}$$

Proof Step 1 First we recall that the theorem is known for B^s_{pq} and F^s_{pq}, that is, in the unweighted case $w(x) = 1$. We refer for a special case to Theorem 5.2.3 and to the references given in Remark 5.2.3/1, which also cover this more general situation. We take these results for granted and extend them in the following steps from the unweighted to the weighted case.

Step 2 We start with a preparation. Let $f \in F^s_{pq}$ and suppose that

$$\mathrm{supp} f \subset K = \left\{ y \in \mathbf{R}^n : |y| < \sqrt{n} \right\}. \tag{4}$$

Let $k \in \mathbf{N}$ and temporarily let f be sufficiently smooth. Then we have by partial integration,

$$\langle x \rangle^{2k} a(x, D) f(x) = \int \left[(\mathrm{id} - \Delta)_\xi^k e^{ix\xi} \right] a(x, \xi) \hat{f}(\xi) \, d\xi$$

$$= \int e^{ix\xi} \sum_{|\alpha| + |\beta| \le 2k} b_{\alpha\beta} D_\xi^\alpha a(x, \xi) \left(y^\beta f(y) \right)^\wedge (\xi) \, d\xi \tag{5}$$

where the $b_{\alpha\beta}$ are appropriate constants. Here we used the fact that $\left(y^\beta f(y) \right)^\wedge (\xi) = i^{|\beta|} D^\beta \hat{f}(\xi)$. Since $D_\xi^\alpha a(x, \xi) \in S_{1,\gamma}^{-|\alpha|} \subset S_{1,\gamma}^0$ we have by Step 1,

$$\left\| \langle x \rangle^{2k} a(x, D) f \mid F_{pq}^s \right\| \le c \sum_{|\beta| \le 2k} \left\| x^\beta f \mid F_{pq}^s \right\| \le c' \left\| f \mid F_{pq}^s \right\|, \tag{6}$$

where the latter estimate comes from (4) and the pointwise multiplier property for the spaces F_{pq}^s; see [Triγ], 4.2.2, p. 203. Let $f \in F_{pq}^s$ and suppose that (4) holds. By Theorem 2.2.3 we may assume that we have an optimal atomic decomposition in which all atoms have supports near the origin. Then (6) holds for all these atoms and their finite linear combinations f^l, and we have

$$\left\| \langle x \rangle^{2k} a(x, D) f^l \mid F_{pq}^s \right\| \le c \left\| f \mid F_{pq}^s \right\|, \tag{7}$$

where c is independent of l. We may assume that $f^l \to f$ in $\mathscr{S}'$. By the Fatou property discussed in 2.4.2 we obtain

$$\left\| \langle x \rangle^{2k} a(x, D) f \mid F_{pq}^s \right\| \le c \left\| f \mid F_{pq}^s \right\| \tag{8}$$

for all $f \in F_{pq}^s$ with property (4).

Step 3 Let K_k be a ball in $\mathbf{R}^n$ of radius $\sqrt{n}$ and centred at $k \in \mathbf{Z}^n$. Let $\psi_k(x) = \psi(x - k)$ be a related resolution of unity. We use the localisation principle for F_{pq}^s which may be found in [Triγ], 2.4.7, p. 124. Let $f \in F_{pq}^s(w(\cdot))$ and let $k, l \in \mathbf{Z}^n$. By Theorem 4.2.2, the pointwise multiplier property for F_{pq}^s spaces and (8) it follows that

$$
\begin{aligned}
\left\| \psi_l a(\cdot, D) \psi_{k+l} f \mid F_{pq}^s(w(\cdot)) \right\| &\le c_1 w(l) \left\| \psi_l a(\cdot, D) \psi_{k+l} f \mid F_{pq}^s \right\| \\
&\le c_2 w(l) \langle k \rangle^{\cdot} \left\| \psi_{l+k} f \mid F_{pq}^s \right\| \\
&\le c_3 \frac{w(l)}{w(l+k)} \langle k \rangle^{-\nu} \left\| \psi_{l+k} f \mid F_{pq}^s(w(\cdot)) \right\| \\
&\le c_4 \langle k \rangle^{-\mu} \left\| \psi_{l+k} f \mid F_{pq}^s(w(\cdot)) \right\|, \tag{9}
\end{aligned}
$$

where we used (4.2.1/2) in the last estimate. In (9), $\mu > 0$ is at our disposal, while the constants c_2, c_3, c_4 depend on v and μ. Let $\lambda = \min(1, p, q)$. Then (9), $p < \infty$, and the λ triangle inequality for F_{pq}^s yield

$$
\begin{aligned}
\left\| \psi_l a(\cdot, D) f \mid F_{pq}^s(w(\cdot)) \right\|^p &\leq \left\| \psi_l a(\cdot, D) \sum_k \psi_{k+l} f \mid F_{pq}^s(w(\cdot)) \right\|^p \\
&\leq \left(\sum_k \left\| \psi_l a(\cdot, D) \psi_{k+l} f \mid F_{pq}^s(w(\cdot)) \right\|^\lambda \right)^{\frac{p}{\lambda}} \\
&\leq c_1 \left(\sum_k \langle k \rangle^{-\mu} \left\| \psi_{k+l} f \mid F_{pq}^s(w(\cdot)) \right\|^\lambda \right)^{\frac{p}{\lambda}} \\
&\leq c_2 \sum_k \langle k \rangle^{-v} \left\| \psi_{k+l} f \mid F_{pq}^s(w(\cdot)) \right\|^p, \qquad (10)
\end{aligned}
$$

where again $v > 0$ is at our disposal. Summation over $l \in \mathbf{Z}^n$ and the above-mentioned localisation principle for F_{pq}^s prove (3):

$$
\left\| a(\cdot, D) f \mid F_{pq}^s(w(\cdot)) \right\| \leq c \left\| f \mid F_{pq}^s(w(\cdot)) \right\|. \qquad (11)
$$

Step 4 By the same argument we have

$$
\left\| a(\cdot, D) f \mid B_{\infty\infty}^s(w(\cdot)) \right\| \leq c \left\| f \mid B_{\infty\infty}^s(w(\cdot)) \right\|. \qquad (12)
$$

We use the real interpolation formula

$$
B_{pq}^s(w(\cdot)) = \left(F_{pq_0}^{s_0}(w(\cdot)), F_{pq_1}^{s_1}(w(\cdot)) \right)_{\theta, q} \qquad (13)
$$

with $s_0 < s < s_1$, $s = (1 - \theta)s_0 + \theta s_1$, $0 < p < \infty$, $0 < q_0 \leq \infty$, $0 < q_1 \leq \infty$ and $0 < q \leq \infty$. If $p = \infty$ then we have (13) with $B_{\infty\infty}^{s_0}(w(\cdot))$ and $B_{\infty\infty}^{s_1}(w(\cdot))$ in place of $F_{pq_0}^{s_0}(w(\cdot))$ and $F_{pq_1}^{s_1}(w(\cdot))$, respectively. The unweighted case of (13) (including the indicated modifications if $p = \infty$) may be found in [Triβ], 2.4.2, p. 64. Its weighted generalisation follows from the unweighted case and Theorem 4.2.2(ii). Now (2) is a consequence of (11), complemented by (12) and the interpolation property.

Remark 1 In the unweighted case one needs only finitely many of the assumptions (5.2.3/1) and all estimates depend only on these related numbers $c_{\alpha,\beta}$ in (5.2.3/1). The above proof shows that one needs, in the weighted case also, only finitely many of the conditions (5.2.3/1). Recently Dintelmann [Di] has given a direct proof of the weighted case which does not require knowledge of the unweighted case.

Remark 2 Let $a(x,D) \in \Psi_{1,\gamma}^\kappa$ for some $\kappa \in \mathbf{R}$ and $0 \le \gamma \le 1$. By (1) and the above theorem it follows that for $0 < p < \infty$, $0 < q \le \infty$ and $s \in \mathbf{R}$ (with $s - \kappa > \sigma_{pq}$ in the exotic case $\gamma = 1$),

$$a(\cdot, D) \text{ is a bounded map from } F_{pq}^s(w(\cdot)) \text{ into } F_{pq}^{s-\kappa}(w(\cdot)). \tag{14}$$

The B spaces can be handled in a similar manner.

5.4.3 Pseudodifferential operators: spectral properties

As in 5.2.6, 5.3.3 we study degenerate pseudodifferential operators of type $b_2(x) b(x,D) b_1(x)$ by looking at mapping properties of b_1, b_2 and $b(\cdot, D)$ between function spaces of the above type, preferably $H_p^s(w(\cdot))$; see (4.2.1/10). Hence we have to handle several types of these spaces simultaneously. This makes it desirable to have a closer look at the dependence of the spectral properties especially of the operators of the class $\Psi_{1,\gamma}^\kappa$ on the underlying spaces. By (5.4.2/1) and the related remarks we may restrict our attention to the spectral properties of the zero class $\Psi_{1,\gamma}^0$. By Theorem 5.4.2 the operator $A = a(\cdot, D) \in \Psi_{1,\gamma}^0$ might be considered as a bounded linear mapping in $B_{pq}^s(w(\cdot))$ or $F_{pq}^s(w(\cdot))$. As usual

$$\rho\left(A, B_{pq}^s(w(\cdot))\right)$$
$$= \left\{\lambda \in \mathbf{C} : (A - \lambda \mathrm{id})^{-1} \text{ exists as a bounded map in } B_{pq}^s(w(\cdot))\right\} \tag{1}$$

is the related resolvent set. Similarly $\rho\left(A, F_{pq}^s(w(\cdot))\right)$ is defined. For shortness, we temporarily let $\rho(A) = \rho(A, L_2)$.

Theorem *Let* $0 \le \gamma < 1$, $A = a(\cdot, D) \in \Psi_{1,\gamma}^0$, $s \in \mathbf{R}$, $0 < q \le \infty$ *and* $w \in W^n$.

(i) Suppose that $0 < p \le \infty$. *Then*

$$\rho(A) \subset \rho\left(A, B_{pq}^s(w(\cdot))\right). \tag{2}$$

(ii) Let $0 < p < \infty$. *Then*

$$\rho(A) \subset \rho\left(A, F_{pq}^s(w(\cdot))\right). \tag{3}$$

Proof Suppose that $\lambda \in \rho(A)$; then

$$(A - \lambda \mathrm{id})^{-1} \in \Psi_{1,\gamma}^0. \tag{4}$$

This remarkable result is due to R. Beals and J. Überberg, see [Bea] and [Übe]. Then $(A - \lambda \mathrm{id})^{-1}$ is also the inverse on $\mathscr{S}$. Since the exotic case

$\gamma = 1$ is excluded, it follows by duality arguments that $(A - \lambda \text{id})^{-1}$ is also the inverse on $\mathscr{S}'$. Now the theorem follows from Theorem 5.4.2.

Remark 1 We emphasise that the exotic case $\gamma = 1$ is now excluded.

Remark 2 It seems to be natural to expect that L_2 is in some sense the best space; in particular L_2 should be the space with the largest resolvent set. Hence, by (2) and (3) the following conjecture seems to be natural.

Conjecture Under the assumptions of the theorem we have

$$\rho(A) = \rho\left(A, B_{pq}^s(w(\cdot))\right), \ 0 < p \le \infty, \tag{5}$$

and

$$\rho(A) = \rho\left(A, F_{pq}^s(w(\cdot))\right), \ 0 < p < \infty. \tag{6}$$

Remark 3 As usual, the complement of the resolvent set in $\mathbf{C}$ is called the spectrum. Hence the above conjecture is the problem of spectral invariance. Some affirmative answers are known: (6) for $H^s = F_{2,2}^s$ may be found in [Bea]. This was extended in [Übe] to $H_p^s = F_{p,2}^s$ with $1 < p < \infty$. Further extensions to B_{pq}^s and F_{pq}^s with $1 < p < \infty, 0 < q < \infty$ were given in [LeS]. In [Schr] some (anisotropic) weights are included. In [LeT] we proved the above conjecture for those $A \in \Psi_{1,\gamma}^0$ which are self-dual and self-adjoint, so that $A = A' = A^*$. This subclass is of interest especially in connection with the problem of the "negative spectrum" treated in subsections 5.4.7–5.4.9.

5.4.4 Degenerate pseudodifferential operators: eigenvalue distributions

Sections 5.2 and 5.3 dealt with the eigenvalue distributions of degenerate elliptic operators in bounded smooth domains with a preference for fractional powers of regular elliptic operators as one of the main ingredients. We now use the same technique as in those sections, but this time we are on $\mathbf{R}^n$ and concentrate on operators of type

$$Bf = b_2 b(\cdot, D) b_1 f, \tag{1}$$

where b_1 and b_2 belong to some spaces $L_{r_j}(\alpha_j) = L_{r_j}(\langle x \rangle^{\alpha_j})$ or $H_{r_j}^s(\alpha_j)$ and $b(\cdot, D) \in \Psi_{1,\gamma}^{-\kappa}$ with $\kappa > 0$ and $0 \le \gamma \le 1$. For the notation we refer to 4.2.1 and especially the notational agreement at the end of that subsection, 4.2.5 and the beginning of 5.4.2. To simplify the situation we confine discussion in this subsection to the "ground level" $s = 0$, so that we are looking for weighted L_p spaces, $1 < p < \infty$, for which B becomes compact. In that case we denote by $\{\mu_k\}$ the sequence of the non-vanishing eigenvalues of B, counted with respect to their algebraic multiplicity and ordered by decreasing modulus, so that $|\mu_1| \geq |\mu_2| \geq ... > 0$. The collection of all associated eigenvectors of a given eigenvalue is the corresponding root space.

The phrase "the root spaces coincide" in the theorem below means that for the different basic spaces in question the eigenvalues coincide and that for any given eigenvalue the corresponding root spaces coincide. We use the notation introduced in 4.2.1, especially the weight class W^n, $L_p(w(\cdot))$, $H_p^s(w(\cdot)\langle\cdot\rangle^\alpha)$, etc. Tacitly we always assume $\mu_k \neq 0$, especially in the assertions about the root spaces. Furthermore, (7) and (9) do not exclude the possibility that there may be only finitely many eigenvalues $\mu_k \neq 0$.

Theorem *Suppose* $\kappa > 0$, $0 \leq \gamma \leq 1$ *and*

$$b(\cdot,D) \in \Psi_{1,\gamma}^{-\kappa}, \tag{2}$$

$$\alpha_1, \alpha_2 \in \mathbf{R} \text{ with } \alpha = \alpha_1 + \alpha_2 > 0, \tag{3}$$

$$r_1, r_2 \in [1,\infty] \text{ with } \frac{1}{r_1} + \frac{1}{r_2} < \min\left(1, \frac{\kappa}{n}\right), \tag{4}$$

$$b_1 \in L_{r_1}(\alpha_1), \quad b_2 \in L_{r_2}(\alpha_2). \tag{5}$$

(i) For any p with

$$\frac{1}{r_2} < \frac{1}{p} < 1 - \frac{1}{r_1} \tag{6}$$

and any $w \in W^n$, *the operator* B *given by (1) is compact in* $L_p(w(\cdot))$. *Furthermore, for some* $c > 0$,

$$|\mu_k| \leq c \, \|b_1 \mid L_{r_1}(\alpha_1)\| \, \|b_2 \mid L_{r_2}(\alpha_2)\| \, k^{-\kappa/n}, \quad k \in \mathbf{N}, \tag{7}$$

if

$$\frac{1}{r_1} + \frac{1}{r_2} > \frac{\kappa - \alpha}{n} \tag{8}$$

and

$$|\mu_k| \leq c_\varepsilon \, \|b_1 \mid L_{r_1}(\alpha_1)\| \, \|b_2 \mid L_{r_2}(\alpha_2)\| \, k^{-\frac{\alpha}{n} - \left(\frac{1}{r_1} + \frac{1}{r_2}\right)} (\log k)^{\varepsilon + \frac{1}{r_1} + \frac{1}{r_2}}, \tag{9}$$

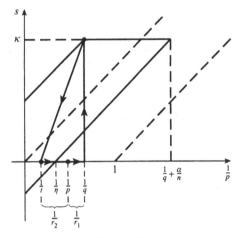

Fig. 5.14.

for all $k \in \mathbf{N}$, $k > 1$, all $\varepsilon > 0$ (with $\varepsilon = 0$ if $r_1 = r_2 = \infty$) and suitably chosen constants $c_\varepsilon > 0$ if

$$\frac{1}{r_1} + \frac{1}{r_2} < \frac{\kappa - \alpha}{n}. \tag{10}$$

(ii) For fixed p satisfying (6) the root spaces coincide for all $w \in W^n$.

Proof Step 1 We use the same technique as in 5.2.4, 5.2.6 and 5.3.3 and decompose B as $B = b_2 \circ \mathrm{id} \circ b(\cdot, D) \circ b_1$ as indicated in Fig. 5.14, where

$$\left.\begin{aligned}
b_1 \quad &: L_p\left(w(\cdot)\right) \to L_q\left(w(\cdot)\langle\cdot\rangle^{\alpha_1}\right), \; \tfrac{1}{q} = \tfrac{1}{p} + \tfrac{1}{r_1}, \\[4pt]
b(x,D) \quad &: L_q\left(w(\cdot)\langle\cdot\rangle^{\alpha_1}\right) \to H_q^\kappa\left(w(\cdot)\langle\cdot\rangle^{\alpha_1}\right), \\[4pt]
\mathrm{id} \quad &: H_q^\kappa\left(w(\cdot)\langle\cdot\rangle^{\alpha_1}\right) \to L_t\left(w(\cdot)\langle\cdot\rangle^{-\alpha_2}\right), \; \tfrac{1}{t} = \tfrac{1}{p} - \tfrac{1}{r_2}, \\[4pt]
b_2 \quad &: L_t\left(w(\cdot)\langle\cdot\rangle^{-\alpha_2}\right) \to L_p\left(w(\cdot)\right).
\end{aligned}\right\} \tag{11}$$

The first line in (11) is simply Hölder's inequality; see (6). The second line is covered by (4.2.1/10) and Remark 5.4.2/2. As for the third line in (11) we have

$$\frac{1}{r_1} + \frac{1}{r_2} < \frac{\kappa}{n} \tag{12}$$

and hence the corresponding line in Fig. 5.14 has a slope steeper than n. The quotient of the two weights involved is $\langle x \rangle^\alpha$. By our remarks at the beginning of 4.3.2 we can apply Theorem 4.3.2. Hence id is a compact embedding and we have for the corresponding entropy numbers

$$e_k \, (\text{id}) \sim k^{-\kappa/n} \text{ if } \frac{1}{t} < \frac{1}{\eta} \tag{13}$$

and

$$e_k \, (\text{id}) \le c_\varepsilon k^{-\frac{\kappa}{n} - \left(\frac{1}{r_1} + \frac{1}{r_2}\right)} (\log k)^{\varepsilon + \frac{1}{r_1} + \frac{1}{r_2}} \text{ if } \frac{1}{t} > \frac{1}{\eta}, \tag{14}$$

where $k > 1$ and

$$\frac{1}{\eta} = \frac{\alpha}{n} + \frac{1}{q} - \frac{\kappa}{n} \tag{15}$$

has the meaning indicated in Fig. 5.14, so that the slope of the line connecting $\left(\frac{1}{\eta}, 0\right)$ and $\left(\frac{\alpha}{n} + \frac{1}{q}, \kappa\right)$ is n. The conditions in (13) and (14) coincide with those in (8) and (10), respectively. Furthermore, (13) corresponds to region I and (4.3.2/10), and (14) to region III and (4.3.2/12) in Fig. 4.2 on p. 170 and Theorem 4.3.2. The last line in (11) is again simply a consequence of Hölder's inequality. Now part (i) of the theorem follows from (13), (14) and Corollary 1.3.4.

Step 2 The embedding id also maps $H_q^\kappa \left(w(\cdot) \, \langle \cdot \rangle^{\alpha_1}\right)$ into $L_t \left(w(\cdot) \, \langle \cdot \rangle^{\alpha_1}\right)$. This follows from Theorem 4.2.2 and the unweighted embedding $H_q^\kappa \subset L_t$ which is covered by (12) and (2.3.3/8). Application of b_2 shows that B also maps $L_p \left(w(\cdot)\right)$ into $L_p \left(w(\cdot) \, \langle \cdot \rangle^\alpha\right)$ with $\alpha = \alpha_1 + \alpha_2 > 0$. Now one can start again with $w(\cdot) \, \langle \cdot \rangle^\alpha$ instead of $w(\cdot)$. By iteration the root spaces of all the basic spaces $L_p \left(w(\cdot) \, \langle \cdot \rangle^\beta\right)$, where β is arbitrary and positive, coincide. But any weight belonging to W^n can be dominated by $w(\cdot) \, \langle \cdot \rangle^\beta$ for some $\beta > 0$. This completes the proof of part (ii).

Remark 1 The root spaces are independent of $w \in W^n$ if p, satisfying (6), is fixed. As we shall see in the following subsection they are also independent of all admissible values of p. A first step in this direction can be taken rather easily. Suppose that

$$\frac{1}{r_2} < \frac{1}{p} \le \frac{1}{u} < 1 - \frac{1}{r_1}. \tag{16}$$

Then (6) is satisfied for both p and u. By part (ii) of the theorem and Hölder's inequality it follows that any associated eigenvector for $L_p \left(w(\cdot)\right)$ is also an associated eigenvector for $L_u \left(w(\cdot)\right)$. As we shall see, the opposite assertion is also true.

Remark 2 By Remark 5.4.2/1 and the above proof it follows that the constants c and c_ε in (7) and (9), respectively, depend only on finitely many constants $c_{\alpha,\beta}$ in (5.2.3/1).

5.4.5 Degenerate pseudodifferential operators: smoothness theory

Just as in 5.2.4 and 5.3.3 one might expect that Theorem 5.4.4 could be complemented by a smoothness theory: if b_1 and b_2 are smoother than stated in (5.4.4/5), then the associated eigenvectors should also be smoother. We treated this problem in some detail in [HaT2] and repeat here some of the main results obtained there, again closely following that paper. A natural replacement of, say, the hypothesis that $b_2 \in L_{r_2}(\alpha_2)$ in (5.4.4/5) is the seemingly very closely related assumption that $b_2 \in H_{r_2^\nu}^\nu(\alpha_2)$ for some $\nu \geq 0$. For notation we refer to 4.2.1, including the notational agreement at the end of 4.2.1, and to 4.2.5; recall in particular that $\frac{1}{r_2^\nu} = \frac{1}{r_2} + \frac{\nu}{n}$. Then we have the limiting embedding

$$H_{r_2^\nu}^\nu(\alpha_2) \subset L_{r_2}(\alpha_2) ; \tag{1}$$

see Theorem 4.2.2 and (2.3.3/8).

Theorem *Let B be given by (5.4.4/1) under the same general assumptions as in Theorem 5.4.4; in particular $w \in W^n$, and (5.4.4/2)–(5.4.4/4) and (5.4.4/6) are assumed to be satisfied. Let (5.4.4/5) be strengthened by*

$$b_1 \in L_{r_1}(\alpha_1) \text{ and } b_2 \in H_{r_2^\nu}^\nu(\alpha_2) \tag{2}$$

for some $\nu \geq 0$ with

$$\nu < n\left(\frac{1}{p} - \frac{1}{r_2}\right) \text{ and } \nu \leq \kappa - n\left(\frac{1}{r_1} + \frac{1}{r_2}\right). \tag{3}$$

(i) Then B is compact in $L_p(w(\cdot))$ and

$$\operatorname{Im} B \subset H_p^\nu\left(w(\cdot)\langle\cdot\rangle^\alpha\right). \tag{4}$$

(ii) The root spaces for all admissible p according to (5.4.4/6) and all $w \in W^n$ coincide.

Proof Step 1 The compactness of B follows from (1) and Theorem 5.4.4. We prove (4) by travelling around in $\left(\frac{1}{p}, s\right)$-diagrams, now employing our improved knowledge (2) rather than (5.4.4/5); see (1). There are two typical situations, characterised by Fig. 5.14 and by Fig. 5.15 in

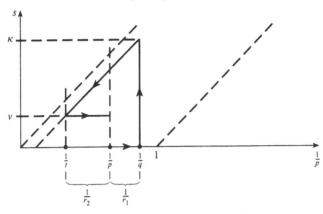

Fig. 5.15.

this subsection. Of course, we always have (5.4.4/11). We deal with the limiting situation

$$0 < v = \kappa - n\left(\frac{1}{r_1} + \frac{1}{r_2}\right) < \frac{n}{t} = n\left(\frac{1}{p} - \frac{1}{r_2}\right) \qquad (5)$$

which corresponds with that in Fig. 5.15. Then (5.4.4/11) can be complemented by

$$
\left.
\begin{aligned}
b_1 \quad &: \ L_p\left(w(\cdot)\right) \qquad\qquad \to \ L_q\left(w(\cdot)\langle\cdot\rangle^{\alpha_1}\right), \\[4pt]
b(x,D) \quad &: \ L_q\left(w(\cdot)\langle\cdot\rangle^{\alpha_1}\right) \ \to \ H_q^\kappa\left(w(\cdot)\langle\cdot\rangle^{\alpha_1}\right), \\[4pt]
\mathrm{id} \quad &: \ H_q^\kappa\left(w(\cdot)\langle\cdot\rangle^{\alpha_1}\right) \ \to \ H_t^v\left(w(\cdot)\langle\cdot\rangle^{\alpha_1}\right), \\[4pt]
b_2 \quad &: \ H_t^v\left(w(\cdot)\langle\cdot\rangle^{\alpha_1}\right) \ \to \ H_p^v\left(w(\cdot)\langle\cdot\rangle^{\alpha}\right),
\end{aligned}
\right\}
\qquad (6)
$$

where q and t have the same meaning as there. The first and the second line coincide with (5.4.4/11). The third line follows from (5), Theorem 4.2.2 and the limiting embedding

$$H_q^\kappa \subset H_t^v, \ \kappa - \frac{n}{q} = v - \frac{n}{t}, \qquad (7)$$

see (2.3.3/8). Finally, the fourth line is a consequence of Theorem 4.2.5 and

$$\frac{1}{t} - \frac{v}{n} + \frac{1}{r_2} = \frac{1}{p} - \frac{v}{n}, \qquad (8)$$

under the assumption (5). Hence we have (4). Obviously, the situation in the above figure is even better if $0 \le v < \kappa - n\left(\frac{1}{r_1} + \frac{1}{r_2}\right)$. Also in the case of Fig. 5.14 the necessary modifications are clear.

Step 2 We prove (ii). By Remark 5.4.4/1 and (5.4.4/16) we have to deal with the case

$$\frac{1}{r_2} < \frac{1}{u} < \frac{1}{p} < 1 - \frac{1}{r_1}. \tag{9}$$

We modify (5.4.4/4) by

$$\frac{1}{r_1} + \frac{1}{r_2} + \frac{1}{p} - \frac{1}{u} < \min\left(1, \frac{\kappa}{n}\right). \tag{10}$$

Then the arguments related to Fig. 5.14 remain valid if one replaces $\frac{1}{t}$ by $\frac{1}{t} - \left(\frac{1}{p} - \frac{1}{u}\right)$. This proves that any associated eigenvector for $L_p(w(\cdot))$ is also an associated eigenvector for $L_u(w(\cdot))$. Iteration and Theorem 5.4.4 now prove part (ii) of the above theorem.

Remark 1 Of course, part (ii) applies only to eigenvalues $\mu_k \ne 0$.

Remark 2 Both in Theorem 5.4.4 and in the above theorem we remained at the ground level $s = 0$. But this is not necessary. By Theorems 4.2.5 and 5.4.2 one can start with any space $H_p^s(w(\cdot))$ if the conditions are satisfied. In particular, the independence of the root spaces can be extended to more general basic spaces. We refer for some results in this direction to [HaT2], especially 4.2, Theorem 2. The theorem above coincides with Theorem 1 in 4.2 of that paper.

5.4.6 Degenerate pseudodifferential operators of positive order

Up to now we have dealt with the operator

$$Bf = b_2 b(\cdot, D) b_1 f, \quad b(\cdot, D) \in \Psi_{1,\gamma}^{-\kappa}, \ \kappa > 0. \tag{1}$$

Now we are interested in degenerate pseudodifferential operators of positive order given by, at least formally,

$$Af = a_1 a(\cdot, D) a_2 f \tag{2}$$

with

$$a(\cdot, D) \in \Psi_{1,\gamma}^{\kappa}, \ \kappa > 0, \ 0 \le \gamma < 1. \tag{3}$$

One may look at A as a bounded mapping between suitable pairs of function spaces or as an unbounded operator in a fixed basic function space, say, of type $H_p^s(w(\cdot))$. We adopt the latter point of view and ask for the spectrum of A and its root spaces. Assume that $a(\cdot,D)$ is invertible in L_2 and that the origin belongs to the resolvent set of $a(\cdot,D)$, which means that

$$0 \in \rho(a(\cdot,D)) \tag{4}$$

according to 1.2 and 5.4.3 with $\mathrm{dom}(a(\cdot,D)) = H_2^\kappa$. Since the exotic case $\gamma = 1$ is now excluded, it follows by Theorem 5.4.3 and what has been said before (see also (5.4.2/1)) that (4) also holds for any other space of interest in our context, and in particular for $H_p^s(w(\cdot))$ with $s \in \mathbf{R}$, $0 < p < \infty$, $w \in W^n$ without indicating that basic space in (4). Of course one has to choose $\mathrm{dom}(a(\cdot,D)) = H_p^{s+\kappa}(w(\cdot))$ in that case and we have

$$b(\cdot,D) = a^{-1}(\cdot,D) \in \Psi_{1,\gamma}^{-\kappa}. \tag{5}$$

Assume that $a_1(x) \neq 0$ and $a_2(x) \neq 0$ a.e. in $\mathbf{R}^n$ and that

$$b_1(x) = a_1^{-1}(x) \text{ and } b_2(x) = a_2^{-1}(x) \tag{6}$$

satisfy, together with (5), the hypotheses (5.4.4/2)–(5.4.4/6) of Theorem 5.4.4. Then it can easily be seen that B, given by (1), is invertible in $L_p(w(\cdot))$, where p is restricted by (5.4.4/6) and $w \in W^n$. Its inverse $A = B^{-1}$ can be written, at least formally, by (2). Since $\gamma < 1$, the dual of $b(\cdot,D)$ also belongs to $\Psi_{1,\gamma}^{-\kappa}$; see [Tay2], p. 13. But then it follows readily that Theorem 5.4.4 can also be applied to the dual of B, say, in $L_{p'}(w^{-1}(\cdot))$, and this dual is invertible. This has the consequence that by Theorem 5.4.4 the range of B considered in $L_p(w(\cdot))$ is dense in $L_p(w(\cdot))$. Hence its inverse A is densely defined in $L_p(w(\cdot))$. We now apply the spectral theory developed in 1.2. By (1.2/2), (1.2/4), Proposition 1.2/3 and Proposition 1.2/2 it follows that A is an unbounded operator in $L_p(w(\cdot))$ with pure point spectrum. Let $\{\lambda_k\}$ be the sequence of its eigenvalues, counted according to algebraic multiplicity and ordered by increasing modulus; then $\lambda_k = \mu_k^{-1}$ where the μ_k are the eigenvalues of B according to Theorem 5.4.4. That theorem proves the following assertion.

Theorem *Let the above hypotheses be satisfied. Then for some* $c > 0$,

$$|\lambda_k| \geq c \, \|b_1 \mid L_{r_1}(\alpha_1)\|^{-1} \|b_2 \mid L_{r_2}(\alpha_2)\|^{-1} k^{\frac{\kappa}{n}}, \quad k \in \mathbf{N}, \tag{7}$$

if

$$\frac{1}{r_1} + \frac{1}{r_2} > \frac{\kappa - \alpha}{n};$$ (8)

and

$$|\lambda_k| \ge c_\varepsilon \, \|b_1 \mid L_{r_1}(\alpha_1)\|^{-1} \, \|b_2 \mid L_{r_2}(\alpha_2)\|^{-1} \, k^{\frac{\alpha}{n}+\frac{1}{r_1}+\frac{1}{r_2}} \, (\log k)^{-\varepsilon - \frac{1}{r_1} - \frac{1}{r_2}},$$ (9)

for all $k \in \mathbf{N}$, $k > 1$, *and any* $\varepsilon > 0$ *(with* $\varepsilon = 0$ *if* $r_1 = r_2 = \infty$*) and suitably chosen constants* $c_\varepsilon > 0$ *if*

$$\frac{1}{r_1} + \frac{1}{r_2} < \frac{\kappa - \alpha}{n}.$$ (10)

Proof This follows from the remarks just before the theorem and Theorem 5.4.4, where now the exotic case $\gamma = 1$ is excluded; see (3).

Remark Of course, part (ii) of Theorem 5.4.4 also has an immediate counterpart. The same holds for the smoothness theory developed in 5.4.5. We again followed [HaT2] and refer to that paper for further information.

5.4.7 The negative spectrum: basic results

We described in 5.4.1 the problem of the negative spectrum. Here we wish to combine Proposition 5.4.1 and (5.4.1/5) on the one hand with the results and techniques developed in the preceding subsections on the other hand. Our basic space in this subsection and the following ones is L_2, always on $\mathbf{R}^n$ although any other Hilbert space $H_2^s(w(\cdot))$ can be taken. In [HaT2], 5.2 we studied in detail the behaviour of the negative spectrum of the self-adjoint unbounded operator

$$A_\beta = a(\cdot, D) - \beta b v(\cdot, D) b \text{ as } \beta \to \infty,$$ (1)

where

$$a(\cdot, D) \in \Psi_{1,\gamma}^\kappa, \text{ with } \kappa > 0 \text{ and } 0 \le \gamma < 1,$$ (2)

is assumed to be positive-definite and self-adjoint in L_2, whereas

$$v(\cdot, D) \in \Psi_{1,\gamma}^\eta \text{ with } -\infty < \eta < \kappa, \, 0 \le \gamma < 1,$$ (3)

is symmetric and b is a real function. For that purpose we used in [HaT2] both parts of Proposition 5.4.1 and (5.4.1/5). Here we restrict ourselves to the technically simple case $v(\cdot, D) = \mathrm{id}$. As we said in 5.4.1 we shall not bother about the domains of definition: everything can be done on $\mathscr{S}$, and the rest may be considered as a matter of completion.

Theorem *Let* $\kappa > 0$, $0 \le \gamma < 1$, $\alpha > 0$ *and*

$$\infty \ge r > \max\left(2, \frac{2n}{\kappa}\right). \tag{4}$$

Suppose that

$$a(\cdot, D) \in \Psi_{1,\gamma}^{\kappa} \text{ is positive-definite and self-adjoint in } L_2, \tag{5}$$

$$b \in L_r\left(\alpha/2\right), \quad b \text{ real-valued}, \tag{6}$$

and

$$H_\beta = a(\cdot, D) - \beta b^2, \quad \beta \ge 1. \tag{7}$$

Then H_β *is self-adjoint in* L_2. *Furthermore, for some* $c > 0$,

$$\#\left\{\sigma\left(H_\beta\right) \cap (-\infty, 0]\right\} \le c \left(\beta \left\|b \mid L_r\left(\alpha/2\right)\right\|^2\right)^{\frac{n}{\kappa}} \tag{8}$$

if $\kappa - \alpha < \frac{2n}{r}$; *and*

$$\left[\#\left\{\sigma\left(H_\beta\right) \cap (-\infty, 0]\right\}\right]^{\frac{\alpha}{n} + \frac{2}{r}} \le c_\varepsilon \beta_0 \left(\log(1 + \beta_0)\right)^{\varepsilon + \frac{2}{r}} \tag{9}$$

with $\beta_0 = \beta \left\|b \mid L_r\left(\alpha/2\right)\right\|^2$ *for any* $\varepsilon > 0$ *(with* $\varepsilon = 0$ *if* $r = \infty$*) and suitably chosen constants* $c_\varepsilon > 0$ *if* $\kappa - \alpha > \frac{2n}{r}$.

Proof We apply (5.4.1/5). Then

$$\#\left\{\sigma\left(H_\beta\right) \cap (-\infty, 0]\right\} \le \#\left\{k \in \mathbf{N} : \sqrt{2}e_k \ge \beta^{-1}\right\}, \tag{10}$$

where $\sqrt{2}e_k$ can be replaced by $|\mu_k|$ and μ_k are the eigenvalues of the operator

$$bb(\cdot, D)b \text{ with } b(\cdot, D) = a^{-1}(\cdot, D) \in \Psi_{1,\gamma}^{-\kappa}. \tag{11}$$

By Theorem 5.4.4 with $r_1 = r_2 = r$, $p = 2$ and $\alpha_1 = \alpha_2 = \frac{\alpha}{2}$ we have

$$ck^{\kappa/n} \le \beta \left\|b \mid L_r\left(\alpha/2\right)\right\|^2 \tag{12}$$

if $\frac{\kappa - \alpha}{n} < \frac{2n}{r}$; *and*

$$c_\varepsilon k^{\frac{\alpha}{n} + \frac{2}{r}} (\log k)^{-\varepsilon - \frac{2}{r}} \le \beta \left\|b \mid L_r\left(\alpha/2\right)\right\|^2 = \beta_0 \tag{13}$$

if $\kappa - \alpha > \frac{2n}{r}$. In the case of (13) we have

$$\log k \le c_\varepsilon \log(1 + \beta_0), \tag{14}$$

which shows that (13) can be reformulated as

$$k^{\frac{\alpha}{n}+\frac{2}{r}} \le c_\varepsilon \beta_0 \left(\log(1 + \beta_0)\right)^{\varepsilon + \frac{2}{r}}. \tag{15}$$

Now (8) follows from (12) and (9) from (15).

Remark 1 Of course, c in (8) and c_ε in (9) are independent of b and $k \in \mathbf{N}$. In [BiS6] the exponent $\frac{n}{\kappa}$ in (8) is called "semiclassical"; it gives the expected behaviour. In that paper sharper results are obtained, based on specific Hilbert space methods and where $a(x, D) = (-\Delta)^l$.

Remark 2 The advantage of (7) compared with (1) is not only its simplicity, but also that one can use the usual symmetric version of the Birman–Schwinger principle formulated in (5.4.1/5). In the case of (1) one has to rely on the unsymmetric version in Proposition 5.4.1. But otherwise the method is rather clear: just as in 5.4.4 and 5.4.5 one travels in $\left(\frac{1}{p}, s\right)$-diagrams using mapping properties for pseudodifferential operators, Hölder inequalities for $H_p^s(w(\cdot))$ spaces, and estimates for entropy numbers of compact embeddings between weighted function spaces. We refer to [HaT2], 5.2, where this programme was followed in some detail. In particular, if H_β is given by (1)–(3) and if one has a suitable replacement of (6) then one obtains a counterpart of (8) with the semiclassical exponent $\frac{n}{\kappa - \eta}$ in place of $\frac{n}{\kappa}$.

5.4.8 The negative spectrum: splitting techniques

The interest in studying the "negative spectrum" (bound states) comes from quantum mechanics, generalising the classical hydrogen operator,

$$H = -\Delta - \frac{c}{|x|}, \quad c > 0, \tag{1}$$

in $L_2\left(\mathbf{R}^3\right)$. Compared with (5.4.7/7) the potential $b(x)$ is of type $|x|^{-\alpha}$, $\alpha > 0$, having local singularities and some decay properties at infinity. Conditions of type (5.4.7/6) may be not very well adapted to such a situation. However, one should compare this remark with the results in [BiS6] which generalise the famous Cwikel–Rosenbljum–Lieb inequality. In our context it seems to be reasonable to split the potential b into two parts,

$$b = b_1 + b_2, \tag{2}$$

where b_1 is compactly supported, collecting the local singularities; and, say, $\langle \cdot \rangle^\alpha b_2 \in L_\infty$ for some $\alpha > 0$.

Theorem *Let* $\kappa > 0$, $0 \leq \gamma < 1$, $\alpha > 0$, $\kappa \neq \alpha$,

$$\infty \geq r > \max\left(2, \frac{2n}{\kappa}\right), \tag{3}$$

and suppose that

$$b = b_1 + b_2, \quad b_1 \text{ and } b_2 \text{ real}, \tag{4}$$

where b_1 *has compact support and*

$$b_1 \in L_r, \quad \langle \cdot \rangle^{\frac{\alpha}{2}} b_2 \in L_\infty. \tag{5}$$

Let

$$a(\cdot, D) \in \Psi^\kappa_{1,\gamma} \text{ be positive-definite and self-adjoint in } L_2 \tag{6}$$

and put

$$H_\beta = a(\cdot, D) - \beta b^2, \quad \beta \geq 1. \tag{7}$$

Then H_β *is self-adjoint in* L_2. *Furthermore, for some* $c > 0$,

$$\left[\#\left\{\sigma\left(H_\beta\right) \cap (-\infty, 0]\right\}\right]^{\frac{\min(\kappa, \alpha)}{n}} \leq c\beta \left(\|b_1 \mid L_r\| + \left\|\langle \cdot \rangle^{\frac{\alpha}{2}} b_2 \mid L_\infty\right\|\right)^2. \tag{8}$$

Proof We use the same arguments as in the proof of Theorem 5.4.7 and put

$$B = bb(\cdot, D)b = B^{1,1} + B^{1,2} + B^{2,1} + B^{2,2}, \tag{9}$$

where again $b(\cdot, D) = a^{-1}(\cdot, D) \in \Psi^{-\kappa}_{1,\gamma}$ and

$$B^{j,k} = b_j b(\cdot, D) b_k. \tag{10}$$

Since b_1 has compact support we have $b_1 \in L_r(\gamma)$ for any $\gamma > 0$. Then it follows from the method in the proof of Theorem 5.4.4 that the entropy numbers of the compact operators $B^{1,1}$, $B^{1,2}$ and $B^{2,1}$ in L_2 can be estimated from above by $e_k \leq ck^{-\kappa/n}$, $k \in \mathbf{N}$. In the case of $B^{2,2}$ we have $r_1 = r_2 = r = \infty$ and hence $\varepsilon = 0$ in (5.4.4/14). Then the corresponding entropy numbers can be estimated from above by $ck^{-\frac{\min(\kappa,\alpha)}{n}}$, $k \in \mathbf{N}$. Hence by (9) the entropy numbers of B can also be estimated from above in that way. Now we obtain (8) by the same arguments as in 5.4.7.

Remark There are two interesting limiting cases. If $\kappa = \alpha$, then one cannot expect an assertion of type (8). By the underlying Theorem 4.3.2 an additional log-term should occur. Secondly, the limiting case $\frac{2n}{\kappa} = r \geq 2$ is of peculiar interest. Here we refer to 5.3.2 where we needed in these cases some limiting embeddings in Orlicz spaces. See also [HaT2], 5.3.

5.4.9 The negative spectrum: homogeneity arguments

Hitherto we have adopted the traditional point of view, asking for the behaviour of

$$\# \{ \sigma (H_\beta) \cap (-\infty, 0] \} \text{ as } \beta \to \infty, \tag{1}$$

where H_β has the same meaning as in 5.4.7 and 5.4.8, namely

$$H_\beta = a(\cdot, D) - \beta b^2, \tag{2}$$

satisfying the conditions detailed there. Looking at the classical origin of the problem described briefly at the beginning of 5.4.8, the following point of view seems to be even more natural. Let

$$H = a(\cdot, D) - b^2, \tag{3}$$

where $a(\cdot, D) \in \Psi^\kappa_{1,\gamma}$, with $\kappa > 0$ and $0 \leq \gamma < 1$, is a self-adjoint positive operator in L_2 with 0 in the essential spectrum $\sigma_e(a(\cdot, D))$; see 1.2. Then, of course, 0 is the bottom point of the spectrum of $a(\cdot, D)$. As before suppose that b is a real function and let $b^2 (\mathrm{id} + a(\cdot, D))^{-1}$ be compact; then H with $\mathrm{dom}(H) = \mathrm{dom}(a(\cdot, D))$ is self-adjoint and

$$\sigma_e(H) = \sigma_e(a(\cdot, D)). \tag{4}$$

Our interest lies in the behaviour of

$$\# \{ \sigma(H) \cap (-\infty, -\varepsilon] \} \text{ as } \varepsilon \downarrow 0. \tag{5}$$

The attempt to reduce (5) to (1) causes some problems, in general. But the situation is much better when both $a(\cdot, D)$ and b satisfy some homogeneity conditions. To be as close as possible to the hydrogen operator H in (5.4.8/1) we restrict ourselves to a comparatively simple but typical example. Let

$$a(D) = (-1)^m \sum_{|\gamma|=m} a_\gamma D^{2\gamma}, \tag{6}$$

with

$$a_\gamma \in \mathbf{R} \text{ and } \sum_{|\gamma|=m} a_\gamma \xi^{2\gamma} \geq c\,|\xi|^{2m}\,, \, \xi \in \mathbf{R}^n, \tag{7}$$

for some $c > 0$, be an elliptic differential operator of order $2m$ with constant coefficients. Then

$$\sigma\,(a\,(D)) = \sigma_e\,(a\,(D)) = [0,\infty). \tag{8}$$

See for background material [EEv], IX. 6, p. 429. Let

$$H = a\,(D) - |x|^{-\eta} \text{ with } 0 < \eta < \min\,(n, 2m). \tag{9}$$

Then we have $b(x) = |x|^{-\eta/2}$ in (3). By (9) there is a number r such that

$$\frac{2n}{\eta} > r > \max\left(2, \frac{n}{m}\right). \tag{10}$$

Since $\kappa = 2m$ in our case, the right-hand side coincides with (5.4.8/3), whereas the left-hand side covers (5.4.8/5) with $\alpha = \eta$ and $b_1(x) = b(x)$ near the origin. Hence we can apply Theorem 5.4.8 to $H_\beta + \mathrm{id}$ with

$$H_\beta = a\,(D) - \beta\,|x|^{-\eta}. \tag{11}$$

We obtain

$$\#\left\{\sigma\,(H_\beta) \cap (-\infty, -1]\right\} \leq c\beta^{n/\eta}. \tag{12}$$

However, (12) is the basis for the estimation of (5) for the operator H given by (9).

Theorem *Let H be given by (9) with (6) and (7). There exists a number $c > 0$ such that for all $\varepsilon > 0$,*

$$\#\left\{\sigma\,(H) \cap (-\infty, -\varepsilon]\right\} \leq c\varepsilon^{-n\left(\frac{1}{\eta} - \frac{1}{2m}\right)}. \tag{13}$$

Proof Let $\lambda \in (-\infty, -1]$ be an eigenvalue of H_β given by (11) and let f be a related eigenfunction. Then we have for any $c > 0$ and with $f(x) = g(cx)$,

$$c^{2m}\,(-1)^m \sum_{|\gamma|=m} a_\gamma\,(D^{2\gamma}g)\,(cx) - \beta\,|x|^{-\eta}\,g\,(cx) = \lambda g\,(cx) \tag{14}$$

and

$$(-1)^m \sum_{|\gamma|=m} a_\gamma D^{2\gamma}g\,(x) - \beta c^{\eta-2m}\,|x|^{-\eta}\,g\,(x) = \lambda c^{-2m}g\,(x). \tag{15}$$

We choose $c^{2m-\eta} = \beta$ and obtain

$$Hg = \lambda\beta^{-\frac{2m}{2m-\eta}}g = \lambda\varepsilon g. \tag{16}$$

The left-hand side of (13) coincides with that of (12) if we choose $\beta = \varepsilon^{-\frac{2m-\eta}{2m}}$. Now (13) follows from (12).

Remark 1 In the case of the hydrogen atom (5.4.8/1) we have $n = 3$, $\eta = 1$ and $2m = 2$. Hence

$$\#\{\sigma(H) \cap (-\infty, -\varepsilon]\} \leq c\varepsilon^{-3/2}, \tag{17}$$

as it should be, since the number of eigenvalues less than or equal to $-\varepsilon = -\frac{1}{4}N^{-2}$ with $N \in \mathbf{N}$, is $\sum_{j=1}^{N} j^2 \sim N^3$; see, for example, [Tri1], 7.3 and 7.3.4.

Remark 2 We again followed closely [HaT2], where one finds further results and comments.

References

[Ag1]   Agmon, S. On the eigenfunctions and on the eigenvalues of general elliptic boundary value problems. *Comm. Pure Appl. Math.* **15** (1962), 119–147.

[Ag2]   Agmon, S. *Lectures on elliptic boundary value problems.* New York: Van Nostrand, 1965.

[ADN]   Agmon, S., Douglis, A. & Nirenberg, L. Estimates near the boundary for solutions of elliptic partial differential equations satisfying general boundary conditions. I. *Comm. Pure Appl. Math.* **12** (1959), 623–727.

[Aub]   Aubin, T. *Nonlinear analysis on manifolds. Monge–Ampère equations.* New York: Springer, 1982.

[Bea]   Beals, R. Characterization of pseudodifferential operators and applications. *Duke Math. J.* **44** (1977), 45–57. Correction, *Duke Math. J.* **46** (1979), 215.

[BeL]   Bergh, J. & Löfström, J. *Interpolation spaces, an introduction.* Berlin: Springer, 1976.

[BeR]   Bennett, C. & Rudnick, K. On Lorentz–Zygmund spaces. *Dissert. Math.* **175** (1980), 1–72.

[BeS]   Bennett, C. & Sharpley, R. *Interpolation of operators.* New York: Academic Press, 1988.

[Bi]   Birman, M.S. On the number of eigenvalues in a quantum scattering problem (Russian). *Vest. LSU* **16** (13) (1961), 163–166.

[BKS]   Birman, M.S., Karadzhov, G.E. & Solomyak, M.Z. Boundedness conditions and spectrum estimates for the operators $b(X)a(D)$ and their analogs. *Adv. in Sov. Math.* **7** (AMS) (1991), 85–106.

[BiS1]   Birman, M.S. & Solomyak, M.Z. Piecewise polynomial approximations of functions of the classes W_p^α. *Mat. Sb.* **73** (115) (1967), 331–355 (Russian). Engl. transl. *Math. USSR Sb.* (1967), 295–317.

[BiS2]   Birman, M.S. & Solomyak, M.Z. Spectral asymptotics of non-smooth elliptic operators. I. *Trans. Moscow Math. Soc.* **27** (1972), 1–52.

[BiS3]   Birman, M.S. & Solomyak, M.Z. Spectral asymptotics of non-smooth elliptic operators. II. *Trans. Moscow Math. Soc.* **28** (1973), 1–32.

[BiS4]   Birman, M.S. & Solomyak, M.Z. The asymptotics of the spectrum of pseudodifferential operators with anisotropically homogeneous symbols, I, II (Russian). *Vestnik Leningrad Univ. Mat. Mekh. Astronom.* **13** (1977), 13–21; **15** (1979), 5–10.

[BiS5]   Birman, M.S. & Solomyak, M.Z. Quantitative analysis in Sobolev imbedding theorems and applications to spectral theory. *Amer. Math. Soc. Transl.* (2) **114** (1980), 1–132.

[BiS6]   Birman, M.S. & Solomyak, M.Z. Estimates for the number of negative eigenvalues of the Schrödinger operator and its generalizations. *Adv. in Sov. Math.* **7** (AMS) (1991), 1–55.

[BPST]   Bourgain, J., Pajor, A., Szarek, S.J. & Tomczak-Jaegermann, N. On the duality problem for entropy numbers of operators. In: *Geometric aspects of function analysis (1987–88)*, Lecture Notes in Mathematics, Vol. 1376. Berlin: Springer, 1989, pp. 50–63.

[Carl1]   Carl, B. Entropy numbers, s-numbers and eigenvalue problems. *J. Functional Anal.* **41** (1981), 290–306.

[Carl2]   Carl, B. Entropy numbers of embedding maps between Besov spaces with an application to eigenvalue problems. *Proc. Roy. Soc. Edinburgh, Sect. A* **90** (1981), 63–70.

[Carl3]   Carl, B. Inequalities of Bernstein–Jackson type and the degree of compactness of operators in Banach spaces. *Ann. Inst. Fourier* **35** (1985), 79–118.

[CaS]   Carl, B. & Stephani, I. *Entropy, compactness and the approximation of operators*. Cambridge: Cambridge University Press, 1990.

[CaT]   Carl, B. & Triebel, H. Inequalities between eigenvalues, entropy numbers, and related quantities of compact operators in Banach spaces. *Math. Ann.* **251** (1980), 129–133.

[CoE]   Cobos, F. & Edmunds, D.E. Clarkson's inequalities, Besov spaces and Triebel–Sobolev spaces. *Zeit. Anal. Anwendungen* **7** (1988), 229–232.

[CoH]   Courant, R. & Hilbert, D. *Methods of mathematical physics, Vol. I.* New York: Interscience, 1953.

[DeVL]   DeVore, R.A. & Lorentz, G.G. *Constructive approximation*. Berlin: Springer-Verlag, 1993.

[Di]   Dintelmann, P. On the boundedness of pseudo-differential operators on weighted Besov–Triebel spaces. *Math. Nachr.* (to appear).

[EE]   Edmunds, D.E. & Edmunds, R.M. Entropy and approximation numbers of embeddings in Orlicz spaces. *J. London Math. Soc.* **32** (1985), 528–538.

[EEv]   Edmunds, D.E. & Evans, W.D. *Spectral theory and differential operators*. Oxford: Oxford University Press, 1987.

[EET]   Edmunds, D.E., Edmunds, R.M. & Triebel, H. Entropy numbers of embeddings of fractional Besov–Sobolev spaces in Orlicz spaces. *J. London Math. Soc.* (2) **35** (1987), 121–134.

[EM]   Edmunds, D.E. & Moscatelli, V.B. Fourier approximations and embeddings of Sobolev spaces. *Diss. Math.* **145** (1977), 1–50.

[ET1]   Edmunds, D.E. & Triebel, H. Entropy numbers and approximation numbers in function spaces. *Proc. London Math. Soc.* **58** (1989), 137–152.

[ET2]   Edmunds, D.E. & Triebel, H. Entropy numbers and approximation numbers in function spaces, II. *Proc. London Math. Soc.* **64** (1992), 153–169.

[ET3]   Edmunds, D.E. & Triebel, H. Eigenvalue distributions of some degenerate elliptic operators : an approach via entropy numbers. *Math. Ann.* **299** (1994), 311–340.

[ET4]   Edmunds, D.E. & Triebel, H. Logarithmic Sobolev spaces and their applications to spectral theory. *Proc. London Math. Soc.* **71** (1995), 333–371.

[ETy]   Edmunds, D.E. & Tylli, H.-O. On the entropy numbers of an operator and its adjoint. *Math. Nachr.* **126** (1986), 231–239.

[Fra1]   Franke, J. Fourier Multiplikatoren, Littlewood–Paley Theoreme und Approximation durch ganze analytische Funktionen in gewichten Funktionenräunen. Forschungsergebnisse Univ. Jena N/86/8, 1986.

[Fra2]   Franke, J. On the spaces F_{pq}^s of Triebel–Lizorkin type: Pointwise multipliers and spaces on domains. *Math. Nachr.* **125** (1986), 29–68.

[FraR]   Franke, J. & Runst, Th. Regular elliptic boundary value problems in Besov–Triebel–Lizorkin spaces. *Math. Nachr.* **174** (1995), 113–149.

[FrJ1]   Frazier, M. & Jawerth, B. Decomposition of Besov spaces. *Indiana Univ. Math. J.* **34** (1985), 777–799.

[FrJ2]   Frazier, M. & Jawerth, B. A discrete transform and decomposition of distribution spaces. *J. Functional Anal.* **93** (1990), 34–170.

[FrJW]   Frazier, M., Jawerth, B. & Weiss, G. Littlewood–Paley theory and the study of function spaces. CBMS-AMS Regional Conf. Ser. 79, 1991.

[Gar]   Garnett, J.B. *Bounded analytic functions*. New York: Academic Press, 1981.

[GiT]   Gilbarg, D. & Trudinger, N.S. *Elliptic partial differential equations of second order*. Berlin: Springer-Verlag, 1977.

[Glu]   Gluskin, E.D. Norms of random matrices and diameters of finite-dimensional sets (Russian). *Mat. Sb.* **120** (1983), 180–189. Engl. transl.: *Math. USSR Sb.* **48** (1984), 173–182.

[GKS]   Gordon, Y., König, H. & Schütt, C. Geometric and probabilistic estimates for entropy and approximation numbers of operators. *J. Approx. Theory* **49** (1987), 219–239.

[Har1]   Haroske, D. Entropy numbers and approximation numbers in weighted function spaces of type $B_{p,q}^s$ and $F_{p,q}^s$, eigenvalue distributions of some degenerate pseudodifferential operators. Thesis, Jena, 1995.

[Har2]   Haroske, D. Approximation numbers in some weighted function spaces. *J. Approx. Theory* **83** (1995), 104–136.

[HaT1]   Haroske, D. & Triebel, H. Entropy numbers in weighted function spaces and eigenvalue distributions of some degenerate pseudodifferential operators I. *Math. Nachr.* **167** (1994), 131–156.

[HaT2]   Haroske, D. & Triebel, H. Entropy numbers in weighted function spaces and eigenvalue distributions of some degenerate pseudodifferential operators II. *Math. Nachr.* **168** (1994), 109–137.

[Hör]   Hörmander, L. *The analysis of linear partial differential operators III*. Berlin: Springer, 1985.

[Jaw]   Jawerth, B. Some observations on Besov and Lizorkin–Triebel spaces. *Math. Scand.* **40** (1977), 94–104.

[JawM]   Jawerth, B. & Milman, M. Extrapolation theory and applications. *Memoirs Amer. Math. Soc.* **89** (1991), 440.

[JK]   Jerison, D.S. & Kenig, C.E. Boundary behavior of harmonic functions in non-tangentially accessible domains. *Adv. in Math.* **46** (1982), 80–147.

[Jo]   Johnsen, J. Elliptic boundary problems and the Boutet de Monvel calculus in Besov and Triebel–Lizorkin spaces. Preprint, Copenhagen 1994.

[JoW] Jonsson, A. & Wallin, H. Function spaces on subsets of $\mathbf{R}^n$. *Math. Reports* **2**, 1 (1984), 1–221.

[Ka] Kato, T. *Perturbation theory for linear operators*, 2nd edition. Berlin: Springer-Verlag, 1976.

[KaT] Kato, M. & Takahashi, Y. Type, cotype constants and Clarkson's inequalities for Banach spaces. *Math. Nachr.* (to appear).

[Kön] König, H. *Eigenvalue distribution of compact operators*. Basel: Birkhäuser, 1986.

[KrR] Krasnosel'skii, M.A. & Rutickii, Ya. B. *Convex functions and Orlicz spaces*. Groningen: Noordhoff, 1961.

[KuJF] Kufner, A., John, O. & Fučik, S. *Function spaces*. Prague: Academia, 1977.

[LeS] Leopold, H.-G. & Schrohe, E. Spectral invariance for algebras of pseudodifferential operators on Besov–Triebel–Lizorkin spaces. *Manuscripta Math.* **78** (1993), 99–110.

[LeT] Leopold, H.-G. & Triebel, H. Spectral invariance for pseudodifferential operators on weighted function spaces. *Manuscripta Math.* **83** (1994), 315–325.

[Lie] Lieb, E.H. Bounds on the eigenvalues of the Laplace and Schrödinger operators. *Bull. Amer. Math. Soc.* **82** (1976), 751–753.

[LiM] Lions, J.-L. & Magenes, E. *Non-homogeneous boundary value problems and applications. I.* Berlin: Springer, 1972.

[Lin] Linde, R. *s*-numbers of diagonal operators and Besov embeddings. *Proc. 13th Winter school, Suppl. Rend. Circ. Mat. Palermo* (2) (1986), 83–110.

[MaVä] Martio, O. & Väisälä, J. Global L^p-integrability of the derivative of a quasiconformal mapping. *Complex Variable Theory Appl.* **9** (1988), 309–319.

[Mil] Milman, M. *Extrapolation and optimal decompositions with applications to analysis*. Lecture Notes in Mathematics, Vol. 1580. Berlin: Springer, 1994.

[MyOt] Mynbaev, K.T. & Otelbaev, M.O. *Weighted function spaces and the spectrum of differential operators* (Russian). Moscow: Nauka, 1988.

[Päi] Päivärinta, L. Pseudo differential operators in Hardy–Triebel spaces. *Z. Anal. Anwendungen* **2** (1983), 235–242.

[Peet] Peetre, J. *New thoughts on Besov spaces*. Duke Univ. Math. Series. Durham, NC: Duke Univ., 1976.

[Pi1] Pietsch, A. *Operator ideals*. Amsterdam: North-Holland, 1980.

[Pi2] Pietsch, A. *Eigenvalues and s-numbers*. Cambridge: Cambridge University Press, 1987.

[RaR] Rao, M.M. & Ren, Z.D. *Theory of Orlicz spaces*. New York: Marcel Dekker, 1991.

[Ro1] Rosenbljum, G.V. The distribution of the discrete spectrum for singular differential operators. *Sov. Math. Dokl.* **13** (1972), 245–249.

[Ro2] Rosenbljum, G.V. Distribution of the discrete spectrum of singular differential operators (Russian). *Izv. Vyssh. Uchebn. Zaved., Mat.* **1** (1976), 75–86.

[Rud] Rudin, W. *Functional analysis*, 2nd edition. New York: McGraw-Hill, 1991.

[Run] Runst, Th. Pseudo-differential operators of the "exotic" class $L^0_{1,1}$ in spaces of Besov and Triebel–Lizorkin type. *Ann. Global Anal. Geom.* **3** (1985), 13–28.

[Sche] Schechter, M. *Spectra of partial differential operators*, 2nd edition. Amsterdam: North-Holland, 1986.

[SchmT] Schmeisser, H.-J. & Triebel, H. *Topics in Fourier analysis and function spaces*. Leipzig: Geest & Portig, 1987; Chichester: Wiley, 1987.

[Schr] Schrohe, E. Boundedness and spectral invariance for standard pseudodifferential operators on anisotropically weighted L^p-Sobolev spaces. *Integral Equations Oper. Theory* **13** (1990), 271–284.

[Schü] Schütt, C. Entropy numbers of diagonal operators between symmetric Banach spaces. *J. Approx. Theory* **40** (1984), 121–128.

[Schw] Schwinger, J. On the bound states of a given potential. *Proc. Nat. Acad. Sci. USA* **47** (1961), 122–129.

[Sic] Sickel, W. On pointwise multipliers in Besov–Triebel–Lizorkin spaces. In: *Seminar Analysis Karl-Weierstrass Institute 1985/86*, Teubner-Texte Math. **96**. Leipzig. Teubner, 1987, 45–103.

[SicT] Sickel, W. & Triebel, H. Hölder inequalities and sharp embeddings in function spaces of B^s_{pq} and F^s_{pq} type. *Z. Anal. Anwendungen* **14** (1995), 105–140.

[Sim1] Simon, B. Analysis with weak trace ideals and the number of bound states of Schrödinger operators. *Trans. Amer. Math. Soc.* **224** (1976), 367–380.

[Sim2] Simon, B. *Trace ideals and their applications*. Cambridge: Cambridge University Press, 1979.

[Sol] Solomyak, M.Z. Piecewise polynomial approximation of functions from $H^\ell\left((0,1)^d\right)$, $2\ell = d$, and applications to the spectral theory of the Schrödinger operator. *Israel J. Math.* **86** (1994), 253–275.

[Ste] Stein, E.M. *Singular integrals and differentiability properties of functions*. Princeton, NJ: Princeton University Press, 1970.

[Str] Strichartz, R.S. A note on Trudinger's extension of Sobolev's inequalities. *Indiana Univ. Math. J.* **21** (1972), 841–842.

[Tal] Talenti, G. Best constant in Sobolev inequality. *Ann. Mat. Pura Appl.* **110** (1976), 353–372.

[Tas] Tashkian, G.M. The classical formula of the asymptotics of the spectrum of elliptic equations, degenerate at the boundary of the domain (Russian). *Mat. Zametki* **30** (6) (1981), 871–880.

[Tay1] Taylor, M.E. *Pseudodifferential operators*. Princeton, NJ: Princeton University Press, 1981.

[Tay2] Taylor, M.E. *Pseudodifferential operators and nonlinear PDE*. Boston, Mass.: Birkhäuser, 1991.

[Tor] Torchinsky, A. *Real-variable methods in harmonic analysis*. San Diego: Academic Press, 1986.

[Torr1] Torres, R.H. Continuity properties of pseudodifferential operators of type 1,1. *Comm. Part. Diff. Equations* **15** (9) (1990), 1313–1328.

[Torr2] Torres, R.H. Boundedness results for operators with singular kernels on distribution spaces. *Memoirs Amer. Math. Soc.* **90** (1991), 442.

[Triα] Triebel, H. *Interpolation theory, function spaces, differential operators*. Berlin: VEB Deutsch. Verl. Wissenschaften, 1978; Amsterdam: North-Holland, 1978. 2nd (revised) edition Leipzig: Barth 1995.

[Triβ] Triebel, H. *Theory of function spaces*. Leipzig: Geest & Portig, 1983; Basel: Birkhäuser, 1983.

[Triγ]   Triebel, H. *Theory of function spaces II*. Basel: Birkhäuser, 1992.

[Tri1]   Triebel, H. *Higher analysis*. Leipzig: Barth, 1992. Translated (and revised) from the 1972 German version.

[Tri2]   Triebel, H. *Spaces of Besov–Hardy–Sobolev type*. Teubner-Texte Math 15. Leipzig: Teubner, 1977.

[Tri3]   Triebel, H. Approximation numbers and entropy numbers of embeddings of fractional Besov–Sobolev spaces in Orlicz spaces. *Proc. London Math. Soc.* (3) **66** (1993), 589–618.

[Tri4]   Triebel, H. A localization property for B_{pq}^s and F_{pq}^s spaces. *Studia Math.* **109** (2) (1994), 183–195.

[Tri5]   Triebel, H. Relations between approximation numbers and entropy numbers, *J. Approx. Theory* **78** (1994), 112–116.

[TriW1]   Triebel, H. & Winkelvoss, H. Intrinsic atomic characterizations of function spaces on domains. *Math. Z.* **222** (1996) (to appear).

[TriW2]   Triebel, H. & Winkelvoss, H. The dimension of a closed subset of $\mathbf{R}^n$ and related function spaces. *Acta Math. Hungarica* **68** (1995), 117–133.

[Tru]   Trudinger, N.S. On imbeddings into Orlicz spaces and some applications. *J. Math. Mach.* **17** (1967), 473–483.

[Übe]   Überberg, J. Zur Spektralinvarianz von Algebren von Pseudodifferentialoperatoren in der L^p-Theorie. *Manuscripta Math.* **61** (1988), 459–475.

[WiW]   Whittaker, E.T. & Watson, G.N. *Modern analysis*. Cambridge: Cambridge University Press 1952.

[Yam]   Yamazaki, M. A quasi-homogeneous version of para-differential operators. I. Boundedness on spaces of Besov type. *J. Fac. Sci. Univ. Tokyo,* Sect. IA, Math. **33** (1986), 131–174.

[Zem]   Zemánek, J. The essential spectral radius and the Riesz part of the spectrum. In: *Functions, series, operators. Proc. Int. Conf. Budapest 1980. Colloq. Math. Soc. Janos Bolyai* **35**, 1275–1289.

Index of Symbols

Page numbers are given in parentheses.

249

Index

Page numbers are given in parentheses.